Comprehensive Remedial Mathematics for B. Pharmacy

Second Edition

Comprehensive Remedial Mathematics for B. Pharmacy

Second Edition

Dr. Shyam Patkar
M.Sc., Ph.D.
Professor & Head

Ramakant Bhardwaj
M.Sc., Ph.D.

Sarvesh Agrawal
M.Sc., M.Phil.

Truba Institute of Engineering & Information Technology,
Bhopal (India)

PharmaMed Press
An imprint of BSP Books Pvt. Ltd.
4-4-309/316, Giriraj Lane,
Sultan Bazar, Hyderabad - 500 095.

Comprehensive Remedial Mathematics for B. Pharmacy Second Edition
***by* Shyam Patkar, Ramakant Bhardwaj, Sarvesh Agrawal**

Sixth Reprint 2026

Published by:

PharmaMed Press
An imprint BSP Books Pvt. Ltd.
4-4-309/316, Giriraj Lane, Sultan Bazar, Hyderabad - 500 095.
Phone: 040-23445688; Fax: 91+40-23445611
E-mail: info@pharmamedpress.com
www.pharmamedpress.com/pharmamedpress.net

Printed at
Repro India Limited.

Price: Rs. 995.00
ISBN: 978-93-81075-69-2 (Paperback)

Dedicated to

Our Parents

Preface

Comprehensive Remedial Mathematics for B. Pharmacy is specially designed for undergraduates. The authors have evolved this from lecture notes and related material developed while teaching various courses in Remedial Mathematics during the last sixteen years.

The authors have given special stress on basic fundamentals, solved examples and oriented problems of various universities in the book. This structure will provide students a platform for building confidence and sharpen their skills in Pharmaceutical Mathematics. The comprehensive edition will surely serve its purpose to help to achieve handsome marks in the examinations.

We hope the students and teachers of Mathematics of various universities will appreciate our efforts and receive the book with enthusiasm.

We sincerely thank our publishers for their untiring efforts in bringing out this book excellently.

- Authors

Contents

CHAPTER 1

EQUATIONS REDUCIBLE TO QUADRATICS EQUATIONS

Numbers

The numbers 1, 2, 3 … ∞ are called natural *numbers* or positive integers.

$\frac{a}{b}$ is called a *fraction* where a and b are any two positive integers.

The number system consisting of positive and negative fractions is called *rational number*. The rational number is of the form $\frac{a}{b}$ where a and b are integers and $b \neq 0$. That is –4, –2, –1, 0, 1, 2, 3, 4 … are called *integers*.

The number $\sqrt{2}$, $\sqrt{3}$, $\sqrt{5}$ … which cannot be expressed as $\frac{a}{b}$ (i.e. ratio of two integers), are called *irrational numbers*.

The rational numbers, positive and negative, irrational numbers and zero constitute the *real number*.

Types of Numbers

Natural number or N

1, 2, 3, 4, …..

Real number or R

The set of rational and irrational numbers i.e., $R = Q \cup I_r$, where I_r denotes the Irrational numbers.

Rational number or Q

$$\ldots\ldots -\frac{3}{4}, -\frac{1}{2}, 0, \frac{1}{2}, \frac{1}{3} \ldots\ldots$$

Primes or P

2, 3, 5, 7, 11, 13 …..

Composite numbers

4, 6, 8, 9, 10, 12 …..

Even numbers

2, 4, 6, 8, 10, 12 …..

Odd numbers

1, 3, 5, 7, 9, 11 …..

Complex Numbers

The numbers of the form x + iy where $i = \sqrt{-1}$ and a and b are real numbers are called complex numbers.

Useful Formulae

$$(a + b)^2 = a^2 + b^2 + 2ab$$
$$(a - b)^2 = a^2 + b^2 - 2ab$$
$$(a + b)^2 + (a - b)^2 = 2a^2 + 2b^2$$
$$(a + b)^2 - (a - b)^2 = 4ab$$
$$(a + b)(a - b) = a^2 - b^2$$

$$(a + b)^3 = a^3 + b^3 + 3a^2b + 3ab^2$$
$$(a + b)^3 = a^3 + b^3 + 3ab(a + b)$$

$$(a - b)^3 = a^3 - b^3 - 3a^2b + 3ab^2$$
$$(a - b)^3 = a^3 - b^3 - 3ab(a - b)$$

$$(a^3 + b^3) = (a + b)^3 - 3ab(a + b)$$
$$a^3 + b^3 = (a + b)(a^2 - ab + b^2)$$

$$a^3 - b^3 = (a - b)^3 + 3ab(a - b)$$
$$a^3 - b^3 = (a - b)(a^2 + ab + b^2)$$
$$(x + a)(x + b) = x^2 + (a + b)x + ab$$
$$(x - a)(x - b) = x^2 - (a + b)x + ab$$

Quadratics Equation

Definition

The equation of the form $ax^2 + bx + c = 0$ is called a quadratic equation, where a, b, c are real numbers and $a \neq 0$.

Example: (i) $5x^2 - 8x + 3 = 0$

(ii) $x^2 - 3x + 2 = 0$

Quadratic Equation of a Roots

It has only two roots, the value of x the given equation is called a root.

Such as $5x^2 - 8x + 3 = 0$

Solution: $5x^2 - 5x - 3x + 3 = 0$ using factorization

$5x(x - 1) - 3(x - 1) = 0$

$(5x - 3)(x - 1) = 0$

$x = \frac{3}{5}$ and $x = 1$

Hence roots are $\left\{\frac{3}{5}, 1\right\}$

Working Rule

Given quadratic equation $ax^2 + bx + c = 0$ can be solved as

$$x = \frac{-b \pm \sqrt{b^2 - 4ac}}{2a}$$

Let $$\alpha = \frac{-b + \sqrt{b^2 - 4ac}}{2a}, \beta = \frac{-b - \sqrt{b^2 - 4ac}}{2a}$$

***Example* 1:** Solve the quadratic equation

$$x^2 - 9x + 20 = 0$$

Solution: Let $f(x) = x^2 - 9x + 20$

$\therefore$ $x^2 - 9x + 20 = 0$ using factionzation

$x^2 - 5x - 4x + 20 = 0$

$x(x - 5) - 4(x - 5) = 0$

$(x - 5)(x - 4) = 0$

Hence solutions, x = 5, 4

***Example* 2:** Solve the quadratic equation $x^2 - 25 = 0$

***Solution*:** we know that $a^2 - b^2 = (a - b)(a + b)$

According to problem $x^2 - 25 = 0$

$$(x)^2 - (5)^2 = 0$$

$$(x - 5)(x + 5) = 0$$

Hence roots are $x = 5, -5$

***Example* 3:** Find the value of x given quadratic equation is $x^2 - 5x + 2 = 0$

***Solution*:** Given equation $x^2 - 5x + 2 = 0$

by quadratic equation $ax^2 + bx + c = 0$

Now, $a = 1, \quad b = -5, \quad c = 2,$ put values

$$x = \frac{-b \pm \sqrt{b^2 - 4ac}}{2a}$$

$$x = \frac{-(-5) \pm \sqrt{(-5)^2 - 4(1)(2)}}{2(1)}$$

$$x = \frac{5 \pm \sqrt{25 - 8}}{2}$$

$$x = \frac{5 \pm \sqrt{17}}{2}$$

Hence $x = \dfrac{5 + \sqrt{17}}{2}$ and $x = \dfrac{5 - \sqrt{17}}{2}$

***Example* 4:** Solve the quadratic equation

$$3x^2 + 14x - 5 = 0$$

***Solution*:** we know that formula $x = \dfrac{-b \pm \sqrt{b^2 - 4ac}}{2a}$

given values $a = 3, \quad b = 14, \quad c = -5$

$$x = \frac{-14 \pm \sqrt{196 - 4(3)(-5)}}{2(3)}$$

$$= \frac{-14 \pm \sqrt{196 + 60}}{6} = \frac{-14 \pm \sqrt{256}}{6}$$

$$x = \frac{-14 \pm 16}{6}$$

$$x = \frac{-14+16}{6} = \frac{2}{6} = \frac{1}{3}$$

$$x = \frac{-14-16}{6} = -\frac{30}{6} = -5$$

Hence roots are $\left(\frac{1}{3}, -5\right)$

Problem 1.1

1. $3x^2 + 11x + 6 = 0$

Ans: $\left(-3, \frac{-2}{3}\right)$

2. $3x^2 - 2x - 21 = 0$

Ans: $\left(\frac{-7}{3}, 3\right)$

3. $x^2 - 16 = 0$

Ans: $(4, -4)$

4. $\frac{1}{x+1} + \frac{2}{x+2} = \frac{4}{x+4}$

Ans: $(2 + 2\sqrt{3}, 2 - 2\sqrt{3})$

Sum of the Roots: If $ax^2 + bx + c = 0$ then sum of the roots $\alpha + \beta = -\frac{b}{a}$

Product of the Roots: If $ax^2 + bx + c = 0$ then product of the roots $(\alpha\,\beta) = \frac{c}{a}$

where α, β are two roots and a, b, c are real numbers

***Example* 1:** Find the value of x and find the sum and product of the roots of given equation $5x^2 - 5\sqrt{3}\,x + 3 = 0$

***Solution*:** According to problem given equation is

$$5x^2 - 5\sqrt{3}\,x + 3 = 0$$

using formula

$$x = \frac{-b \pm \sqrt{b^2 - 4ac}}{2a} \quad .....(1)$$

$a = 5, b = -5\sqrt{3}, c = 3$ put values in equation (1)

$$x = \frac{-\left(-5\sqrt{3}\right) \pm \sqrt{\left(-5\sqrt{3}\right)^2 - 4(5)(3)}}{2(5)}$$

$$x = \frac{5\sqrt{3} \pm \sqrt{75-60}}{2(5)} = \frac{5\sqrt{3} \pm \sqrt{+15}}{10}$$

Now sum of roots, $\alpha + \beta = -\frac{b}{a}$

$$= -\left(\frac{-5\sqrt{3}}{5}\right) = \sqrt{3}$$

Product of the roots $\alpha\ \beta = \frac{c}{a}$

$$= \frac{3}{5}$$

***Example* 2:** Find the sum of the roots and product of the roots $x^2 + 15x + 20 = 0$

***Solution*:** $a = 1, \quad b = 15, \quad c = 20$

Sum of the roots, $\alpha + \beta = -\frac{b}{a}$

$$= -\frac{15}{1} = -15$$

product of the root $(\alpha\ \beta) = \frac{c}{a}$

$$= \frac{20}{1} = 20$$

Hence required sum of the roots –15 and product of the roots 20

***Example* 3:** Find the set of values of λ for given equation has real roots.

$$\lambda x^2 - 6x - 2 = 0$$

***Solution*:** We know that quadratic equation

$$ax^2 + bx + c = 0$$

$$x = \frac{-b \pm \sqrt{b^2 - 4ac}}{2a} \qquad .....(1)$$

$a = \lambda$, $b = -6$ and $c = -2$, put values in equation (1)

$$x = \frac{-(-6) \pm \sqrt{(-6)^2 - 4(\lambda)(-2)}}{2(\lambda)}$$

$$x = \frac{6 \pm \sqrt{36 + 8\lambda}}{2\lambda}$$

the given equation will have real roots if $\delta \geq 0$.

$$\delta = b^2 - 4ac = 36 + 8\lambda$$

i.e. $36 + 8\lambda \geq 0$

i.e. $8\lambda \geq -36$

i.e. $\lambda \geq \frac{-9}{2}$

$\therefore$ the required set of values $= \left\{\lambda \in R : \lambda \geq \frac{-9}{2}\right\}$

Where R is the set of real numbers.

Similar proof or solution method same.

(i) $9x^2 + 3\lambda x + 4 = 0$

(ii) $3x^2 + \lambda x + 4 = 0$

Nature of roots

Some important cases of finding roots of the quadratic equations. $ax^2 + bx + c = 0$

1. The roots α and β are real and distinct, if $\delta > 0$ i.e. if $b^2 - 4ac > 0$.
2. The roots α and β are real and equal ($\alpha = \beta$), if $\delta = 0$ i.e. if $b^2 - 4ac = 0$
3. The roots α and β are imaginary and distinct, if $\delta < 0$ i.e. $b^2 - 4ac < 0$
4. The roots are rational and distinct, if $b^2 - 4ac$ is a perfect square

***Example* 4:** If α, β be the roots of the equation $2x^2 - 5x + 2 = 0$, find the value of $\frac{\alpha}{\beta} + \frac{\beta}{\alpha}$

***Solution*:** Given equation $2x^2 - 5x + 2 = 0$

factorization $2x^2 - 4x - x + 2 = 0$

$2x(x-2) - 1(x-2) = 0$

$(2x-1)(x-2) = 0$

$2x - 1 = 0$ and $x - 2 = 0$

$x = \frac{1}{2}$ and $x = 2$

or $\alpha = \frac{1}{2}$ and $\beta = 2$

According to problem

$$\therefore \quad \frac{\alpha}{\beta} + \frac{\beta}{\alpha} = \frac{\frac{1}{2}}{2} + \frac{\frac{2}{1}}{\frac{1}{2}}$$

$$= \frac{1}{4} + \frac{4}{1}$$

$$= \frac{1+16}{4} = \frac{17}{4}$$

***Example* 5:** solve $x - \sqrt{1-x} = -5$

***Solution*:** the given equation can be written as

$$x + 5 = \sqrt{1-x}$$

squaring the both sides

$$(x+5)^2 = \left(\sqrt{1-x}\right)^2$$

$$x^2 + 25 + 10x = 1 - x$$

$$x^2 + 10x + x + 25 - 1 = 0$$

$$x^2 + 11x + 24 = 0$$

$$x^2 + 8x + 3x + 24 = 0$$

$$x(x+8) + 3(x+8) = 0$$

$$(x+3)(x+8) = 0$$

$$x + 3 = 0 \text{ and } x + 8 = 0$$

$$x = -3 \text{ and } x = -8$$

Hence roots are $\{-3, -8\}$

Formation of Quadratic Equation with Given Roots

If α, β be the given roots, then the required quadratic equation is

$$x^2 - (\alpha + \beta)x + \alpha\beta = 0$$

or x^2 – (sum of the roots)x + product of the roots = 0

***Example 6*:** Find the equation whose roots are $2+\sqrt{3}$ and $2-\sqrt{3}$

***Solution*:** Let $\alpha = 2+\sqrt{3}$ and $\beta = 2-\sqrt{3}$

Now, if α, β are any root of complex number then equation will be

$$x^2 - (\alpha + \beta)x + \alpha\beta = 0$$

$$x^2 - [2+\sqrt{3}+2-\sqrt{3}]\,x + (2+\sqrt{3})(2-\sqrt{3}) = 0$$

$$x^2 - 4x + (2)^2 - (\sqrt{3})^2 = 0 \qquad \because a^2 - b^2 = (a-b)(a+b)$$

$$x^2 - 4x + 4 - 3 = 0$$

$$x^2 - 4x + 1 = 0$$

Problem 1.2

Solve the following

1. $x^2 = 4x$

   Ans: (0, 4)

2. $(x-3)(x+7) = 0$

   Ans: (3, –7)

3. $\dfrac{x^2 - 5x}{2} = 0$

   Ans: (0, 5)

4. $x^2 + x - 12 = 0$

   Ans: (–4, 3)

5. $10x - \dfrac{1}{x} = 3$

   Ans: $\left(\dfrac{1}{2}, \dfrac{-1}{5}\right)$

6. $2x^2 - 11x + 5 = 0$

   Ans: $\left(5, \dfrac{1}{2}\right)$

7. $2(x^2 + 1) = 5x$

Ans: $\left(2, \frac{1}{2}\right)$

8. $\frac{x}{2} + \frac{6}{x} = 4$

Ans: (2, 6)

9. $\frac{x+3}{x-2} - \frac{1-x}{x} = \frac{17}{4}, x \neq 0, x \neq 2$

Ans: $\left(4, \frac{-2}{9}\right)$

10. $\frac{x}{5} + \frac{28}{x+2} = 5$

Ans: (5, 18)

11. $3x^2 = x + 4$

Ans: $\left(-1, \frac{4}{3}\right)$

12. $21x^2 - 8x - 4 = 0$

Ans: $\left(\frac{2}{3}, \frac{-2}{7}\right)$

13. $x(2x + 5) = 25$

Ans: $\left(-5, \frac{5}{2}\right)$

14. $x(6x - 1) = 35$

Ans: $\left(\frac{-7}{3}, \frac{5}{2}\right)$

15. $x(2x + 1) = 6$

Ans: $\left(-2, \frac{3}{2}\right)$

Equations Reducible to Quadratic Equations of Different Types

Type-I $ax^{2n} + bx^n + c = 0$

Method: In this type of equations, we put $x^n = y$, so that $y^2 = (x^n)^2 = x^{2n}$ and the given equation becomes $ay^2 + by + c = 0$, which is a quadratic equation

***Example* 1:** Solve $x^{\frac{2}{3}} = 7x^{\frac{1}{3}} - 12$

***Solution*:** The given equation can be written as

$$\left(x^{\frac{1}{3}}\right)^2 - 7\left(x^{\frac{1}{3}}\right) + 12 = 0 \qquad .....(1)$$

Putting $x^{\frac{1}{3}} = y$ in (1)

$$y^2 - 7y + 12 = 0$$
$$y^2 - 3y - 4y + 12 = 0$$
$$y(y - 3) - 4(y - 3) = 0$$
$$(y - 3)(y - 4) = 0$$

$\therefore$ y = 3, or y = 4, return value

$y = 3 \Rightarrow x^{\frac{1}{3}} = 3^1 \Rightarrow x = 3^3 = 27$

and $y = 4 \Rightarrow x^{\frac{1}{3}} = 4 \Rightarrow x = 4^3 = 64$

$\therefore$ solution set is {27, 64}

***Example* 2:** Solve $x^{\frac{2}{3}} + x^{\frac{1}{3}} - 2 = 0$ .....(1)

***Solution*:** putting $x^{\frac{1}{3}} = y$ in (1)

$$\left(x^{\frac{1}{3}}\right)^2 + \left(x^{\frac{1}{3}}\right) - 2 = 0$$
$$y^2 + y - 2 = 0$$
$$y^2 + 2y - y - 2 = 0$$
$$y(y + 2) - 1(y + 2) = 0$$
$$(y - 1)(y + 2) = 0$$
$$y - 1 = 0 \text{ or } y + 2 = 0$$

y = 1 or y = –2, return value

then

$y = 1 \Rightarrow x^{\frac{1}{3}} = 1 \Rightarrow x = 1^3 \Rightarrow x = 1$

$y = -2 \Rightarrow x^{\frac{1}{3}} = -2 \Rightarrow x = -2^3 \Rightarrow x = -8$

Hence solution set is {1, –8}.

***Example* 3:** Solve $x^6 - 9x^3 + 8 = 0$ .....(1)

***Solution*:** Putting $x^3 = y$ in (1)

$$y^2 - 9y + 8 = 0$$
$$y^2 - 8y - y + 8 = 0$$
$$y(y - 8) - 1(y - 8) = 0$$
$$(y - 1)(y - 8) = 0$$
$$y - 1 = 0 \text{ or } y - 8 = 0$$
$$y = 1 \text{ or } y = 8, \text{ return value}$$

then

$$y = 1 \Rightarrow x^3 = 1^3 \Rightarrow x = 1$$
$$y = 8 \Rightarrow x^3 = 8 \Rightarrow x^3 = 2^3 \Rightarrow x = 2$$

Hence solution is {1, 2}.

***Example* 4:** Solve for x

$$\left(\frac{x}{x-1}\right)^2 - 3\left(\frac{x}{x-1}\right) - 18 = 0, \quad (x \neq 1) \qquad .....(1)$$

***Solution*:** Putting $\frac{x}{x-1} = y$ in (1)

$$y^2 - 3y - 18 = 0$$
$$y^2 - 6y + 3y - 18 = 0$$
$$y(y - 6) + 3(y - 6) = 0$$
$$(y + 3) = 0 \text{ or } y - 6 = 0$$
$$y = -3 \text{ or } y = 6, \text{ return the values}$$

$$y = -3 \Rightarrow \frac{x}{x-1} = -3$$
$$\Rightarrow x = -3x + 3$$
$$4x = 3 \qquad \therefore x = \frac{3}{4}$$

$$y = 6 \Rightarrow \frac{x}{x-1} = 6$$
$$x = 6x - 6$$
$$-5x = -6 \qquad \therefore x = \frac{6}{5}$$

Hence solution is $\left\{\frac{3}{4}, \frac{6}{5}\right\}$.

***Example* 5:** Solve for x, $\left(\frac{3x-1}{2x+3}\right)^4 - 5\left(\frac{3x-1}{2x+3}\right)^2 = -4$

***Solution*:** Putting $\left(\frac{3x-1}{2x+3}\right)^2 = y$ and squaring the both sides, we get

$$\left(\frac{3x-1}{2x+3}\right)^4 = y^2$$

The given equation becomes

$$y^2 - 5y + 4 = 0$$

$$y^2 - 4y - y + 4 = 0$$

$$y(y-4) - 1(y-4) = 0$$

$$(y-1)(y-4) = 0$$

$$y - 1 = 0 \text{ or } y - 4 = 0$$

y = 1 or y = 4, return the values

when $y = 1 \Rightarrow \left(\frac{3x-1}{2x+3}\right)^2 = 1$

$$\left(\frac{3x-1}{2x+3}\right)^2 = (\pm 1)^2$$

$$\therefore \quad \frac{3x-1}{2x+3} = \pm 1$$

For (+)	For (–)
$3x - 1 = 2x + 3$	$3x - 1 = -2x - 3$
$x = 4$	$5x = -2$
	$\therefore \quad x = \frac{-2}{5}$

and $y = 4 \Rightarrow \left(\frac{3x-1}{2x+3}\right)^2 = 4$

$$\left(\frac{3x-1}{2x+3}\right)^2 = 2^2$$

$$\frac{3x-1}{2x+3} = \pm 2$$

or $\quad 3x - 1 = \pm 2(2x + 3)$

For (+) $\quad$ For (–)

$3x - 1 = 4x + 6 \quad\quad 3x - 1 = -4x - 6$

$-x = 7 \quad\quad 7x = -5$

$\therefore x = -7 \quad\quad x = \dfrac{-5}{7}$

Hence $\left\{4, \dfrac{-2}{5}, -7, \dfrac{-5}{7}\right\}$.

***Example* 6:** Solve $6x + 2 = 7\sqrt{x}$

***Solution*:** The given equation can be written as

$$6\left(\sqrt{x}\right)^2 - 7\sqrt{x} + 2 = 0 \quad \text{.....(1)}$$

Putting $\sqrt{x} = y$ in equation (1)

$6y^2 - 7y + 2 = 0$

$6y^2 - 3y - 4y + 2 = 0$

$3y(2y - 1) - 2(2y - 1) = 0$

$(2y - 1)(3y - 2) = 0$

$2y - 1 = 0$ or $3y - 2 = 0$

$y = \dfrac{1}{2}$ or $y = \dfrac{2}{3}$, return the values

$\Rightarrow y = \dfrac{1}{2} \quad \Rightarrow \sqrt{x} = \dfrac{1}{2}$, squaring the both sides

$\therefore \quad x = \dfrac{1}{4}$

$\Rightarrow y = \dfrac{2}{3} \quad \Rightarrow \sqrt{x} = \dfrac{2}{3}$, squaring the both sides

$\therefore \quad x = \dfrac{4}{9}$

Hence the solution is $\left\{\dfrac{1}{4}, \dfrac{4}{9}\right\}$.

Type-II $ax + \frac{b}{x} = c,\ x \neq 0$ where a, b, c are constants.

***Example* 1:** Solve for x

$$\sqrt{\frac{x}{1-x}} + \sqrt{\frac{1-x}{x}} = 2\frac{1}{6}, \qquad x \neq 0, 1$$

***Solution*:**

$$\sqrt{\frac{x}{1-x}} + \sqrt{\frac{1-x}{x}} = \frac{13}{6}$$

Putting $\sqrt{\frac{x}{1-x}} = y$, the given equation becomes

$$y + \frac{1}{y} = \frac{13}{6}$$

$$6y^2 + 6 = 13y$$

$$6y^2 - 13y + 6 = 0$$

$$6y^2 - 9y - 4y + 6 = 0$$

$$3y\,(2y - 3) - 2(2y - 3) = 0$$

$$(2y - 3)\,(3y - 2) = 0$$

$$3y - 2 = 0 \text{ or } 2y - 3 = 0$$

$y = \frac{2}{3}$ or $y = \frac{3}{2}$, return the values

when $y = \frac{2}{3}$ then $\sqrt{\frac{x}{1-x}} = \frac{2}{3}$, squaring the both sides

$$\frac{x}{1-x} = \frac{4}{9}$$

$$9x = 4 - 4x$$

$$13x = 4$$

$$\therefore \qquad x = \frac{4}{13}$$

and $y = \frac{3}{2}$ then $\sqrt{\frac{x}{1-x}} = \frac{3}{2}$, squaring the both sides

$$\frac{x}{1-x} = \frac{9}{4}$$

$$4x = 9 - 9x$$
$$13x = 9$$
$$x = \frac{9}{13}$$

Hence, the solution set are $\left\{\frac{9}{13}, \frac{4}{13}\right\}$

***Example* 2:** Solve $5x - \frac{35}{x} = 18, \quad x \neq 0$

***Solution*:** The given equation can be written as

$$5x^2 - 35 = 18x$$
$$5x^2 - 18x - 35 = 0$$
$$5x^2 - 25x + 7x - 35 = 0$$
$$5x(x - 5) + 7(x - 5) = 0$$
$$(5x + 7)(x - 5) = 0$$
$$5x + 7 = 0, \quad x - 5 = 0$$
$$x = \frac{-7}{5} \text{ or } x = 5$$

Hence the solution set are $\left\{\frac{-7}{5}, 5\right\}$.

***Example* 3:** Solve $3x + \frac{3}{x} = 10$

***Solution*:** The given equation can be written as

$$3x^2 + 3 = 10x$$
$$3x^2 - 10x + 3 = 0$$
$$3x^2 - 9x - x + 3 = 0$$
$$3x(x - 3) - 1(x - 3) = 0$$
$$(x - 3)(3x - 1) = 0$$
$$x - 3 = 0 \quad \text{and} \quad 3x - 1 = 0$$
$$x = 3 \quad \text{and} \quad 3x = 1$$
$$\therefore \quad x = \frac{1}{3}$$

Hence the solution set are $\left\{\frac{1}{3}, 3\right\}$.

Type-III **Reciprocal Equations**

$$P\left(x^2+\frac{1}{x^2}\right) \pm q\left(x \pm \frac{1}{x}\right) + r = 0$$

Where p, q, r are constants.

Put $\left(x \pm \frac{1}{x}\right) = y$ then the given equation becomes $p\,(y^2 \pm 2) \pm qy + r = 0.$

***Example* 1:** Solve $3\left(x^2+\frac{1}{x^2}\right) - 16\left(x+\frac{1}{x}\right) + 26 = 0, \qquad x \neq 0$

***Solution*:** Put $x+\frac{1}{x} = y \Rightarrow \left(x+\frac{1}{x}\right)^2 = y^2$

$$x^2 + \frac{1}{x^2} + 2.\,x.\,\frac{1}{x} = y^2$$

$$x^2 + \frac{1}{x^2} + 2 = y^2$$

$$x^2 + \frac{1}{x^2} = y^2 - 2$$

Putting these value in given equation

$$3(y^2 - 2) - 16\,(y) + 26 = 0$$

$$3y^2 - 6 - 16y + 26 = 0$$

$$3y^2 - 16y + 20 = 0$$

$$3y^2 - 10y - 6y + 20 = 0$$

$$y\,(3y - 10) - 2\,(3y - 10) = 0$$

$$(3y - 10)\,(y - 2) = 0$$

$$3y - 10 = 0 \text{ and } y - 2 = 0$$

$$3y = 10$$

$$y = \frac{10}{3}, \qquad y = 2$$

when $y = \frac{10}{3}$ | when $y = 2$

then $x + \frac{1}{x} = \frac{10}{3}$ | then $x + \frac{1}{x} = 2$

$\frac{x^2+1}{x} = \frac{10}{3}$ | $\frac{x^2+1}{x} = 2$

by cross multiplication | $x^2 - 2x + 1 = 0$

$3x^2 - 10x + 3 = 0$ | $(x-1)^2 = 0$

$3x^2 - 9x - x + 3 = 0$ | $x - 1 = 0, (x-1) = 0$

$3x(x-3) - 1(x-3) = 0$ | $x = 1, 1$

$(3x-1)(x-3) = 0$

$3x - 1 = 0, \; x - 3 = 0$

$x = \frac{1}{3}$ and $x = 3$

Hence $x = \left\{3, \frac{1}{3}, 1, 1\right\}$.

***Example* 2:** Solve $2\left(x^2 + \frac{1}{x^2}\right) - 9\left(x + \frac{1}{x}\right) + 14 = 0, \quad x \neq 0$

***Solution*:** the given equation can be written as

$$2\left\{\left(x + \frac{1}{x}\right)^2 - 2\right\} - 9\left(x + \frac{1}{x}\right) + 14 = 0$$

Put $x + \frac{1}{x} = y$, then given equation becomes

$2(y^2 - 2) - 9y + 14 = 0$

$2y^2 - 9y + 10 = 0$

$2y^2 - 5y - 4y + 10 = 0$

$y(2y-5) - 2(2y-5) = 0$

$(2y-5)(y-2) = 0$

$2y - 5 = 0$ or $y - 2 = 0$

$y = \frac{5}{2}, \quad y = 2$

when $y = \frac{5}{2}$

then $x + \frac{1}{x} = \frac{5}{2}$

$\frac{x^2+1}{x} = \frac{5}{2}$

$2x^2 + 2 = 5x$

$2x^2 - 5x + 2 = 0$

$2x^2 - 4x - x + 2 = 0$

$2x(x-2) - 1(x-2) = 0$

$(x-2)(2x-1) = 0$

$x = 2,\ x = \frac{1}{2}$

when $y = 2$

then $x + \frac{1}{x} = 2$

$\frac{x^2+1}{x} = 2$

$x^2 - 2x + 1 = 0$

$x^2 - x - x + 1 = 0$

$x(x-1) - 1(x-1) = 0$

$(x-1)(x-1) = 0$

$x = 1, 1$

Hence $x = \left\{1, 1, 2, \frac{1}{2}\right\}$

Problem 1.3

Solve

1. $10\left(x^2 + \frac{1}{x^2}\right) - 63\left(x - \frac{1}{x}\right) + 52 = 0$

Ans: $\left(2, \frac{-1}{2}, 5, \frac{-1}{5}\right)$

2. $\left(x^2 + \frac{1}{x^2}\right) + \left(x + \frac{1}{x}\right) - 4 = 0$

Ans: $\left(1, 1, \frac{-3+\sqrt{5}}{2}, \frac{-3-\sqrt{5}}{2}\right)$

3. $8\left(x^2 + \frac{1}{x^2}\right) - 42\left(x - \frac{1}{x}\right) + 29 = 0$

Ans: $\left\{\frac{-1}{2}, \frac{-1}{4}, 2, 4\right\}$

Type-IV Equation of the Form $Pm^{2x} + qm^x + r = 0$

Put $m^x = y$, then the given equation becomes

$$py^2 + qy + r = 0$$

or $$ax^2 + bx + c = 0 \text{ which is quadratic}$$

Type-V Equation of the Form $(x + a)(x + b)(x + c)(x + d) + K = 0$

where a, b, c, d, k are real constants.

Type-VI Irrational equations reducible to quadratics as

$$\sqrt{(ax + b)} = k \text{ or } \sqrt{(ax + b)} + \sqrt{(cx + d)} = k$$

or $$\sqrt{(ax + b)} + \sqrt{(cx + d)} = \sqrt{(ex + f)}$$

Type-VII Simultaneous equations in two variables

***Problem* 1:** Solve for x

$$9^{x+2} - 6.\ 3^{x+1} + 1 = 0$$

***Solution*:** The given equation can be written as

$$(3^2)^{x+2} - 6.\ 3^x.\ 3^1 + 1 = 0$$
$$3^{2x+4} - 18.\ 3^x + 1 = 0$$
$$3^{2x}.\ 3^4 - 18.\ 3^x + 1 = 0$$
$$81.\ 3^{2x} - 18.\ 3^x + 1 = 0$$

Let $3^x = y$ and squaring the both sides

$$3^{2x} = y^2$$
$$81y^2 - 18y + 1 = 0$$
$$\Rightarrow \quad 9y(9y - 1) - 1(9y - 1) = 0$$
$$(9y - 1)(9y - 1) = 0$$
$$9y - 1 = 0 \text{ or } 9y - 1 = 0$$
$$y = \frac{1}{9} \text{ or } y = \frac{1}{9}$$
$$3^x = \frac{1}{9} \text{ or } 3^x = \frac{1}{9}$$
$$3^x = 3^{-2} \text{ or } 3^x = 3^{-2} \quad \text{Base same then power equal}$$
$$x = -2 \text{ or } x = -2$$

Hence $x = -2$

***Problem* 2:** Solve $3^{x+2} + 3^{-x} = 10$

***Solution*:** The given equation can be written as

$$3^x . 3^2 + 3^{-x} = 10 \quad \text{by exponent rule} \quad .....(1)$$

Let $3^x = y$, put in equation (1)

$$y. 3^2 + y^{-1} = 10$$

or $$9y + \frac{1}{y} = 10$$

$$9y^2 - 10y + 1 = 0$$

$$9y^2 - 9y - y + 1 = 0$$

$$9y (y - 1) -1 (y - 1) = 0$$

$$(9y - 1) (y - 1) = 0$$

$$9y - 1 = 0 \text{ and } y - 1 = 0$$

$$y = \frac{1}{9} \text{ and } y = 1$$

$$3^x = \frac{1}{9} \text{ and } 3^x = 1$$

$$3^x = 3^{-2} \quad 3^x = 1^o$$

$$x = -2 \text{ and } x = 0$$

Hence $x = 0, -2$

***Problem* 3:** Solve for x

$$(x + 1) (x + 2) (x + 3) (x + 4) + 1 = 0 \quad \text{...Type - V}$$

***Solution*:**

$$[(x + 1) (x + 2)] [(x + 3) (x + 4)] + 1 = 0$$

$$[x^2 + 2x + x + 2] [x^2 + 4x + 3x + 12] + 1 = 0$$

$$(x^2 + 3x + 12) (x^2 + 7x + 12) + 1 = 0$$

Which is not similar, so the given equation can be written as

$$[(x + 1) (x + 4)] [(x + 2) (x + 3)] + 1 = 0$$

$$(x^2 + 5x + 4) (x^2 + 5x + 6) + 1 = 0 \quad .....(1)$$

Let $x^2 + 5x = y$, put in (1)

$$(y +4) (y + 6) + 1 = 0$$

$$y^2 + 4y + 6y + 24 + 1 = 0$$

$$y^2 + 10y + 25 = 0$$

$y^2 + 5y + 5y + 25 = 0$

$y(y + 5) + 5(y + 5) = 0$

$(y + 5)(y + 5) = 0$

$(y + 5)^2 = 0$

$\therefore \quad y = -5 \qquad \because x^2 + 5x = y$

Now $\quad x^2 + 5x = -5$

$x^2 + 5x + 5 = 0$

$$x = \frac{-5 \pm \sqrt{25 - 4(1)(5)}}{2(1)} = \frac{-5 \pm \sqrt{5}}{2}$$

Problem 4: Solve $\sqrt{3x^2 - 4x - 6} + 3x^2 - 4x = 18$ .....(1)

Solution: Suppose $\sqrt{3x^2 - 4x - 6} = y$ so that

$3x^2 - 4x - 6 = y^2$ and so

$3x^2 - 4x = y^2 + 6$, putting these values in (1)

$y + y^2 + 6 = 18$

$y^2 + y - 12 = 0$

$y^2 + 4y - 3y - 12 = 0$

$y(y + 4) - 3(y + 4) = 0$

$(y - 3)(y + 4) = 0$

$y - 3 = 0$ and $y + 4 = 0$

$y = 3$ and $y = -4$

$\sqrt{3x^2 - 4x - 6} = 3$ and $\sqrt{3x^2 - 4x - 6} = -4$

squaring the both sides

$3x^2 - 4x - 6 = 9$ and $3x^2 - 4x - 6 = 16$

$3x^2 - 4x - 15 = 0$ and $3x^2 - 4x - 22 = 0$

$3x^2 - 9x + 5x - 15 = 0$ and $x = \dfrac{4 \pm \sqrt{(-4)^2 - 4(3)(-22)}}{2(3)}$

$3x(x - 3) + 5(x - 3) = 0$ and $x = \dfrac{4 \pm \sqrt{16 + 264}}{6}$

$(x - 3)(3x + 5) = 0$ and $x = \dfrac{4 \pm \sqrt{280}}{6}$

$x = 3$ and $x = \dfrac{-5}{3}$ $\qquad x = \dfrac{4 \pm \sqrt{4 \times 70}}{2 \times 3}$

$= \dfrac{2 \pm \sqrt{70}}{3}$

$\therefore$ The solution set is $\left\{3, \dfrac{-5}{3}, \dfrac{2+\sqrt{70}}{3}, \dfrac{2-\sqrt{70}}{3}\right\}$.

***Problem* 5:** Solve $\sqrt{(x^2 + 5x + 1)} - 1 = 2x$ ...Type-VI

***Solution*:** We have $\sqrt{(x^2 + 5x + 1)} = 2x + 1$, squaring both the sides.

$x^2 + 5x + 1 = 4x^2 + 1 + 4x$

$-3x^2 + x = 0$

$3x^2 - x = 0$

$x(3x - 1) = 0$

$x = 0, 3x - 1 = 0$

$x = 0$ or $x = \dfrac{1}{3}$

Hence the solution set $\left\{0, \dfrac{1}{3}\right\}$

***Problem* 6:** $\sqrt{4 - x} + \sqrt{9 + x} = 5$

***Solution*:** The given equation can be written as

$\sqrt{9 + x} = 5 - \sqrt{4 - x}$, squaring the both sides

$9 + x = 25 + 4 - x - 2(5)\sqrt{4 - x}$

$2x - 20 = -10\sqrt{4 - x}$

$2(x - 10) = -10\sqrt{4 - x}$

$x - 10 = -5\sqrt{4 - x}$, squaring the both sides

$x^2 + 100 - 20x = 25(4 - x)$

$$x^2 + 100 - 20x = 100 - 25x$$
$$x^2 + 5x = 0$$
$$x(x + 5) = 0$$
$$x = 0, x = -5$$

Hence the solution set $\{0, -5\}$

Problem 7: $x(2x + 1)(x - 2)(2x - 3) = 63$

Solution: The given equation can be written as

$$(x + 0)\left(x + \frac{1}{2}\right)(x - 2)\left(x - \frac{3}{2}\right) = \frac{63}{4}$$

$$0 + \left(\frac{-3}{2}\right) = \frac{1}{2} + (-2)$$

Therefore taking together first and fourth and second and third factors, we get

$$\left\{x\left(x - \frac{3}{2}\right)\right\}\left\{\left(x + \frac{1}{2}\right)(x - 2)\right\} = \frac{63}{4}$$

$$\left(x^2 - \frac{3}{2}x\right)\left(x^2 - \frac{3}{2}x - 1\right) = \frac{63}{4} \qquad \text{.....(A)}$$

Putting $x^2 - \frac{3}{2}x = y$, put in (A)

$$y(y - 1) = \frac{63}{4}$$
$$4y^2 - 4y - 63 = 0$$
$$4y^2 - 18y + 14y - 63 = 0$$
$$2y(2y - 9) + 7(2y - 9) = 0$$
$$(2y - 9)(2y + 7) = 0$$
$$y = \frac{9}{2} \text{ or } y = \frac{-7}{2}$$

Now, when $y = \frac{9}{2}$

Then $x^2 - \frac{3x}{2} = \frac{9}{2}$

$$2x^2 - 3x - 9 = 0$$

$$2x^2 - 6x + 3x - 9 = 0$$

$$2x(x-3) + 3(x-3) = 0$$

$$(x-3)(2x+3) = 0$$

$$x = 3 \text{ or } x = \frac{-3}{2}$$

Again, When

$$y = \frac{-7}{2}$$

Then $\quad x^2 - \frac{3}{2}x = \frac{-7}{2}$

$$2x^2 - 3x + 7 = 0 \qquad \text{.....(B)}$$

Its discriminant = $b^2 - 4ac$

$$= (-3)^2 - 4 \times 2 \times 7$$

$$= -47 < 0$$

$\therefore$ Equation (B) has no real roots.

Hence real roots the given equation are 3, $\frac{-3}{2}$

$\therefore$ solution set $\left\{3, \frac{-3}{2}\right\}$

Problem 6: Solve $\sqrt{5x^2 - 6x + 8} - \sqrt{5x^2 - 6x - 7} = 1$ ...Type VI

Solution: Let $A = \sqrt{5x^2 - 6x + 8}$, $B = \sqrt{5x^2 - 6x - 7}$

then $\quad A - B = 1 \qquad$ .... (ii)

$$A^2 - B^2 = 15 \qquad \text{..... (iii)}$$

Dividing (iii) by (ii)

$$A + B = 15$$

then A = 8

$$\sqrt{5x^2 - 6x + 8} = 8$$

Squaring the both sides

$$5x^2 - 6x + 8 = 64$$

$$5x^2 - 6x - 56 = 0$$

$$5x^2 - 20x + 14x - 56 = 0$$

$$5x(x - 4) + 14(x - 4) = 0$$

$$(x - 4)(5x + 14) = 0$$

$$x = 4 \quad \text{or} \quad x = \frac{-14}{5}$$

$$\therefore \text{Solution set} = \left(4, \frac{-14}{5}\right)$$

Problem 1.4

1. $\frac{x+2}{x-2} - \frac{x-2}{x+2} = \frac{5}{6}$

   Ans: $\left(\frac{-2}{5}, 10\right)$

2. $2x^4 + 9x^3 + 8x^2 + 9x + 2 = 0$

   Ans: $\left(-2+\sqrt{3}, -2-\sqrt{3}\right)$

3. $3x^4 - 5x^2 - 2 = 0$

   Ans: $\left(-\sqrt{2}, \sqrt{2}\right)$ imaginary

4. $7^{X+1} + 7^{1-X} = 7^2 + 1$

   Ans: (–1, 1)

5. $4\left(x^2 + \frac{1}{x^2}\right) - 16\left(x - \frac{1}{x}\right) + 7 = 0$

   Ans: $\left(2, -\frac{1}{2}, \frac{5+\sqrt{41}}{4}, \frac{5-\sqrt{41}}{4}\right)$

6. $6\left(x^2 + \frac{1}{x^2}\right) - 35\left(x + \frac{1}{x}\right) + 62 = 0, x \neq 0$

   Ans: $\left(\frac{1}{3}, \frac{1}{2}, 2, 3\right)$

7. $(x + 1)(x + 2)(x - 5)(x - 6) = 144$

Ans: (2, 7, –3)

8. $x(x + 1)(x + 2)(x + 3) = \frac{9}{16}$

Ans: $\left(-\frac{3}{2}, \frac{-3+\sqrt{10}}{2}, \frac{-3-\sqrt{10}}{2}\right)$

9. $x^2 + 6x + 2\sqrt{x^2 + 6x} = 24$

Ans: (2, –8, $-3 + 3\sqrt{5}$, $-3 - 3\sqrt{5}$)

10. $\sqrt{3x^2 + 7x - 1} + \sqrt{3x^2 + 7x - 10} = 9$

Ans: $\left(2, -\frac{13}{3}\right)$

11. $\sqrt{2y + 8} = 20 - y$

Ans: (14)

12. $\sqrt{4x^2 + x - 2} = 1\ 2x - 1$

Ans: $\left(\frac{3}{5}\right)$

13. $(2x - 1) = \sqrt{3x^2 - 2}$

Ans: (1, 3)

14. $\sqrt{3x + 10} + \sqrt{6 - x} = 6$

Ans: (2, 5)

15. $\sqrt{x + 5} + \sqrt{x + 21} = \sqrt{6x + 40}$

Ans: (4)

Solution of Word Problems involving Quadratic Equations

We consider such word problems in the examples given below.

***Example* 1:** The product of Ramu's age (in years) five years ago with age 9 years later is 15. Find Ramu's present age.

***Solution*:** Let Ramu's present age = x

His age 5 years ago = x – 5

His age 9 years later = x + 9

According to the problem

Therefore (x – 5) (x + 9) = 15

$\therefore \quad x^2 + 4x - 45 = 15$

i.e., $\quad x^2 + 4x - 60 = 0$

Solving $\quad x^2 + 10x - 6x - 60 = 0$

$$x(x + 10) - 6(x + 10) = 0$$

$$(x - 6)(x + 10) = 0$$

$$x = 6 \quad \text{or} \quad x = -10$$

Since x is the present age of Ramu, it cannot be negative.

Therefore, we reject the solution x = –10.

Thus Ramu's present age is 6 years

(check – Ramu's age 5 years ago = 6 – 5 = 1 year

Ramu's age 9 years later = 15 years

Their product = 15, as required in the problem

***Example* 2:** Find two natural numbers which differ by 3 and whose squares have the sum 117.

***Solution*:** Let the larger of the two numbers be x

Then the other number = x – 3

Sum of their squares = 117

$\therefore \quad x^2 + (x - 3)^2 = 117$

$$2x^2 - 6x - 108 = 0$$

Dividing both sides by 2,

$$x^2 - 3x - 54 = 0$$

$$x^2 - 9x + 6x - 54 = 0$$

$$x(x - 9) + 6(x - 9) = 0$$

$$(x - 9)(x + 6) = 0$$

The solution of this quadratic are x = 9 or x = –6

Since x must be natural number, it cannot be negative and so we reject the solution $x = -6$

Therefore $x = 9$

Therefore, the larger number = 9

The other number = $x - 3 = 6$

Therefore, the numbers required are 6 and 9.

Check: Sum of their squares = 36 + 81 = 117 as required and they differ by 3.

***Example* 3:** The product of two successive multiples of 5 is 300. Determine the multipliers.

Solution: Let the two required numbers be 5x and 5(x + 1)

According to the problem

$$5x \times 5(x + 1) = 300$$

$$x(x + 1) = 12$$

$$x^2 + x - 12 = 0$$

$$(x - 3)(x + 4) = 0$$

$$x - 3 = 0 \quad \therefore \quad x = 3$$

$$x + 4 = 0 \quad \therefore \quad x = -4 \text{ which is rejected as } -4$$

Thus, the numbers are 5×3, $5(3 + 1)$

i.e., 15, 20.

***Example* 4:** Divide 16 into two parts such that twice the square of the larger part exceeds the square of the smaller part by 164.

Solution: Let the larger part be x

So the smaller part is $16 - x$ $\quad 0 < x < 16$

According to the problem

$$2(x)^2 = (16 - x)^2 + 164$$

$$2x^2 = 256 - 32x + x^2 + 164$$

$$x^2 + 32x - 420 = 0$$

$$x^2 + 42x - 10x - 420 = 0$$

$$(x + 42)(x - 10) = 0$$

$$\therefore \quad x = -42 \quad \text{or} \quad x = 10$$

But $x = -42$ does not satisfy $0 < x < 16$ so it is rejected.

$\therefore$ $x = 10$, which is the largest part

$\therefore$ smaller part $= 16 - 10 = 6$

***Example* 5:** A fast train 3 hours less than a slow train for a journey of 600 km. If the speed of the slow train is 10 km/hr less than that of the fast train, find the speed of two trains.

Solution: Let the speed of fast train = x km/hr

Then the speed of slow train = (x – 10) km/hr
Time taken for journey of 600 km by fast train

$$= \frac{600}{x} \text{ hours} \qquad \left(\because \text{Time} = \frac{\text{dis tan ce}}{\text{speed}}\right)$$

And time taken for journey of 600 km by slow train

$$= \frac{600}{x-10} \text{ hours}$$

Since the fast train takes 3 hours less than the slow train

$$\frac{600}{x} = \frac{600}{x-10} - 3$$

$$600(x - 10) = 600x - 3x(x - 10) \qquad \text{(Multiplying by } x(x - 3)$$

$$600x - 6000 = 600x - 3x^2 + 30x$$

$$3x^2 - 30x - 6000 = 0$$

$$x^2 - 10x - 2000 = 0$$

$$(x - 50)(x + 40) = 0$$

$$\therefore \qquad x = 50 \text{ or } \quad x = -40$$

Rejecting $x = -40$, as the speed of train cannot be negative.

$\therefore$ speed of fast train $= x = 50$ km/hr

Speed of slow train $= x - 10 = 40$ km/hr

***Example* 6:** A motor-boat whose speed is 15 km/hr in still water goes 30 km downstream and comes back in a total of 4 hours 30 minutes. Determine the speed of water.

Solution: Let x km/hr be the speed of water.

The speed of motor boat in still water is 15 km/hr

Therefore, its speed downstream is (15 + x) km/hr
and the speed upstream is (15 – x) km/hr

Time taken for going 30 km downstream = $\frac{30}{15+x}$ hours

Time taken for going 30 km upstream = $\frac{30}{15-x}$ hours

Since the total time is given to be 4 hours 30 minutes i.e., $\frac{9}{2}$ hours

$$\frac{30}{15+x}+\frac{30}{15-x}=\frac{9}{2} \text{ hours}$$

$$\frac{30(15-x)+30(15+x)}{(15+x)(15-x)}=\frac{9}{2}$$

$$\frac{900}{225-x^2}=\frac{9}{2}$$

$$1800 = 9\,(225 - x^2)$$

$$200 = 225 - x^2$$

$$x^2 = 25$$

$$x = 5 \text{ or } x = -5 \qquad [\because x \neq -5 \text{ as } x > 0]$$

$$\therefore \qquad x = 5$$

Hence the speed of water = 5 km/hr.

***Example* 7:** The hypotenuse of a right triangle is 1m less than twice the shortest side. If the third side is 1 m more than the shortest side, find the sides of the triangle.

***Solution*:** Let x m be the shortest side of the right triangle

then hypotenuse = (2x – 1)m

and third side = (x + 1)m.

using pythagorus theorem

$$(2x-1)^2 = x^2 + (x+1)^2$$

$$4x^2 - 4x + 1 = x^2 + x^2 + 2x + 1$$

$$2x^2 - 6x = 0$$

$$2x\,(x-3) = 0$$

$x = 0$ or $x = 3$ but $x = 0$ m is in inadmissible as x is the length of a side. Thus $x = 3$ m is the length of the shortest side of the triangle.

$\therefore$ Length of hypotenuse $= 2x - 1$

$= 2 \times 3 - 1 = 5m$

Length of third side $= x + 1 = 4m$

***Example* 8:** The sides (in cm) of a right triangle containing the right angle are 5x and $3x - 1$. If the area of the triangle is 60 cm^2, find the sides of the triangle.

Solution: Given that area of right triangle ABC = 60 cm^2

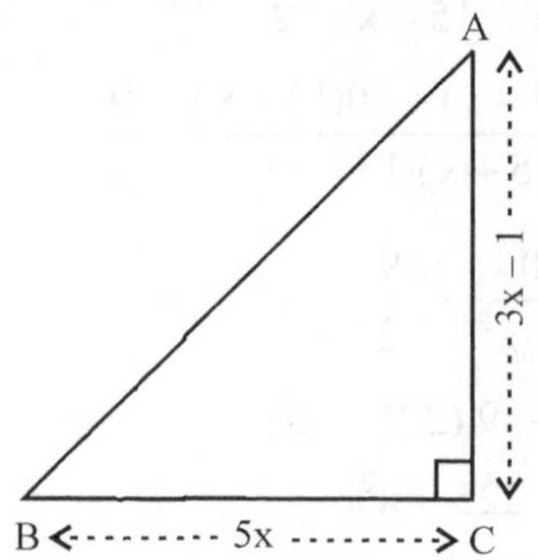

$$\frac{1}{2} \text{ Base} \times \text{length of perpendicular} = 60$$

$$\frac{1}{2} BC \times AC = 60$$

$$\frac{1}{2} 5x \times (3x - 1) = 60$$

$$15x^2 - 5x = 120$$

$$3x^2 - x - 24 = 0$$

$$3x^2 - 9x + 8x - 24 = 0$$

$$3x(x - 3) + 8(x - 3) = 0$$

$$(x - 3)(3x + 8) = 0$$

$$x = 3 \quad \text{or} \quad x = \frac{-8}{3}$$

Since side cannot be negative,

So taking $x = 3$

The sides of the triangle are

$5x = 5 \times 3 = 15$ cm and $3x - 1 = 3 \times 3 - 1 = 8$ cm

***Example* 9:** Find two consecutive positive odd integers whose squares have the sum 290.

Solution: Let the two consecutive positive odd integers be $2n + 1$ and $2n + 3$, where n is a whole no.

According to the problem

$$(2n + 1)^2 + (2n + 3)^2 = 290$$

$$(4n^2 + 4n + 1) + (4n^2 + 12n + 9) = 290$$

$$8n^2 + 16n - 280 = 0$$

$$n^2 + 2n - 35 = 0$$

$$n^2 + 7n - 5n - 35 = 0$$

$$n(n + 7) - 5(n + 7) = 0$$

$$(n + 7)(n - 5) = 0$$

$\therefore$ $n = -7$ or $n = 5$

Since n is a whole number

$n = -7$ is rejected

$\therefore$ $n = 5$

Hence, the two consecutive positive odd integers are $2 \times 5 + 1$ and $2 \times 5 + 3$

$\therefore$ i.e., 11 and 13.

***Example* 10:** The Product of two consecutive integers is 72. Find the integers.

Solution: Let one of the integers be = x

$\therefore$ The second integer = $x + 1$

Now according to the problem

$$x(x + 1) = 72$$

$$x^2 + x = 72$$

$$x^2 + x - 72 = 0$$

$$x^2 + 9x - 8x - 72 = 0$$

$$x(x + 9) - 8(x + 9) = 0$$

$(x + 9) = 0$ or $x - 8 = 0$

$x = -9$ or $x = 8$

Hence the required integers are 8 and 9 or –8 and –9.

Problem 1.5

1. The sum of the squares of there consecutive integers is 50. Find the integers.

   Ans: –5, –4, –3, or 3, 4, 5

2. A two digit number is such that product of the digits is 12. When 36 is added to this number the digits interchange their places. Determine the number.

   Ans: 26

3. The Perimeter of a rectangle is 82 meter and its area is 400 m^2. Find the breadth of the rectangle.

   Ans: 16 m

4. The sides in cm of a right triangle are (x – 1), x, x + 1. Find the sides of the triangle.

   Ans: 3 cm, 4 cm, 5 cm

5. An integer when added to its square, equals 90. Find all possible values of the integer.

   Ans: (9, –10)

Chapter 2

SIMULTANEOUS EQUATIONS

Definition

Let two equations

$$a_1x + b_1y + c_1 = 0$$

$$a_2x + b_2y + c_2 = 0$$

These equations containing two or more unknowns are said to be simultaneous equation. If these equations are satisfied by the same values of the unknowns. These equations are called linear simultaneous equations if the unknowns, say x and y;

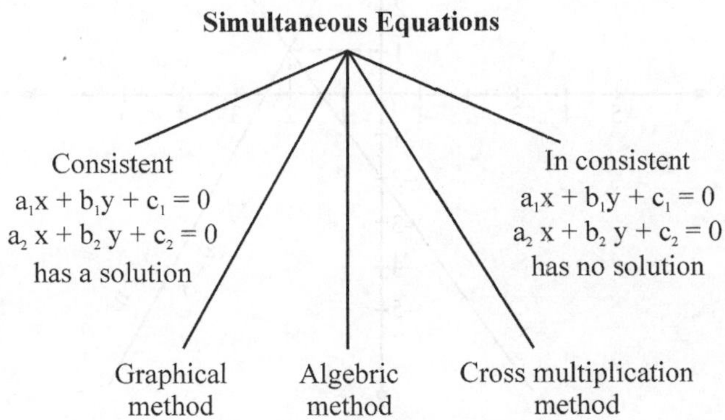

Simultaneous Equation of Graphical Method

There are infinitely many solutions of a linear equation in two variables, each pair of the value of x, y satisfies the given linear equation corresponds to the co-ordinates of a point on the line representing the graph of the equation.

Let us consider two linear equations

$$3x - 2y = 4 \qquad .....(1)$$

$$2x + y = 5 \qquad .....(2)$$

We know that the graphs of (1) and (2) are lines which we denote respectively by *l* and m. It is clear that the co-ordinates of the point of intersection i.e. x = 2, y = 1 of *l* and m will satisfy both (1) and (2) simultaneously. Since *l*, m are intersecting in a unique point with co-ordinates (2, 1); x = 2, y = 1 is the unique solution of the two equations.

It can be seen from the graph of equations (1) and (2) (Fig. 2.1), that the lines *l* and m are neither coincident nor parallel.

Draw the lines *l* and m representing the graphs of the two equations respectively. The following three possibilities arise.

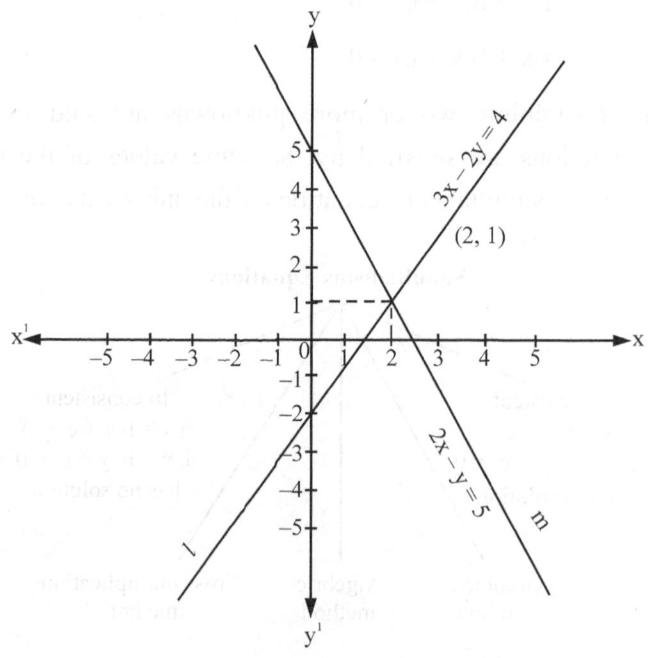

Fig. 2.1

(a) If l and m are coincident then any solution of any of the two equations will be common solution of the two and so there are infinitely many solution of the system.

(b) If l and m are parallel there are no common solutions and we say that the system has no solutions.

(c) If l and m intersect and the co-ordinates of the point of intersection are (h, k) we say that x = h, y = k is the only solution of the system.

Note that two simultaneous linear equations in two variables have (i) Infinitely many solutions, or (ii) no solutions, (iii) exactly one solution.

***Example* 1:** Solve graphically the system

$$2x - 3y = 1 \quad .....(1)$$

$$3x - 4y = 1 \quad .....(2)$$

***Solution*:** It is easily seen that the points (2, 1) and (1, 3) lie on the graph of equation (i) joining these points we get the line which is the graph of (1).

The graph of two lines are drawn in Fig. 2.2 they are seen to intersect in the points (–1, –1) so, that the solution is x = –1, y = –1.

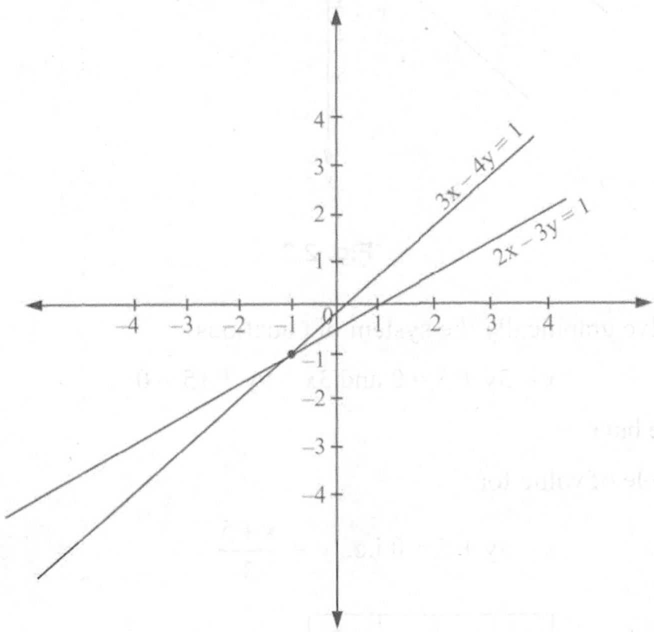

Fig. 2.2

***Example* 2:** Show graphically that the system of equations

$$3x - 4y = 1 \text{ and } -2x + \frac{8}{3}y = 5 \text{ has no solution}$$

***Solution*:** As seen in Fig. 2.3, the graphs of the two equations are parallel lines. As they have no common point, there is no common solution. Hence the system of equations has no solution.

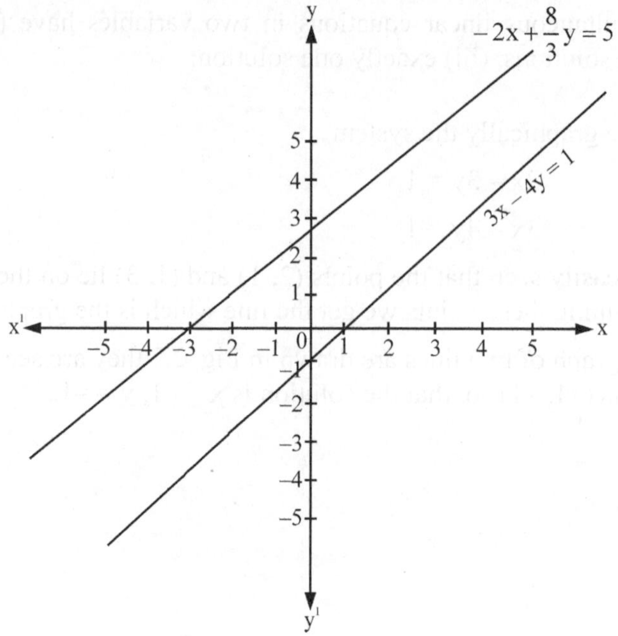

Fig. 2.3

***Example* 3:** Solve graphically the system of equations

$$x - 3y + 5 = 0 \text{ and } 3x - 9y + 15 = 0$$

***Solution*:** We have

Table of value for

$$x - 3y + 5 = 0 \text{ i.e., } y = \frac{x+5}{3}$$

x	1	4	–2
y	2	3	1

Table of values for

$$3x - 9y + 15 = 0 \text{ i.e., } y = \frac{3x+15}{9}$$

x	1	4	–2
y	2	3	1

Now we plot the points (1, 2), (4, 3) and (–2, 1) and then joining them we get the straight line l_1 as the graph of the equation $x - 3y + 5 = 0$. Again, we plot the points (–5, 0), (7, 4) and (–8, –1). We see these points lie on the straight line l_1. Consequently, the two lines coincide with each other. Hence the given system of equations is consistent and has infinite number of solutions.

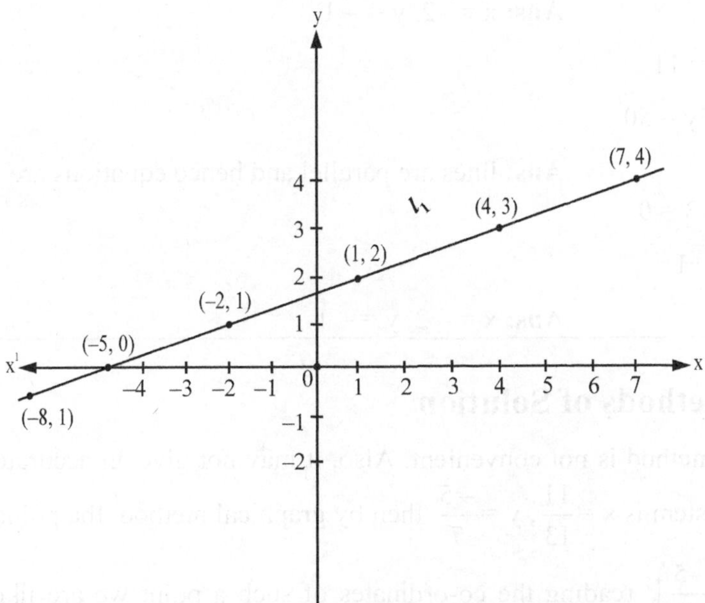

Fig. 2.4

Problem 2.1

Solve the following system of simultaneous linear equations graphically.

1. $4x + 6y = 9$

$5x + 7y = 10$

Ans: $x = \frac{-3}{2}, y = \frac{5}{2}$

2. $3x + 2y = 14$

$-x + 4y = 7$

Ans: $x = 3, y = \frac{5}{2}$

3. $3x - 5y = -1$

$2x - y = 3$

Ans: $x = -2, y = -1$

4. $2x + 7y = 11$

$10x + 35y = 50$

Ans: lines are parallel and hence equations are inconsistent.

5. $2x - y + 3 = 0$

$5y = 3x + 1$

Ans: $x = -2, y = -1$

Algebric Methods of Solution

The graphical method is not convenient. Also, it may not give an accurate answer. If the solution to a system is $x = \frac{11}{13}, y = \frac{-5}{7}$ then by graphical method, the point of intersection would be $\left(\frac{11}{13}, \frac{-5}{7}\right)$, reading the co-ordinates of such a point we are likely to make an error. This point is so close to the point (0.8, – 0.7) that x = 0.8 and y = –0.7 are likely to be mistaken to be the solution of the given system of simultaneous equations. Hence it is necessary to use a precise method to solve two simultaneous linear equations. The two algebric methods described below achieve the purpose.

First Method: Elimination by Substitution

We first find the value of one variable say y in terms of the other, i.e., x, form either equation. We get an equation in only one variable x, which can easily be solved. Substituting the value of x thus obtained, in the expression for y in terms of x found earlier, the value of y is obtained.

***Example* 1:** Solve $2x + 3y = 11$

$$2x - y = -1$$

***Solution*:** $2x - y = -1$

$$\therefore \quad y = 2x + 1$$

Putting value y in $2x + 3y = 11$

$$2x + 3(2x + 1) = 11$$

$$2x + 6x + 3 = 11$$

$$8x = 11 - 3$$

i.e., $8x = 8$ or $x = 1$

putting this value in $y = 2x + 1$

$$y = 2(1) + 1 = 3$$

So the solution is $x = 1$, $y = 3$.

Alternatively, we could have eliminated x, as follows. From $2x + 3y = 11$, we have $x = \frac{11-3y}{2}$, substituting this in $2x - y = -1$, we get

$$2\left(\frac{11-3y}{2}\right) - y = -1$$

$$11 - 3y - y = -1$$

$$4y = 1 + 11$$

$$y = 3$$

Putting this value of y in $x = \frac{11-3y}{2}$ we get $x = 1$.

So the solution is $x = 1$, $y = 3$

Second Method: Elimination by Equating the Coefficients

We multiply both equations by suitable non zero constants so that the coefficients of the variable to be eliminated become equal. We then add or subtract one equation from the other so that this variable with equal coefficient gets eliminated.

***Example* 1:** Solve

$$3x + 2y = 11 \qquad .....(1)$$

$$2x + 3y = 4 \qquad .....(2)$$

***Solution*:** To eliminate y, we must make the coefficients of y in (1) and (2) equal. The coefficients of y are 2 and 3 respectively. The least common multiple of 2 and 3 is 6. So it will be convenient to make both the coefficients equal to 6. Multiplying (1) by 3 and (2) by 2, we get.

$$9x + 6y = 33 \qquad .....(3)$$

$$4x + 6y = 8 \qquad .....(4)$$

If we subtract (4) from (3), we get

$$5x = 25$$

$\therefore$ $x = 5$ put this value in (1)

$$15 + 2y = 11$$

Which gives $y = -2$

Hence the solution is $x = 5$, $y = -2$

If we had eliminated x instead of y we would have multiplied (1) by 2 and (2) by 3

$$6x + 4y = 22$$

$$6x + 9y = 12$$

Subtraction would yield $-5y = +10$ or $y = -2$, put this value in (1) $3x - 4 = 11$ i.e., $x = 5$

***Example* 2:** Solve

$$8v - 3u = 5uv \qquad .....(1)$$

$$6v - 5u = -2uv \qquad .....(2)$$

***Solution*:** Here the equations are not linear in the variables u and v, but we can convert them into linear equations by an appropriate substitution. By setting u = 0 in either of the two equations, it is easily found that v = 0. Thus u =0 and v = 0 form one solution of (1) and (2).

To look for other solution (if any), we assume that $u \neq 0, v \neq 0$. Then dividing both equations by uv, we get

$$\frac{8}{u} - \frac{3}{v} = 5$$

$$\frac{6}{u} - \frac{5}{v} = -2$$

Let us set $x = \frac{1}{u}$, $y = \frac{1}{v}$ in these equations. Then the equations reduce to

$$8x - 3y = 5 \qquad .....(3)$$

$$6x - 5y = -2 \qquad .....(4)$$

The equations (3) and (4) are linear equations in x and y and can be solved using the elimination method. In order to eliminate 'y' multiply (3) by 5 and (4) by 3. This gives us

$$40x - 15y = 25 \qquad .....(5)$$

$$18x - 15y = -6 \qquad .....(6)$$

Subtracting (6) from (5), we get 22x = 31 so that $x = \frac{31}{22}$, put in equation (3), we get $\frac{124}{11} - 3y = 5$ i.e., $3y = \frac{124}{11} - 5 = \frac{69}{11}$

Therefore $y = \frac{23}{11}$

Thus $u = \frac{22}{31}$, and $v = \frac{11}{23}$

Hence, the given system of equations has the two solutions (i) u = 0, v = 0, (ii) $u = \frac{22}{31}$, $v = \frac{11}{23}$.

Problem 2.2

1. $x + y = 7$

$5x + 12y = 7$

Ans: $x = 11$ and $y = -4$

2. $5x - 3y = 1$

$2x + 5y = 19$

Ans: $x = 2$ and $y = 3$

3. $-6x + 5y = 2$

$-5x + 6y = 9$

Ans: $x = 3$ and $y = 4$

4. $7x + 11y = 1$

$8x + 13y = 2$

Ans: $x = -3$ and $y = 2$

5. $8x + 5y = 9$

$3x + 2y = 4$

Ans: $x = -2$ and $y = 5$

6. $2x + 3y = 31$

$17x - 11y = 8$

Ans: $x = 5$ and $y = 7$

Common Solution and Conditions for Solvability

In general equations

$$a_1x + b_1y = c_1 \quad \text{.....(i)}$$

$$a_2x + b_2y = c_2 \quad \text{.....(ii)}$$

of two simultaneous linear equations. If we eliminate y,

$$(a_1b_2 - a_2b_1)x = c_1b_2 - c_2b_1 \quad \text{.....(iii)}$$

Multiplying (i) by b_2 and (ii) by b_1, we get

$$a_1b_2\,x + b_1b_2\,y = c_1b_2$$

$$a_2b_1\,x + b_1b_2\,y = c_2b_1$$

By subtraction, we get (iii).

Similarly if we eliminate x, we get

$$(a_1b_2 - a_2b_1)y = a_1c_2 - a_2c_1 \qquad \text{.....(iv)}$$

Now two cases

(A) First suppose that $a_1b_2 - a_2b_1 \neq 0$. Then clearly from (iii) and (iv)

$$x = \frac{c_1b_2 - c_2b_1}{a_1b_2 - a_2b_1}, \qquad y = \frac{a_1c_2 - a_2c_1}{a_1b_2 - a_2b_1}$$

From the one and only one solution of the given equation. So we conclude that, if $a_1b_2 - a_2b_1 \neq 0$ i.e., $\frac{a_1}{a_2} \neq \frac{b_1}{b_2}$, then the system of the equation is consistent and has a unique solution.

The solution can be written as

$$\frac{x}{b_1c_2 - b_2c_1} = \frac{y}{c_1a_2 - c_2a_1} = \frac{-1}{a_1b_2 - a_2b_1} \qquad \text{.....(A)}$$

Equation (A) is called rule of cross multiplication. The following point may be written as

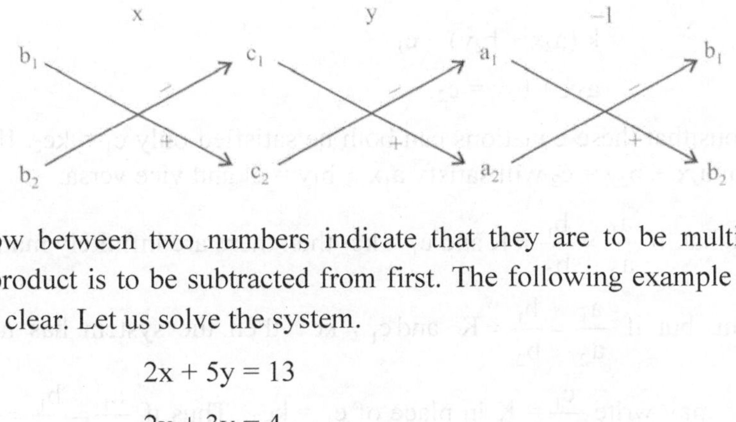

The arrow between two numbers indicate that they are to be multiplied and the second product is to be subtracted from first. The following example will make the meaning clear. Let us solve the system.

$$2x + 5y = 13$$

$$2x + 3y = 4$$

Given coefficients may be written as

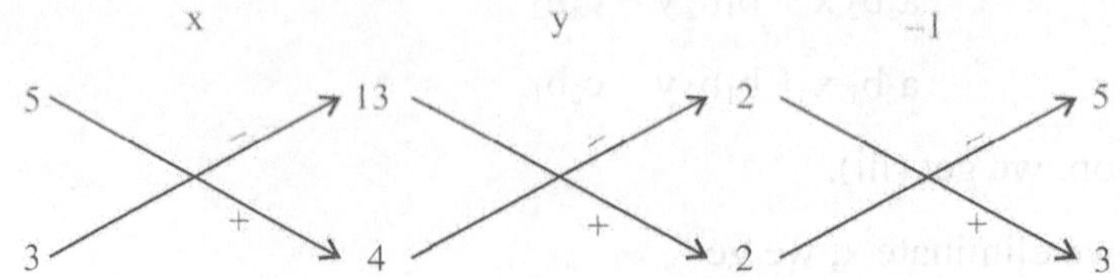

and

$$\frac{x}{5(4)-3(13)} = \frac{y}{13(2)-4(2)} = \frac{-1}{2(3)-2(5)}$$

This gives

$$\frac{x}{-19} = \frac{y}{18} = \frac{1}{+4}$$

$$\therefore \qquad x = \frac{-19}{4},\ y = \frac{+18}{4} = \frac{+9}{2}$$

(B) Next suppose that $a_1b_2 - a_2b_1 = 0$

$$\text{i.e., } \frac{a_1}{a_2} = \frac{b_1}{b_2}$$

If we write $\frac{a_1}{a_2} = \frac{b_1}{b_2} = K$ then

$$a_1 = Ka_2, \qquad b_1 = Kb_2$$

So the given equations are

$$k\,(a_2x + b_2y) = c_1$$

$$a_2x + b_2y = c_2$$

It is obvious that these equations can both be satisfied only $c_1 = kc_2$. If $c_1 = kc_2$, any solution of $a_2x + b_2y = c_2$ will satisfy $a_1x + b_1y = c_1$ and vice versa.

It follows that, if $\frac{a_1}{a_2} = \frac{b_1}{b_2} = K$ and $c_1 = kc_2$ then there are infinitely many solutions of the system, but if $\frac{a_1}{a_2} = \frac{b_1}{b_2} = K$ and $c_1 \neq kc_2$ then the system has no solution. If $c_2 \neq 0$ we may write $\frac{c_1}{c_2} = K$ in place of $c_1 = kc_2$. Thus if $\frac{a_1}{a_2} = \frac{b_1}{b_2} = \frac{c_1}{c_2}$ then there are infinitely many solutions and for $\frac{a_1}{a_2} = \frac{b_1}{b_2} \neq \frac{c_1}{c_2}$ there are no solution. When a system of equations has a solution, we say that the system is consistent and when it has no solution, we say that the system is inconsistent.

Conclusion: the system of simultaneous equations

$a_1x + b_1y + c_1 = 0$ and

$a_2x + b_2y + c_2 = 0$

1. Has unique solution (only one solution) if $\frac{a_1}{a_2} \neq \frac{b_1}{b_2}$ and the solution given by

$$\frac{x}{b_1c_2 - b_2c_1} = \frac{y}{c_1a_2 - c_2a_1} = \frac{+1}{a_1b_2 - a_2b_1}$$

2. Has infinitely many solution if $\frac{a_1}{a_2} = \frac{b_1}{b_2} = \frac{c_1}{c_2}$

3. Has no solution if $\frac{a_1}{a_2} = \frac{b_1}{b_2} \neq \frac{c_1}{c_2}$

***Example* 1:** Solve $2x + 3y = 17$, $x + y = 7$

***Solution*:** we have $2x + 3y = 17$ .....(i)

$x + y = 7$ .....(ii)

From (ii), $y = 7 - x$ this value put in (i)

$2x + 3(7-x) = 17 \Rightarrow 2x + 21 - 3x = 17$

$-x = -4 \Rightarrow x = 4$

Put the value $x = 4$ in (ii), we have

$4 + y = 7 \quad \therefore y = 3$

Hence required solution is (4, 3)

***Example* 2:** solve $8x + 3y = 17$ and $2x + 3y = 5$

***Solution*:** $8x + 3y = 17$ .....(i)

$2x + 3y = 5$ .....(ii)

Let equation (ii)

$2x + 3y = 5$ i.e., $y = \frac{5 - 2x}{3}$

Put y value in (i)

$$8x + \frac{3(5-2x)}{3} = 17$$

$$8x + 5 - 2x = 17$$

$$6x = 12 \quad \therefore x = 2$$

Put $x = 2$ in (ii)

$$2(2) + 3y = 5$$

$$3y = 5 - 4$$

$$y = \frac{1}{3}$$

Hence required solution is $\left(2, \frac{1}{3}\right)$

***Example* 3:** solve $x - y = 3$, $xy = 18$

***Solution*:**

$$x - y = 3 \quad \text{.....(i)}$$

$$xy = 18 \quad \text{.....(ii)}$$

$$x = y + 3 \quad \text{.....(iii)}$$

put x value in (ii)

$$(y + 3)y = 18$$

$$y^2 + 3y = 18$$

$$y^2 + 3y - 18 = 0$$

$$y^2 + 6y - 3y - 18 = 0$$

$$y(y + 6) - 3(y + 6) = 0$$

$$(y + 6)(y - 3) = 0$$

$$y = 3, y = -6$$

when $y = 3$ then from (iii) $x = y + 3$ i.e., 6 and

when $y = -6$ then from (iii) $x = y + 3$ i.e., $x = -6 + 3 \therefore x = -3$

Hence the required results $x = -3, y = -6$ and

$$x = 6, y = 3$$

***Example* 4:** A fraction reduces to 1/4 when 2 is subtracted from the numerator and 3 is added to the denominator. But it reduces to 2/3 if 6 is added to the numerator and the denominator is multiplied by 3. Find the fraction.

***Solution*:** Let the fraction be x/y. Then

$$\frac{x-2}{y+3} = \frac{1}{4} \text{ and } \frac{x+6}{3y} = \frac{2}{3}$$

Thus

$$4(x-2) = y+3 \text{ and } x+6 = 2y$$

i.e.,

$$4x - y = 11 \text{ and } x - 2y = -6$$

giving x = 4, y = 5 so that the fraction is 4/5.

***Example* 5:** The sum of two digit number and the number obtained by reversing the order of its digits is 121 and the two digits differ by 3. Find the number.

***Solution*:** Suppose x is the digit in the ten's place and y is the digit in the unit's place of the given number. Then the given number is 10x + y. If we reverse the order of the digits, the new number becomes 10y + x.

Hence $(10x + y) + (10y + x) = 121$

also $x - y = \pm 3$

that is $11(x + y) = 121$

and $x - y = \pm 3$

so we have either

$$x + y = 11 \qquad\qquad x + y = 11$$

$$x - y = 3 \qquad \text{or} \qquad -x + y = 3$$

Solving we get x = 7, y = 4 or x = 4, y = 7

Hence the number is 74 or 47

***Example* 6:** In a ΔABC, $\angle A = x^o$, $\angle B = 3x^o$ and $\angle C = y^o$. If $3y - 5x = 30$ prove that the triangle is right angled.

***Solution*:** Since $\angle A + \angle B + \angle C = 180^o$ we have $x + 3x + y = 180$ or $4x + y = 180$

Thus the system of equations is

$$-5x + 3y = 30$$

$$4x + y = 180$$

Solving we get x = 30, y = 60. Hence $\angle B = 3x^o = 90^o$

Thus ΔABC is right angled.

Problem 2.3

1. Seven times a two digits number is equal to four times the number obtained by reversing the order of digits, and the sum of digits of number is 3. Find the number.

Ans: 12

2. Five years ago I was thrice as old as my son and ten years later I shall be twice as old as my son. How old are we now?

Ans: My age = 50 years, Son's age = 20 years

3. In a ΔABC, $\angle C = 3$, $\angle B = 2(\angle A + \angle B)$. Find the three angles.

Ans: $\angle A = 20^{o}$, $\angle B = 40^{o}$, $\angle C = 120^{o}$

4. I am three times as old as my son. Five years later I shall be two and a half times as old as my son. How old am I and how old is my son.

Ans: My age = 45 years, Son's age = 15 years

CHAPTER 3

DETERMINANTS

Determinants

Consider equations $a_1x + b_1y = 0$ .....(i)

and $a_2x + b_2y = 0$ .....(ii)

From equation (i) $\frac{x}{y} = \frac{-b_1}{a_1}$

(ii) $\frac{x}{y} = \frac{-b_2}{a_2}$

$\therefore \quad \frac{b_1}{a_1} = \frac{b_2}{a_2}$

or $a_2b_1 - a_1b_2 = 0$

The expression $a_1b_2 - a_2b_1$ is a function a_1, b_1, a_2, b_2. This is usually denoted by $\begin{vmatrix} a_1 & b_1 \\ a_2 & b_2 \end{vmatrix}$ and is called determinant of second order.

a_1, a_2, b_1, b_2 are called constituents or elements of the determinant.

The determinant above has two horizontal rows and two vertical columns.

Determinant of order Three

Let us consider the following equations in three variables x, y and z

$a_1x + b_1y + c_1z = 0$ ..... (i)

$$a_2x + b_2y + c_2z = 0 \qquad .....(ii)$$

$$a_3x + b_3y + c_3z = 0 \qquad .....(iii)$$

by solving these three equations, we have

$$\begin{vmatrix} a_1 & b_1 & c_1 \\ a_2 & b_2 & c_2 \\ a_3 & b_3 & c_3 \end{vmatrix}$$

is determinant of order three. It is denoted by Δ.

Thus

$$\begin{vmatrix} a_1 & b_1 & c_1 \\ a_2 & b_2 & c_2 \\ a_3 & b_3 & c_3 \end{vmatrix}$$

$$= a_1\begin{vmatrix} b_2 & c_2 \\ b_3 & c_3 \end{vmatrix} - b_1\begin{vmatrix} a_2 & c_2 \\ a_3 & c_3 \end{vmatrix} + c_1\begin{vmatrix} a_2 & b_2 \\ a_3 & b_3 \end{vmatrix}$$

$$= a_1(b_2\, c_3 - b_3\, c_2) - b_1(a_2c_3 - c_2\, a_3) + c_1(a_2b_3 - a_3\, b_2)$$

Sarus Rule: To find the value of third order determinant only

$$\text{consider} \quad \begin{vmatrix} a_1 & b_1 & c_1 \\ a_2 & b_2 & c_2 \\ a_3 & b_3 & c_3 \end{vmatrix}$$

Write the three columns and repeat 1 and 2 columns

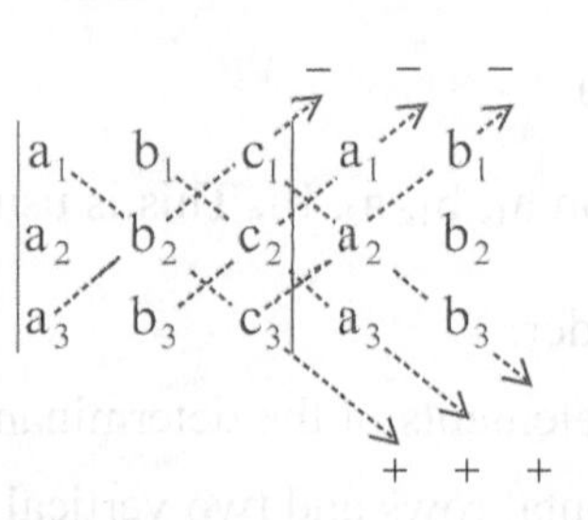

Add the products on the downward moving arrows and subtract the products on the upward moving arrows, we have

$$\Delta = a_1b_2\, c_3 + b_1\, c_2\, a_3 + c_1\, a_2\, b_3 - c_3\, a_2\, b_1 - b_3\, c_2\, a_1 - a_3\, b_2\, c_1$$

***Example* 1:** Evaluate $\begin{vmatrix} 3 & 2 \\ 5 & 7 \end{vmatrix}$

***Solution*:** $\begin{vmatrix} 3 & 2 \\ 5 & 7 \end{vmatrix} = 3 \times 7 - 5 \times 2 = 21 - 10 = 11$

***Example* 2:** Evaluate $\begin{vmatrix} \cos x & \sin x \\ -\sin x & \cos x \end{vmatrix}$

***Solution*:** $\begin{vmatrix} \cos x & \sin x \\ -\sin x & \cos x \end{vmatrix} = \cos x \,.\, \cos x - (-\sin x) \sin x$

$= \cos^2 x + \sin^2 x = 1$

***Example* 3:** Evaluate $\Delta = \begin{vmatrix} 3 & 4 & 5 \\ -6 & 2 & -3 \\ 8 & 1 & 7 \end{vmatrix}$

***Solution*:** $\Delta = 3\begin{vmatrix} 2 & -3 \\ 1 & 7 \end{vmatrix} - 4\begin{vmatrix} -6 & -3 \\ 8 & 7 \end{vmatrix} + 5\begin{vmatrix} -6 & 2 \\ 8 & 1 \end{vmatrix}$

$= 3\,(2 \times 7 - (-3) \times 1) - 4\,((-6) \times 7 - (-3) \times 8) + 5\,((-6) \times 1 - 2 \times 8)$

$= 3\,(14 + 3) - 4\,(-42 + 24) + 5\,(-6 - 16)$

$= 3 \times 17 - 4\,(-18) + 5(-22)$

$= 51 + 72 - 110 = 13$

***Example* 4:** Evaluate $\Delta = \begin{vmatrix} a & h & g \\ h & b & f \\ g & f & c \end{vmatrix}$ by using sarus diagram.

***Solution*:** By Sarus diagram

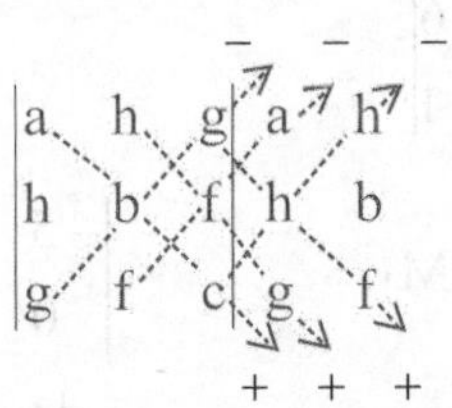

$\Delta = abc + hfg + ghf - ch^2 - af^2 - bg^2$

$= abc + 2hgf - af^2 - bg^2 - ch^2$

Minors and Cofactors

Minor

The determinant obtained by deleting the row and the column in which an element of a determinant lies is called the minor of that element. Let the determinant be

$$\Delta = \begin{vmatrix} a_{11} & a_{12} & a_{13} \\ a_{21} & a_{22} & a_{23} \\ a_{31} & a_{32} & a_{33} \end{vmatrix}$$

and minor of element a_{ij} be denoted by M_{ij}. Then

Minor of a_{11} i.e., $$M_{11} = \begin{vmatrix} a_{22} & a_{23} \\ a_{32} & a_{33} \end{vmatrix} = (a_{22}\,a_{33} - a_{32}\,a_{23})$$

$$M_{12} = \begin{vmatrix} a_{21} & a_{23} \\ a_{31} & a_{33} \end{vmatrix} = (a_{21}\,a_{33} - a_{31}\,a_{23})$$

Cofactor

The cofactor of an element of a determinant is its minor with proper sign. The cofactor C_{ij} of an element a_{ij} which lies in ith row and jth column of a determinant. Then $c_{ij} = (-1)^{i+j}$ M_{ij}

Hence, $$C_{11} = (-1)^{1+1}\,M_{11} = M_{11} = (a_{22}\,a_{33} - a_{32}\,a_{23})$$

$$C_{21} = (-1)^{2+1}\,M_{21} = -M_{21} = -(a_{12}\,a_{33} - a_{32}\,a_{13})$$

***Example* 5:** Find all cofactor of $\begin{vmatrix} 1 & 2 & 3 \\ 4 & -5 & 6 \\ 7 & 0 & -1 \end{vmatrix}$

***Solution*:** $$C_{11} = (-1)^{1+1}\,M_{11} = M_{11} = \begin{vmatrix} -5 & 6 \\ 0 & -1 \end{vmatrix} = 5$$

$$C_{12} = (-1)^{1+2}\,M_{12} = -M_{12} = -\begin{vmatrix} 4 & 6 \\ 7 & -1 \end{vmatrix} = -(-4 - 42) = 46$$

$$C_{13} = (-1)^{1+3}\, M_{13} = M_{13} \quad = \begin{vmatrix} 4 & -5 \\ 7 & 0 \end{vmatrix} = -(-35) = 35$$

$$C_{21} = (-1)^{2+1}\, M_{21} = -M_{21} \quad = -\begin{vmatrix} 2 & 3 \\ 0 & -1 \end{vmatrix} = -(-2) = 2$$

$$C_{22} = (-1)^{2+2}\, M_{22} = M_{22} \quad = \begin{vmatrix} 1 & 3 \\ 7 & -1 \end{vmatrix} = -1 - 21 = -22$$

$$C_{23} = (-1)^{2+3}\, M_{23} = -M_{23} \quad = -\begin{vmatrix} 1 & 2 \\ 7 & 0 \end{vmatrix} = -(-14) = 14$$

$$C_{31} = (-1)^{3+1}\, M_{31} = M_{31} \quad = \begin{vmatrix} 2 & 3 \\ -5 & 6 \end{vmatrix} = 12 + 15 = 27$$

$$C_{32} = (-1)^{3+2}\, M_{32} = -M_{32} \quad = -\begin{vmatrix} 1 & 3 \\ 4 & 6 \end{vmatrix} = -(6 - 12) = 6$$

$$C_{33} = (-1)^{3+3}\, M_{33} = M_{33} \quad = \begin{vmatrix} 1 & 2 \\ 4 & -5 \end{vmatrix} = -5 - 8 = -13$$

Properties of Determinant

***Property* 1:** The value of a determinant is not altered by changing its rows into columns and columns into rows.

***Example*:** Let $\Delta = \begin{vmatrix} a_1 & b_1 & c_1 \\ a_2 & b_2 & c_2 \\ a_3 & b_3 & c_3 \end{vmatrix}$ and $\Delta^1 = \begin{vmatrix} a_1 & a_2 & a_3 \\ b_1 & b_2 & b_3 \\ c_1 & c_2 & c_3 \end{vmatrix}$

Then $\Delta = \Delta^1$, Δ and Δ^1 are transpose of each other

***Property* 2:** If two adjacent rows or columns of a determinant are inter-changed, the determinant changes in sign but its numerical values are unaltered.

***Example*:** Let $\Delta^1 = \begin{vmatrix} a_1 & b_1 & c_1 \\ a_2 & b_2 & c_2 \\ a_3 & b_3 & c_3 \end{vmatrix}$ and $-\Delta = \begin{vmatrix} b_1 & a_1 & c_1 \\ b_2 & a_2 & c_2 \\ b_3 & a_3 & c_3 \end{vmatrix}$

Then $\Delta^1 = -\Delta$

***Property* 3:** If two rows or columns are identical then the value of determinant is zero.

***Example*:**

$$\Delta = \begin{vmatrix} a_1 & a_1 & c_1 \\ a_2 & a_2 & c_2 \\ a_3 & a_3 & c_3 \end{vmatrix}$$

Then $\Delta = 0$

***Property* 4:** If the elements in a row or column of a determinant are respectively equal to k times the corresponding elements in another row or column, the value of determinant is zero.

***Example*:**

$$\Delta = \begin{vmatrix} a_1 & b_1 & c_1 \\ a_2 & b_2 & c_2 \\ a_3 & b_3 & c_3 \end{vmatrix} \text{ and } \Delta^1 = \begin{vmatrix} ka_1 & kb_1 & kc_1 \\ a_2 & b_2 & c_2 \\ a_3 & b_3 & c_3 \end{vmatrix}$$

Then $\Delta^1 = k\,\Delta$

***Property* 5:** If each element in any row or column consists of two terms, then the determinant can be expressed as the sum of two determinants of the same order.

Example:

$$\Delta = \begin{vmatrix} a_1 + x & b_1 & c_1 \\ a_2 + y & b_2 & c_2 \\ a_3 + z & b_3 & c_3 \end{vmatrix} = \begin{vmatrix} a_1 & b_1 & c_1 \\ a_2 & b_2 & c_2 \\ a_3 & b_3 & c_3 \end{vmatrix} + \begin{vmatrix} x & b_1 & c_1 \\ y & b_2 & c_2 \\ z & b_3 & c_3 \end{vmatrix}$$

***Property* 6:** If to each element of a row or column of a determinant be added or subtracted, the equimultiples of the corresponding elements of one or more rows or columns, the determinant remains unaltered.

***Example*:**

$$\Delta = \begin{vmatrix} a_1 & b_1 & c_1 \\ a_2 & b_2 & c_2 \\ a_3 & b_3 & c_3 \end{vmatrix}$$

Then

$$\Delta^1 = \begin{vmatrix} a_1 + pb_1 & b_1 & c_1 \\ a_2 + pb_2 & b_2 & c_2 \\ a_3 + pb_3 & b_3 & c_3 \end{vmatrix}$$

$$= \begin{vmatrix} a_1 & b_1 & c_1 \\ a_2 & b_2 & c_2 \\ a_3 & b_3 & c_3 \end{vmatrix} + \begin{vmatrix} pb_1 & b_1 & c_1 \\ pb_2 & b_2 & c_2 \\ pb_3 & b_3 & c_3 \end{vmatrix} \quad \text{(by property 5)}$$

$$= \begin{vmatrix} a_1 & b_1 & c_1 \\ a_2 & b_2 & c_2 \\ a_3 & b_3 & c_3 \end{vmatrix} + p\begin{vmatrix} b_1 & b_1 & c_1 \\ b_2 & b_2 & c_2 \\ b_3 & b_3 & c_3 \end{vmatrix}$$

$$= \begin{vmatrix} a_1 & b_1 & c_1 \\ a_2 & b_2 & c_2 \\ a_3 & b_3 & c_3 \end{vmatrix} + p(0) \quad \text{(by property 3)}$$

$\therefore \quad \Delta^1 = \Delta$

***Property* 7:** *Factor Theorem:* If the element of determinant Δ involving x are polynomials in x and $\Delta = 0$, where a is put for x.

Then$(x - a)$ is a factor of Δ

***Example*:** Let $\Delta = \begin{vmatrix} 1 & 1 & 1 \\ x & b & c \\ x^2 & b^2 & c^2 \end{vmatrix}$

If we replace x by b then

$$\Delta = \begin{vmatrix} 1 & 1 & 1 \\ b & b & c \\ b^2 & b^2 & c^2 \end{vmatrix}$$

$\Delta = 0$ (by Identical Property)

$\therefore \quad (x - b)$ is a factor of Δ.

To Find the Value of a Determinant

The main purpose behind the simplification of a determinant lies in making maximum zeros in a row or column by using the above properties and then to expand the determinant by that row or column. We denote the I^{st}, 2nd, 3rd rows of a determinant by R_1, R_2, R_3 and the 1st, 2nd, 3rd columns by C_1, C_2, C_3 respectively. We express the addition of k times the elements of jth row with the corresponding elements of ith row by

$$R_i \to R_i + KR_j$$

***Example* 6:** Prove the following results $\begin{vmatrix} a+b & a & b \\ a & a+c & c \\ b & c & b+c \end{vmatrix} = 4abc$

Solution: $\begin{vmatrix} a+b & a & b \\ a & a+c & c \\ b & c & b+c \end{vmatrix}$

$R_1 \rightarrow R_1 + R_2 + R_3$

$$= \begin{vmatrix} 2a+2b & 2a+2c & 2b+2c \\ a & a+c & c \\ b & c & b+c \end{vmatrix}$$

$$= 2\begin{vmatrix} a+b & a+c & b+c \\ a & a+c & c \\ b & c & b+c \end{vmatrix}$$

$R_2 \rightarrow R_2 - R_1$ and $R_3 \rightarrow R_3 - R_1$

$$= 2\begin{vmatrix} a+b & a+c & b+c \\ -b & 0 & -b \\ -a & -a & 0 \end{vmatrix}$$

$R_1 \rightarrow R_1 + R_2 + R_3$

$$= 2\begin{vmatrix} 0 & c & c \\ -b & 0 & -b \\ -a & -a & 0 \end{vmatrix}$$

$$= 2c\,(-b)\,(-a)\begin{vmatrix} 0 & 1 & 1 \\ 1 & 0 & 1 \\ 1 & 1 & 0 \end{vmatrix}$$

$$= 2abc\,[0 - 1\,(-1) + 1(1)]$$

$$= 2acb\,[1 + 1]$$

$$= 4abc$$

***Example* 7:** Show that $\begin{vmatrix} a-b-c & 2a & 2a \\ 2b & b-c-a & 2b \\ 2c & 2c & c-a-b \end{vmatrix} = (a+b+c)^3$

Solution: by $R_1 \to R_1 + R_2 + R_3$

$$= \begin{vmatrix} a-b-c+2b+2c & 2a+b-c-a+2c & 2a+2b+c-a-b \\ 2b & b-c-a & 2b \\ 2c & 2c & c-a-b \end{vmatrix}$$

$$= \begin{vmatrix} a+b+c & a+b+c & a+b+c \\ 2b & b-c-a & 2b \\ 2c & 2c & c-a-b \end{vmatrix}$$

Taking $(a + b + c)$ common from R_1

$$= (a+b+c)\begin{vmatrix} 1 & 1 & 1 \\ 2b & b-c-a & 2b \\ 2c & 2c & c-a-b \end{vmatrix}$$

by $C_2 \to C_2 - C_1$ and $C_3 \to C_3 - C_1$

$$= (a+b+c)\begin{vmatrix} 1 & 0 & 0 \\ 2b & -a-b-c & 0 \\ 2c & 0 & -a-b-c \end{vmatrix}$$

Taking $(a + b + c)$ common from C_2 and C_3

$$= (a+b+c)\,(a+b+c)\,(a+b+c)\begin{vmatrix} 1 & 0 & 0 \\ 2b & -1 & 0 \\ 2c & 0 & -1 \end{vmatrix}$$

$$= (a+b+c)^3\begin{vmatrix} -1 & 0 \\ 0 & -1 \end{vmatrix} \qquad \text{(expanding w.r.t } R_1)$$

$$= (a+b+c)^3\,(-1)\,(-1)$$

$$= (a+b+c)^3$$

Example 8: Prove that $\begin{vmatrix} a+b+2c & a & b \\ c & b+c+2a & b \\ c & a & c+a+2b \end{vmatrix} = 2(a+b+c)^3$

Solution: by $c_1 \to c_1 + c_2 + c_3$

$$= \begin{vmatrix} 2(a+b+c) & a & b \\ 2(a+b+c) & b+c+2a & b \\ 2(a+b+c) & a & c+a+2b \end{vmatrix}$$

Taking 2(a + b + c) common from c_1

$$= 2(a+b+c) \begin{vmatrix} 1 & a & b \\ 1 & b+c+2a & b \\ 1 & a & c+a+2b \end{vmatrix}$$

by $R_2 \to R_2 - R_1$ and $R_3 \to R_3 - R_1$

$$= 2(a+b+c) \begin{vmatrix} 1 & a & b \\ 0 & a+b+c & 0 \\ 0 & 0 & a+b+c \end{vmatrix}$$

Taking (a + b + c) common from R_2 and R_3

$$= 2(a+b+c)\,(a+b+c)\,(a+b+c) \begin{vmatrix} 1 & a & b \\ 0 & 1 & 0 \\ 0 & 0 & 1 \end{vmatrix}$$

$$= 2(a+b+c)^3\, 1(1)$$

$$= 2(a+b+c)^3$$

Example 9: Show that $\begin{vmatrix} 1 & 1 & 1 \\ a & b & c \\ a^3 & b^3 & c^3 \end{vmatrix} = (b-c)\,(c-a)\,(a-b)\,(a+b+c)$

Solution: by $c_2 \to c_2 - c_1$ and $c_3 \to c_3 - c_1$

$$= \begin{vmatrix} 1 & 0 & 0 \\ a & b-a & c-a \\ a^3 & b^3-a^3 & c^3-a^3 \end{vmatrix}$$

$$= \begin{vmatrix} 1 & 0 & 0 \\ a & b-a & c-a \\ a^3 & (b-a)(b^2+ab+a^2) & (c-a)(c^2+ac+a^2) \end{vmatrix}$$

Taking (b – a) (c – a) common from c_2 and c_3

$$= (b-a)(c-a)\begin{vmatrix} 1 & 0 & 0 \\ a & 1 & 1 \\ a^3 & b^2+ab+a^2 & c^2+ac+a^2 \end{vmatrix}$$

by $c_2 \to c_2 - c_3$

$$= (b-a)(c-a)\begin{vmatrix} 1 & 0 & 0 \\ a & 0 & 1 \\ a^3 & b^2-c^2+ab-ac & c^2+ac+a^2 \end{vmatrix}$$

$$= (b-a)(c-a)\,[1\{0-(b^2-c^2+ab-ac)\}]$$

$$= (b-a)(c-a)\,[-\{(b-c)(b+c)+a(b-c)\}]$$

$$= -(b-a)(c-a)(b-c)(b+c+a)$$

$$= (a-b)(b-c)(c-a)(a+b+c)$$

***Example* 10:** Prove that $\begin{vmatrix} 1 & 1 & 1 \\ x & y & z \\ yz & zx & xy \end{vmatrix} = (x-y)(y-z)(z-x)$

Solution: by $c_1 \to c_1 - c_2$ and $c_2 \to c_2 - c_3$

$$= \begin{vmatrix} 0 & 0 & 1 \\ x-y & y-z & z \\ yz-zx & zx-xy & xy \end{vmatrix}$$

$$= \begin{vmatrix} 0 & 0 & 1 \\ x-y & y-z & z \\ -z(x-y) & -x(y-z) & xy \end{vmatrix}$$

Taking (x – y) (y – z) common from c_1 and c_2

$$= (x-y)(y-z)\begin{vmatrix} 0 & 0 & 1 \\ 1 & 1 & z \\ -z & -x & xy \end{vmatrix}$$

$$= (x-y)(y-z)\,[0-0+1(-x+z)]$$

$$= (x-y)(y-z)(z-x)$$

***Example* 11:** Show that $\begin{vmatrix} 1 & 1 & 1 \\ a^2 & b^2 & c^2 \\ a^3 & b^3 & c^3 \end{vmatrix} = (a-b)(b-c)(c-a)(ab+bc+ca)$

Solution: by $c_2 \to c_2 - c_1$ and $c_3 \to c_3 \to c_1$

$$= \begin{vmatrix} 1 & 0 & 0 \\ a^2 & b^2-a^2 & c^2-a^2 \\ a^3 & b^3-a^3 & c^3-a^3 \end{vmatrix}$$

$$= \begin{vmatrix} 1 & 0 & 0 \\ a^2 & (b-a)(b+a) & (c-a)(c+a) \\ a^3 & (b-a)(b^2+ab+a^2) & (c-a)(c^2+ac+a^2) \end{vmatrix}$$

Taking common $(b-a)(c-a)$ from c_2 and c_3

$$= (b-a)(c-a) \begin{vmatrix} 1 & 0 & 0 \\ a^2 & b+a & c+a \\ a^3 & b^2+ab+a^2 & c^2+ac+a^2 \end{vmatrix}$$

On expanding

$$= (b-a)(c-a) \begin{vmatrix} b+a & c+a \\ b^2+ab+a^2 & c^2+ac+a^2 \end{vmatrix}$$

By $c_2 \to c_2 - c_1$

$$= (b-a)(c-a) \begin{vmatrix} b+a & c-b \\ b^2+ab+a^2 & c^2+ac-ab \end{vmatrix}$$

$$= (b-a)(c-a) \begin{vmatrix} b+a & c-b \\ b^2+ab+a^2 & (c-b)(c+b)+a(c-b) \end{vmatrix}$$

$$= (b-a)(c-a) \begin{vmatrix} b+a & c-b \\ b^2+ab+a^2 & (c-b)(c+b+a) \end{vmatrix}$$

Taking $(c-b)$ common from c_2

$$= (b-a)(c-a)(c-b) \begin{vmatrix} b+a & 1 \\ b^2+ab+a^2 & c+b+a \end{vmatrix}$$

$$= (b-a)(c-a)(c-b)[(b+a)(c+b+a) - (b^2+ab+a^2)]$$

$$= (a-b)(b-c)(c-a)[bc+b^2+ab+ac+ab+a^2-b^2-ab-a^2]$$

$$= (a-b)(b-c)(c-a)(ab+bc+ca)$$

***Example* 12:** Prove that $\begin{vmatrix} 1 & a & a^2-bc \\ 1 & b & b^2-ca \\ 1 & c & c^2-ab \end{vmatrix} = 0$

Solution:

$$\Delta = \begin{vmatrix} 1 & a & a^2 \\ 1 & b & b^2 \\ 1 & c & c^2 \end{vmatrix} + \begin{vmatrix} 1 & a & -bc \\ 1 & b & -ca \\ 1 & c & -ab \end{vmatrix}$$

Multiply the rows of 2nd determinant by a, b, c and divide by abc.

$$= \begin{vmatrix} 1 & a & a^2 \\ 1 & b & b^2 \\ 1 & c & c^2 \end{vmatrix} - \frac{1}{abc}\begin{vmatrix} a & a^2 & abc \\ b & b^2 & abc \\ c & c^2 & abc \end{vmatrix}$$

Taking abc common from c_3 by 2nd determinant

$$= \begin{vmatrix} 1 & a & a^2 \\ 1 & b & b^2 \\ 1 & c & c^2 \end{vmatrix} - \begin{vmatrix} a & a^2 & 1 \\ b & b^2 & 1 \\ c & c^2 & 1 \end{vmatrix}$$

Exchanging c_2 and c_3 in 2nd determinant

$$= \begin{vmatrix} 1 & a & a^2 \\ 1 & b & b^2 \\ 1 & c & c^2 \end{vmatrix} - (-1)\begin{vmatrix} a & 1 & a^2 \\ b & 1 & b^2 \\ c & 1 & c^2 \end{vmatrix}$$

Exchanging c_1 and c_2 in 2nd determinant

$$= \begin{vmatrix} 1 & a & a^2 \\ 1 & b & b^2 \\ 1 & c & c^2 \end{vmatrix} - (-1)^2\begin{vmatrix} 1 & a & a^2 \\ 1 & b & b^2 \\ 1 & c & c^2 \end{vmatrix}$$

$$= \begin{vmatrix} 1 & a & a^2 \\ 1 & b & b^2 \\ 1 & c & c^2 \end{vmatrix} - \begin{vmatrix} 1 & a & a^2 \\ 1 & b & b^2 \\ 1 & c & c^2 \end{vmatrix}$$

$$= 0$$

***Example* 13:** Solve the equation $\begin{vmatrix} x+a & b & c \\ a & x+b & c \\ a & b & x+c \end{vmatrix} = 0$

Solution: by $c_1 \to c_1 + c_2 + c_3$

$$\Rightarrow \begin{vmatrix} x+a+b+c & b & c \\ x+a+b+c & x+b & c \\ x+a+b+c & b & x+c \end{vmatrix} = 0$$

Taking $(x + a + b + c)$ common from c_1

$$\Rightarrow (x+a+b+c) \begin{vmatrix} 1 & b & c \\ 1 & x+b & c \\ 1 & b & x+c \end{vmatrix} = 0$$

by $R_2 \to R_2 - R_1$ and $R_3 \to R_3 - R_1$

$$\Rightarrow (x+a+b+c) \begin{vmatrix} 1 & b & c \\ 0 & x & 0 \\ 0 & 0 & x \end{vmatrix} = 0$$

on expanding, $(x + a + b + c)\, x^2 = 0$

$\therefore \quad x^2 = 0 \quad$ or $\quad x + a + b + c = 0$

Hence $\quad x = 0 \quad$ or $\quad x = -(a + b + c)$

Applications of Determinants

1. ***Area of Triangle*:** If the vertices of a triangle are (x_1, y_1) (x_2, y_2) and (x_3, y_3) respectively. Then area of Δ is

$$= \frac{1}{2}\,[x_1\,(y_2 - y_3) + x_2(y_3 - y_1) + x_3(y_1 - y_2)]$$

In determinant form $\Delta = \dfrac{1}{2}\begin{vmatrix} x_1 & y_1 & 1 \\ x_2 & y_2 & 1 \\ x_3 & y_3 & 1 \end{vmatrix}$

Condition for collinearty: Three points $A(x_1, y_1)$, $B(x_2, y_2)$ and $C(x_3, y_3)$ are collinear if area of $\Delta ABC = 0$

$$\text{i.e.} \begin{vmatrix} x_1 & y_1 & 1 \\ x_2 & y_2 & 1 \\ x_3 & y_3 & 1 \end{vmatrix} = 0$$

2. ***Crammer's rule:*** Consider three linear simultaneous equations are

$$a_1x + b_1y + c_1z = d_1 \quad .....(1)$$

$$a_2x + b_2y + c_2z = d_2 \quad .....(2)$$

$$a_3x + b_3y + c_3z = d_3 \quad .....(3)$$

If $$\Delta = \begin{vmatrix} a_1 & b_1 & c_1 \\ a_2 & b_2 & c_2 \\ a_3 & b_3 & c_3 \end{vmatrix} \qquad \Delta_1 = \begin{vmatrix} d_1 & b_1 & c_1 \\ d_2 & b_2 & c_2 \\ d_3 & b_3 & c_3 \end{vmatrix}$$

$$\Delta_2 = \begin{vmatrix} a_1 & d_1 & c_1 \\ a_2 & d_2 & c_2 \\ a_3 & d_3 & c_3 \end{vmatrix} \qquad \Delta_3 = \begin{vmatrix} a_1 & b_1 & d_1 \\ a_2 & b_2 & d_2 \\ a_3 & b_3 & d_3 \end{vmatrix}$$

Then by Crammer's rule, we have

$$x = \frac{\Delta_1}{\Delta}, y = \frac{\Delta_2}{\Delta}, z = \frac{\Delta_3}{\Delta}, (\Delta \neq 0)$$

Case I If $\Delta \neq 0$, then, in this case the given system of equations is called consistent and has unique solution.

Case II If $\Delta = 0$ and any of Δ_1, Δ_2, Δ_3 is non zero, then, the given system of equation is inconsistent i.e., it has no solution.

Case III If $\Delta = 0$ and all of Δ_1, Δ_2, Δ_3 are zero. Then given system of equation is consistent and has infinite number of solutions.

Case IV In case of linear equation of two variables i.e., $a_1x + b_1y = c_1$ and $a_2x + b_2y = c_2$. Then

$$\Delta = \begin{vmatrix} a_1 & b_1 \\ a_2 & b_2 \end{vmatrix}, \qquad \Delta_1 = \begin{vmatrix} c_1 & b_1 \\ c_2 & b_2 \end{vmatrix}, \qquad \Delta_2 = \begin{vmatrix} a_1 & c_1 \\ a_2 & c_2 \end{vmatrix}$$

$$\therefore \quad x = \frac{\Delta_1}{\Delta}, y = \frac{\Delta_2}{\Delta}$$

***Example* 14:** Find the area of the triangle with vertices (–2, –3), (3, 2), (–1, –8)

Solution: Area of $\Delta = \frac{1}{2}\begin{vmatrix} x_1 & y_1 & 1 \\ x_2 & y_2 & 1 \\ x_3 & y_3 & 1 \end{vmatrix}$

$$= \frac{1}{2}\begin{vmatrix} -2 & -3 & 1 \\ 3 & 2 & 1 \\ -1 & -8 & 1 \end{vmatrix}$$

$$= \frac{1}{2}\,[-2\,(2 + 8) + 3\,(3 + 1) + 1(-24 + 2)]$$

$$= \frac{1}{2}\,[-20 + 12 - 22\,]$$

$$= \frac{1}{2} \times (-30) = -15 = 15$$

***Example* 15:** Show that the following points are collinear: A(b, c + a), B(c, a + b), C(a, b + c)

***Solution*:** Area of $\Delta ABC = \frac{1}{2}\begin{vmatrix} b & c+a & 1 \\ c & a+b & 1 \\ a & b+c & 1 \end{vmatrix}$

By $c_2 \rightarrow c_2 + c_1$

$$\Delta = \frac{1}{2}\begin{vmatrix} b & c+a+b & 1 \\ c & a+b+c & 1 \\ a & a+b+c & 1 \end{vmatrix}$$

Taking (a + b + c) common from c_2

$$\Delta = \frac{1}{2}(a+b+c)\begin{vmatrix} b & 1 & 1 \\ c & 1 & 1 \\ a & 1 & 1 \end{vmatrix}$$

$\Delta = 0$ (by identical property)

Since area of ΔABC is zero, then points are collinear.

***Example* 16:** Solve the following equation with the help of determinants.

$$x - 3y + z = 2$$

$$3x + y + z = 6$$

$$5x + y + 3z = 3$$

***Solution*:** Let

$$\Delta = \begin{vmatrix} 1 & -3 & 1 \\ 3 & 1 & 1 \\ 5 & 1 & 3 \end{vmatrix}$$

$$= 1(3-1) + 3\,(9-5) + 1(3-5)$$

$$= 2 + 12 - 2 = 12$$

$$\Delta_1 = \begin{vmatrix} 2 & -3 & 1 \\ 6 & 1 & 1 \\ 3 & 1 & 3 \end{vmatrix}$$

$$= 2(3-1) + 3\,(18-3) + 1(6-3)$$

$$= 4 + 45 + 3 = 52$$

$$\Delta_2 = \begin{vmatrix} 1 & 2 & 1 \\ 3 & 6 & 1 \\ 5 & 3 & 3 \end{vmatrix}$$

$$= 1(18-3) - 2(9-5) + 1(9-30)$$

$$= 15 - 8 - 21$$

$$= -14$$

$$\Delta_3 = \begin{vmatrix} 1 & -3 & 2 \\ 3 & 1 & 6 \\ 5 & 1 & 3 \end{vmatrix}$$

$$= 1(3-6) + 3(9-30) + 2(3-5)$$

$$= -3 - 63 - 4 = -70$$

By Crammer's rule, $x = \dfrac{\Delta_1}{\Delta} = \dfrac{52}{12} = \dfrac{13}{3}$

$$y = \frac{\Delta_2}{\Delta} = \frac{-14}{12} = \frac{-7}{6}$$

$$z = \frac{\Delta_3}{\Delta} = \frac{-70}{12} = \frac{-35}{6}$$

***Example* 17:** Solve the system of equations using Cramer's rule

$$2x + y - z = 3$$
$$x + 2y + 2z = 4$$
$$x + y + z = 2$$

***Solution*:** Let

$$\Delta = \begin{vmatrix} 2 & 1 & -1 \\ 1 & 2 & 2 \\ 1 & 1 & 1 \end{vmatrix}$$

$$= 2(2-2) - 1(1-2) - 1(1-2)$$
$$= 0 + 1 + 1 = 2$$

$$\Delta_1 = \begin{vmatrix} 3 & 1 & -1 \\ 4 & 2 & 2 \\ 2 & 1 & 1 \end{vmatrix}$$

$$= 3(2-2) - 1(4-4) - 1(4-4)$$
$$= 0 - 0 - 0 = 0$$

$$\Delta_2 = \begin{vmatrix} 2 & 3 & -1 \\ 1 & 4 & 2 \\ 1 & 2 & 1 \end{vmatrix}$$

$$= 2(4-4) - 3(1-2) - 1(2-4)$$
$$= 0 + 3 + 2 = 5$$

$$\Delta_3 = \begin{vmatrix} 2 & 1 & 3 \\ 1 & 2 & 4 \\ 1 & 1 & 2 \end{vmatrix}$$

$$= 2(4-4) - 1(2-4) + 3(1-2)$$
$$= 0 + 2 - 3 = -1$$

$$\therefore \quad x = \frac{\Delta_1}{\Delta} = 0$$

$$y = \frac{\Delta_2}{\Delta} = \frac{5}{2}$$

$$z = \frac{\Delta_3}{\Delta} = \frac{-1}{2}$$

***Example* 18:** Solve 2x – 3y + z = 1, x + y + z = 2, 3x – 4z –17 = 0 by using Crammer's rule.

***Solution*:** Given equations are

$$2x - 3y + z = 1$$

$$x + y + z = 2$$

$$3x - 4z = 17$$

$$\therefore \quad \Delta = \begin{vmatrix} 2 & -3 & 1 \\ 1 & 1 & 1 \\ 3 & 0 & -4 \end{vmatrix}$$

$$= 2(-4) + 3(-4 - 3) + 1(-3)$$

$$= -8 - 21 - 3 = -32$$

$$\Delta_1 = \begin{vmatrix} 1 & -3 & 1 \\ 2 & 1 & 1 \\ 17 & 0 & -4 \end{vmatrix}$$

$$= 1(-4) + 3\,(-8 - 17) + 1(-17)$$

$$= -4 - 75 - 17 = -96$$

$$\Delta_2 = \begin{vmatrix} 2 & 1 & 1 \\ 1 & 2 & 1 \\ 3 & 17 & -4 \end{vmatrix}$$

$$= 2(-8 - 17) - 1(-4 - 3) + 1(17 - 6)$$

$$= -50 + 7 + 11 = -32$$

$$\Delta_3 = \begin{vmatrix} 2 & -3 & 1 \\ 1 & 1 & 2 \\ 3 & 0 & 17 \end{vmatrix}$$

$$= 2(17) + 3(17 - 6) + 1(-3)$$

$$= 34 + 33 - 3 = 64$$

$$\therefore \quad x = \frac{\Delta_1}{\Delta} = \frac{-96}{-32} = 3$$

$$y = \frac{\Delta_2}{\Delta} = \frac{-32}{-32} = 1$$

$$z = \frac{\Delta_3}{\Delta} = \frac{64}{-32} = -2$$

***Example* 19:** Solve the equation using Crammer's rule

$$x + y + z = 1$$
$$2x - y + z = 2$$
$$x + 2y + 2z = 3$$

***Solution*:**

$$\Delta = \begin{vmatrix} 1 & 1 & 1 \\ 2 & -1 & 1 \\ 1 & 2 & 2 \end{vmatrix}$$

$$= 1(-2 -2) -1(4 - 1) + 1(4 + 1)$$
$$= -4 -3 + 5 = -2$$

$$\Delta_1 = \begin{vmatrix} 1 & 1 & 1 \\ 2 & -1 & 1 \\ 3 & 2 & 2 \end{vmatrix}$$

$$= 1(-2 -2) -1 (4 - 3) + 1(4 +3)$$
$$= -4 -1 +7 = 2$$

$$\Delta_2 = \begin{vmatrix} 1 & 1 & 1 \\ 2 & 2 & 1 \\ 1 & 3 & 2 \end{vmatrix}$$

$$= 1 (4 - 3) -1 (4 - 1) +1 (6 - 2)$$
$$= 1 - 3 + 4 = 2$$

$$\Delta_3 = \begin{vmatrix} 1 & 1 & 1 \\ 2 & -1 & 2 \\ 1 & 2 & 3 \end{vmatrix}$$

$$= 1(-3 -4) -1(6 -2) +1(4 + 1)$$
$$= -7 -4 + 5 = -6$$

$\therefore$
$$x = \frac{\Delta_1}{\Delta} = \frac{2}{-2} = -1$$

$$y = \frac{\Delta_2}{\Delta} = \frac{2}{-2} = -1$$

$$z = \frac{\Delta_3}{\Delta} = \frac{-6}{-2} = 3$$

***Example* 20:** Solve the system of equations using Cramer's rule

$$2y - 3z = 0$$

$$x + 3y = -4$$

$$3x + 4y = 3$$

***Solution*:** Let

$$\Delta = \begin{vmatrix} 0 & 2 & -3 \\ 1 & 3 & 0 \\ 3 & 4 & 0 \end{vmatrix}$$

$$= 0 - 2\,(0) - 3(4 - 9)$$

$$= 15$$

$$\Delta_1 = \begin{vmatrix} 0 & 2 & -3 \\ -4 & 3 & 0 \\ 3 & 4 & 0 \end{vmatrix}$$

$$= 0 - 2(0) - 3(-16 - 9) = 75$$

$$\Delta_2 = \begin{vmatrix} 0 & 0 & -3 \\ 1 & -4 & 0 \\ 3 & 3 & 0 \end{vmatrix}$$

$$= 0 - 0 - 3\,(3 + 12) = -45$$

$$\Delta_3 = \begin{vmatrix} 0 & 2 & 0 \\ 1 & 3 & -4 \\ 3 & 4 & 3 \end{vmatrix}$$

$$= 0 - 2\,(3 + 12) + 0 = -30$$

$$\therefore \quad x = \frac{\Delta_1}{\Delta} = \frac{75}{15} = 5$$

$$y = \frac{\Delta_2}{\Delta} = \frac{-45}{15} = -3$$

$$z = \frac{\Delta_3}{\Delta} = \frac{-30}{15} = -2$$

Some Practice Questions

***Example* 1:** Evaluate $\begin{vmatrix} 265 & 240 & 219 \\ 240 & 225 & 198 \\ 219 & 198 & 181 \end{vmatrix}$

***Solution*:** By $c_1 \to c_1 - c_2$ and $c_2 \to c_2 - c_3$

$$\Delta = \begin{vmatrix} 25 & 21 & 219 \\ 15 & 27 & 198 \\ 21 & 17 & 181 \end{vmatrix}$$

by $c_1 \to c_1 - c_2$ and $c_3 \to c_3 - 10c_2$

$$= \begin{vmatrix} 4 & 21 & 9 \\ -12 & 27 & -72 \\ 4 & 17 & 11 \end{vmatrix}$$

by $R_2 \to R_2 + 3R_1$ and $R_3 \to R_3 - R_1$

$$= \begin{vmatrix} 4 & 21 & 9 \\ 0 & 90 & -45 \\ 0 & -4 & 2 \end{vmatrix}$$

$$= 4 \begin{vmatrix} 90 & -45 \\ -4 & 2 \end{vmatrix}$$

$$= 4(90 \times 2 - 45 \times 4)$$

$$= 4(180 - 180) = 0$$

***Example* 2:** Evaluate $\begin{vmatrix} a - ib & -c + id \\ c + id & a + ib \end{vmatrix}$

***Solution*:** Let $\Delta = \begin{vmatrix} a - ib & -c + id \\ c + id & a + ib \end{vmatrix}$

$$= (a - ib)\,(a + ib) - (-c + id)\,(c + id)$$

$$= a^2 - i^2 b^2 - (i^2 d^2 - c^2) \qquad (\because i^2 = -1)$$

$$= a^2 + b^2 - (-d^2 - c^2)$$

$$= a^2 + b^2 + c^2 + d^2$$

***Example* 3:** Find all the co-factors of $\begin{vmatrix} 3 & 2 \\ 1 & 4 \end{vmatrix}$

***Solution*:** Co-factor $c_{11} = (-1)^{1+1}. 4 = 4$

$c_{12} = (-1)^{1+2}. 1 = -1$

$c_{21} = (-1)^{2+1}. 2 = -2$

$c_{22} = (-1)^{2+2}. 3 = 3$

***Example* 4:** Show that $\begin{vmatrix} 60 & 12 & 8 \\ 35 & 21 & 4 \\ 17 & 9 & 2 \end{vmatrix} = 0$

***Solution*:** Δ

$$= 60(21 \times 2 - 9 \times 4) - 12(35 \times 2 - 17 \times 4) + 8(35 \times 9 - 21 \times 17)$$

$$= 60 \times (42 - 36) - 12(70 - 68) + 8(315 - 357)$$

$$= 60 \times 6 - 12 \times 2 + 8 \times (-42)$$

$$= 360 - 24 - 336$$

$$= 360 - 360 = 0$$

***Example* 5:** Prove that $\begin{vmatrix} 1 & w & w^2 \\ w & w^2 & 1 \\ w^2 & 1 & w \end{vmatrix} = 0$, where w is a cube root of unity.

***Solution*:** By $c_1 \to c_1 + c_2 + c_3$

$$\Delta = \begin{vmatrix} 1+w+w^2 & w & w^2 \\ 1+w+w^2 & w^2 & 1 \\ 1+w+w^2 & 1 & w \end{vmatrix}$$

$$= \begin{vmatrix} 0 & w & w^2 \\ 0 & w^2 & 1 \\ 0 & 1 & w \end{vmatrix} \qquad (\because 1 + w + w^2 = 0)$$

$$\Delta = 0$$

***Example* 6:** Evaluate $\begin{vmatrix} a-b & b-c & c-a \\ b-c & c-a & a-b \\ c-a & a-b & b-c \end{vmatrix}$

***Solution*:** Apply $R_1 \to R_1 + R_2 + R_3$, we get

$$\Delta = \begin{vmatrix} 0 & 0 & 0 \\ b-c & c-a & a-b \\ c-a & a-b & b-c \end{vmatrix}$$

$$\Delta = 0$$

***Example* 7:** Prove that $\begin{vmatrix} b+c & a & a \\ b & c+a & b \\ c & c & a+b \end{vmatrix} = 4abc$

***Solution*:** Applying $R_1 \to R_1 - R_2 - R_3$, we get

$$\Delta = \begin{vmatrix} 0 & -2c & -2b \\ b & c+a & b \\ c & c & a+b \end{vmatrix}$$

On expanding $= 2c\begin{vmatrix} b & b \\ c & a+b \end{vmatrix} - 2b\begin{vmatrix} b & c+a \\ c & c \end{vmatrix}$

$$= 2c[b(a+b) - bc] - 2b[bc - c(c+a)]$$
$$= 2c[ab + b^2 - bc] - 2b[bc - c^2 - ac]$$
$$= 2abc + 2b^2c - 2bc^2 - 2b^2c + 2bc^2 + 2abc$$
$$= 4abc$$

***Example* 8:** Show that $\begin{vmatrix} x+\lambda & x & x \\ x & x+\lambda & x \\ x & x & x+\lambda \end{vmatrix} = \lambda^2(3x+\lambda)$

***Solution*:** Applying $c_1 \to c_1 + c_2 + c_3$

$$\Delta = \begin{vmatrix} 3x+\lambda & x & x \\ 3x+\lambda & x+\lambda & x \\ 3x+\lambda & x & x+\lambda \end{vmatrix}$$

Taking $(3x + \lambda)$ common from c_1, we get

$$= (3x+\lambda)\begin{vmatrix} 1 & x & x \\ 1 & x+\lambda & x \\ 1 & x & x+\lambda \end{vmatrix}$$

by $R_2 \to R_2 - R_1$ and $R_3 \to R_3 - R_1$, we get

$$= (3x+\lambda)\begin{vmatrix} 1 & x & x \\ 0 & \lambda & 0 \\ 0 & 0 & \lambda \end{vmatrix}$$

On expanding, we get

$$\Delta = (3x+\lambda).\ 1\begin{vmatrix} \lambda & 0 \\ 0 & \lambda \end{vmatrix}$$

$$= (3x+\lambda)\,\lambda^2$$

***Example* 9:** Solve the equation $\begin{vmatrix} 1 & 4 & 4 \\ 1 & -2 & 1 \\ 1 & 2x & x^2 \end{vmatrix} = 0$

***Solution*:** Applying $c_2 \to c_2 - 4c_1$ and $c_3 \to c_3 - 4c_1$

$$\Rightarrow \begin{vmatrix} 1 & 0 & 0 \\ 1 & -6 & -3 \\ 1 & 2x-4 & x^2-4 \end{vmatrix} = 0$$

on expanding, we get, $-6(x^2 - 4) + 3(2x - 4) = 0$

$$-6x^2 + 24 + 6x - 12 = 0$$

$$6x^2 - 6x - 12 = 0$$

$$6(x^2 - x - 2) = 0$$

$$x^2 - x - 2 = 0$$

$$x^2 - 2x + x - 2 = 0$$

$$x(x-2) + 1(x-2) = 0$$

$$(x-2)(x+1) = 0$$

$$x - 2 = 0 \quad \text{or} \quad x + 1 = 0$$

$$x = +2 \quad \text{or} \quad x = -1$$

***Example* 10:** Prove the following results $\begin{vmatrix} 1 & bc & a(b+c) \\ 1 & ca & b(c+a) \\ 1 & ab & c(a+b) \end{vmatrix} = 0$

***Solution*:** Applying $c_2 \to c_2 + c_3$, we get

$$\Delta = \begin{vmatrix} 1 & bc+ab+ac & a(b+c) \\ 1 & ca+bc+ba & b(c+a) \\ 1 & ab+ca+cb & c(a+b) \end{vmatrix}$$

Taking (ab + bc + ca) common from c_2, we get

$$= (ab + bc + ca) \begin{vmatrix} 1 & 1 & a(b+c) \\ 1 & 1 & b(c+a) \\ 1 & 1 & c(a+b) \end{vmatrix}$$

$$= (ab + bc + ca) \times 0 \qquad (\because \text{ by property of identical column})$$

$$= 0$$

***Example* 11:** Solve the following system of equations

$$x + 3y + 3z = 1, \qquad x + 4y + 3z = 0, \qquad x + 3y + 4z = 2$$

***Solution*:** From the given equations

$$\Delta = \begin{vmatrix} 1 & 3 & 3 \\ 1 & 4 & 3 \\ 1 & 3 & 4 \end{vmatrix}$$

$$= 1(16 - 9) - 3(4 - 3) + 3(3 - 4)$$

$$= 7 - 3 - 3 = 1$$

$$\Delta_1 = \begin{vmatrix} 1 & 3 & 3 \\ 0 & 4 & 3 \\ 2 & 3 & 4 \end{vmatrix}$$

$$= 1(16 - 9) - 3(0 - 6) + 3(0 - 8)$$

$$= 7 + 18 - 24 = 1$$

$$\Delta_2 = \begin{vmatrix} 1 & 1 & 3 \\ 1 & 0 & 3 \\ 1 & 2 & 4 \end{vmatrix}$$

$$= 1(0 - 6) - 1(4 - 3) + 3(2 - 0)$$

$$= -6 - 1 + 6 = -1$$

$$\Delta_3 = \begin{vmatrix} 1 & 3 & 1 \\ 1 & 4 & 0 \\ 1 & 3 & 2 \end{vmatrix}$$

$$= 1(8 - 0) - 3(2 - 0) + 1(3 - 4)$$

$$= 8 - 6 - 1 = 1$$

using Crammer's rule

$$x = \frac{\Delta_1}{\Delta} = \frac{1}{1} \qquad \therefore \qquad x = 1$$

$$y = \frac{\Delta_2}{\Delta} = \frac{-1}{1} \qquad\qquad y = -1$$

$$z = \frac{\Delta_3}{\Delta} = \frac{1}{1} \qquad\qquad z = 1$$

***Example* 12:** The total sale (s) in lakhs of rupees of a firm selling two drugs p and q is given by the relationship.

$$s = ap + bq + c$$

Data for the sales of first three months is given:

Months	Total sale	p	q
2	12	2	3
2	13	6	2
3	15	5	3

Using determinant method, determine the sales in next months when it sells 4 units of p and 5 units of q.

***Solution*:** Given the equation $s = ap + bq + c$

For the one month, $s = 12$, $p = 2$, $q = 3$

$$\therefore \qquad 2a + 3b + c = 12 \qquad \text{.....(1)}$$

For the second month, $s = 13$, $p = 6$, $q = 2$

$$\therefore \qquad 6a + 2b + c = 13 \qquad \text{.....(2)}$$

For the third month, $s = 15$, $p = 5$, $q = 3$

$$\therefore \qquad 5a + 3b + c = 15 \qquad \text{.....(3)}$$

From the given equations,

$$\Delta = \begin{vmatrix} 2 & 3 & 1 \\ 6 & 2 & 1 \\ 5 & 3 & 1 \end{vmatrix}$$

$$= 2(2-3) - 3(6-5) + 1(18-10)$$

$$= -2 - 3 + 8 = 3$$

$$\Delta_1 = \begin{vmatrix} 12 & 3 & 1 \\ 13 & 2 & 1 \\ 15 & 3 & 1 \end{vmatrix}$$

$$= 12(2-3) - 3(13-15) + 1(39-30)$$

$$= -12 + 6 + 9 = 3$$

$$\Delta_2 = \begin{vmatrix} 2 & 12 & 1 \\ 6 & 13 & 1 \\ 5 & 15 & 1 \end{vmatrix}$$

$$= 2(13-15) - 12(6-5) + 1(90-65)$$

$$= -4 - 12 + 25 = 9$$

$$\Delta_3 = \begin{vmatrix} 2 & 3 & 12 \\ 6 & 2 & 13 \\ 5 & 3 & 15 \end{vmatrix}$$

$$= 2(30-39) - 3(90-65) + 12(18-10)$$

$$= -18 - 75 + 96 = 3$$

Using Cramer's rule

$$a = \frac{\Delta_1}{\Delta} = \frac{3}{3} = 1$$

$$b = \frac{\Delta_2}{\Delta} = \frac{9}{3} = 3$$

$$c = \frac{\Delta_3}{\Delta} = \frac{3}{3} = 1$$

Now, for the next month, if $p = 4$, $q = 5$

Then total sale, s = ap + bq + c

$= 1 \times 4 + 3 \times 5 + 1$

$= 4 + 15 + 1$

S = 20 lakhs

***Example* 13:** Show that the following system of equations is inconsistent

$$2x + y = 3, \qquad 4x + 2y = 5.$$

***Solution*:** From the given equations

$$\Delta = \begin{vmatrix} 2 & 1 \\ 4 & 2 \end{vmatrix} = 4 - 4 = 0$$

$$\Delta_1 = \begin{vmatrix} 3 & 1 \\ 5 & 2 \end{vmatrix} = 6 - 5 = 1$$

Here $\Delta = 0$ but $\Delta_1 \neq 0$

$\therefore$ The given system is inconsistent

Hence, it has no solutions

***Example* 14:** Solve the system of equations $x + 2y = 3$ and $4x + 8y = 12$ using determinants.

***Solution*:** From given system of equations

$$\Delta = \begin{vmatrix} 1 & 2 \\ 4 & 8 \end{vmatrix} = 8 - 8 = 0$$

$$\Delta_1 = \begin{vmatrix} 3 & 2 \\ 12 & 8 \end{vmatrix} = 24 - 24 = 0$$

$$\Delta_2 = \begin{vmatrix} 1 & 3 \\ 4 & 12 \end{vmatrix} = 12 - 12 = 0$$

Here $\Delta = \Delta_1 = \Delta_2 = 0$

Therefore, the system is consistent and has infinite number of solutions.

***Example* 15:** Show that the points (8 – 1), (4, 7) and (6, 3) are collinear

***Solution*:** we have, Area of $\Delta = \frac{1}{2}\begin{vmatrix} x_1 & y_1 & 1 \\ x_2 & y_2 & 1 \\ x_3 & y_3 & 1 \end{vmatrix}$

$$= \frac{1}{2}\begin{vmatrix} 8 & -1 & 1 \\ 4 & 7 & 1 \\ 6 & 3 & 1 \end{vmatrix}$$

$$= \frac{1}{2}[8(7-3) + 1(4-6) + 1(12-42)]$$

$$= \frac{1}{2}[8 \times 4 - 2 - 30]$$

$$= \frac{1}{2}[32 - 32] = 0$$

Since, area of triangle is zero, hence the given points are collinear

Example based on the Pharmaceutical Applications of Determinants

***Example* 1:** A mixture is to be made of milk, water and sugar. The three materials contains quantities P, Q, R as shown in table. How to form a mixture which will have 8 gms of P, 5 gms of Q and 7 gms of sugar.

Material	P	Q	R
Milk	1	3	4
Water	2	1	2
Sugar	5	1	1

Solve by using Crammer's rule.

***Solution*:** Let milk is x, water is y and sugar is z in a mixture.

Then we have the equation

$$x + 3y + 4z = 8$$

$$2x + y + 2z = 5$$

$$5x + y + z = 7$$

By using Cramer's rule

$$\Delta = \begin{vmatrix} 1 & 3 & 4 \\ 2 & 1 & 2 \\ 5 & 1 & 1 \end{vmatrix}$$

$$= 1(1-2) - 3(2-10) + 4(2-5)$$

$$= -1 + 24 - 12 = 11$$

$$\Delta_1 = \begin{vmatrix} 8 & 3 & 4 \\ 5 & 1 & 2 \\ 7 & 1 & 1 \end{vmatrix}$$

$$= 8(1-2) - 3(5-14) + 4(5-7)$$

$$= -8 + 27 - 8 = 11$$

$$\Delta_2 = \begin{vmatrix} 1 & 8 & 4 \\ 2 & 5 & 2 \\ 5 & 7 & 1 \end{vmatrix}$$

$$= 1(5-14) - 8(2-10) + 4(14-25)$$

$$= -9 + 64 - 44 = 11$$

$$\Delta_3 = \begin{vmatrix} 1 & 3 & 8 \\ 2 & 1 & 5 \\ 5 & 1 & 7 \end{vmatrix}$$

$$= 1(7-5) - 3(14-25) + 8(2-5)$$

$$= 2 + 33 - 24 = 11$$

$$\therefore \quad x = \frac{\Delta_1}{\Delta} = \frac{11}{11} = 1$$

$$y = \frac{\Delta_2}{\Delta} = \frac{11}{11} = 1$$

$$z = \frac{\Delta_3}{\Delta} = \frac{11}{11} = 1$$

***Example* 2:** The monthly expenditure in an Industry for 3 months is given below according to the staff employed.

Month	Employees			Total monthly Salary
	Clerks	**Typists**	**Peons**	
April	4	2	3	4900
March	3	3	2	4500
Feb	4	3	4	5800

Calculate the salary for each type of staff by using determinant

***Solution*:** Let the salary of clerks, typists and peons are a, b, c respectively.

Then
$$4a + 2b + 3c = 4900$$
$$3a + 3b + 2c = 4500$$
$$4a + 3b + 4c = 5800$$

By using Cramer's rule

$$\Delta = \begin{vmatrix} 4 & 2 & 3 \\ 3 & 3 & 2 \\ 4 & 3 & 4 \end{vmatrix}$$

$$= 4(12 - 6) - 2(12 - 8) + 3(9 - 12) = 7$$

$$\Delta_1 = \begin{vmatrix} 4900 & 2 & 3 \\ 4500 & 3 & 2 \\ 5800 & 3 & 4 \end{vmatrix}$$

$$= 4900(12 - 6) - 2(18000 - 11600) + 3\,(13500 - 17400)$$

$$= 4900$$

$$\Delta_2 = \begin{vmatrix} 4 & 4900 & 3 \\ 3 & 4500 & 2 \\ 4 & 5800 & 4 \end{vmatrix}$$

$$= 4(18000 - 11600) - 4900(12 - 8) + 3\,(17400 - 18000)$$

$$= 4200$$

$$\Delta_3 = \begin{vmatrix} 4 & 2 & 4900 \\ 3 & 3 & 4500 \\ 4 & 3 & 5800 \end{vmatrix}$$

$= 4(17400 - 13500) - 2(17400 - 18000) + 4900(9 - 12)$

$= 2100$

$\therefore \quad a = \dfrac{\Delta_1}{\Delta} = \dfrac{4900}{7} = 700$

$b = \dfrac{\Delta_2}{\Delta} = \dfrac{4200}{7} = 600$

$c = \dfrac{\Delta_3}{\Delta} = \dfrac{2100}{7} = 300$

$\therefore$ Salary of clerks, typists and peons are Rs 700, Rs 600, and Rs 300

Exercise 3.1

1. Evaluate $\begin{vmatrix} 5 & 1 & 1 \\ 5 & 2 & 0 \\ 3 & 0 & 6 \end{vmatrix}$

2. Evaluate each of the following determinants.

(i) $\begin{vmatrix} x+1 & x^2 \\ 2 & x-1 \end{vmatrix}$ (ii) $\begin{vmatrix} \operatorname{cosec} x & \cot x \\ \cot x & \operatorname{cosec} x \end{vmatrix}$ (iii) $\begin{vmatrix} 1 & w \\ w & -w \end{vmatrix}$

3. Evaluate $\begin{vmatrix} 31 & 37 & 92 \\ 31 & 58 & 71 \\ 31 & 55 & 74 \end{vmatrix}$

4. Evaluate $\begin{vmatrix} 1^2 & 2^2 & 3^2 \\ 2^2 & 3^2 & 4^2 \\ 3^2 & 4^2 & 5^2 \end{vmatrix}$

5. Show that $\begin{vmatrix} 1 & w^3 & w^2 \\ w^3 & 1 & w \\ w^2 & w & 1 \end{vmatrix} = 3$, w is the cube root of unity

6. If $= \begin{vmatrix} x & -6 & 1 \\ 2x & -3 & 3 \\ 0 & 1 & 2 \end{vmatrix} = 0$, find the value of x.

7. Show that $\begin{vmatrix} 0 & 1 & \sec\alpha \\ \tan\alpha & -\sec\alpha & \tan\alpha \\ 1 & 0 & 1 \end{vmatrix} = \sec^2\alpha$

8. If $\begin{vmatrix} 1 & -3 & 2 \\ 4 & -1 & 2 \\ 3 & 5 & 2 \end{vmatrix}$

find the cofactor of elements 1, –3, 2 and also find the value of determinant.

9. Find the ratio of the cofactor and minor of 4 in determinant

$$\begin{vmatrix} 0 & 1 & -8 \\ -1 & 0 & 4 \\ 2 & -1 & 0 \end{vmatrix}$$

10. Show that $\begin{vmatrix} 0 & ab^2 & ac^2 \\ a^2b & 0 & bc^2 \\ a^2c & cb^2 & 0 \end{vmatrix} = 2a^3b^3c^3$

11. Show that $\begin{vmatrix} bc & a & a^2 \\ ca & b & b^2 \\ ab & c & c^2 \end{vmatrix} = \begin{vmatrix} 1 & a^2 & a^3 \\ 1 & b^2 & b^3 \\ 1 & c^2 & c^3 \end{vmatrix}$

12. Prove that $\begin{vmatrix} \frac{1}{a} & a^2 & bc \\ \frac{1}{b} & b^2 & ca \\ \frac{1}{c} & c^2 & ab \end{vmatrix} = 0$

13. Prove that $\begin{vmatrix} b^2+c^2 & ab & ac \\ ab & c^2+a^2 & bc \\ ac & bc & a^2+b^2 \end{vmatrix} = 4a^2b^2c^2$

14. Without expanding, prove that $\begin{vmatrix} 1 & a & bc \\ 1 & b & ca \\ 1 & c & ab \end{vmatrix} = \begin{vmatrix} 1 & a & a^2 \\ 1 & b & b^2 \\ 1 & c & c^2 \end{vmatrix}$

[Hint: multiply R_1, R_2, R_3 by a, b, c and divide by abc]

15. Solve $\begin{vmatrix} x & 3 & 7 \\ 2 & x & 2 \\ 7 & 6 & x \end{vmatrix} = 0$

[Hint: $R_1 \to R_1 + R_2 + R_3$]

16. Show that x = 8 is a root of the equation $\begin{vmatrix} x & -6 & -1 \\ 2 & -3x & x-3 \\ -3 & 2x & x+2 \end{vmatrix}$ and solve it.

17. Solve the following for x.

(i) $\begin{vmatrix} x+7 & 3x & 3x \\ 3x & x+7 & 3x \\ 3x & 3x & x+7 \end{vmatrix} = 0$ (ii) $\begin{vmatrix} 1 & x & x^3 \\ 1 & b & b^3 \\ 1 & c & c^3 \end{vmatrix} = 0$

18. Find the value of x so that the points A(3, –2), B(x, 2) and C(8, 8) lie on a line.

19. Find k so that the points are collinear (3, –2) (k, 2) and (8, 8).

20. If points (a, 0), (0, b) and (p, q) are collinear, prove that $\frac{p}{a} + \frac{q}{b} = 1$.

21. Solve the following system of linear equations using Cramer's rule :

$$x + 2y + 3z = 6,\ 2x + 4y + z = 7 \text{ and } 3x + 2y + 9z = 14$$

22. Solve the following system of equations using Cramer's rule :

$$5x - 7y + z = 11,\ 6x - 8y - z = 15 \text{ and } 3x + 2y - 6z = 7.$$

23. Does the following system of equations have a unique solution?

$$x - y + z = -1$$

$$3x - y + 2z = 1$$

$$2x + z = 3$$

24. Find the value of k for which the following system is consistent $2x - y + 3 = 0$, $kx - y + 1 = 0$, $5x - y - 3 = 0$

[Hint : solving equations (i) and (ii), we get $x = 2$, $y = 7$ by Cramer's rule and substitute in equation (iii) we get $k = 3$]

25. (i) Can we add two determinants of order (2×3) and (3×2)

(ii) what about the value of determinant when row and columns are interchanged.

26. By Cramer's rule solve the following system of equations.

(i) $3x + y = 19,\ \ 3x - y = 23$

(ii) $x - 2y = 4,\ \ -3x + 5y = -7$

27. By Cramer's rule solve the system of equations

$$x + y = 5,\ y + z = 3,\ z + x = 4$$

28. Solve the following equations using Cramer's rule

$$\frac{4}{x+2} + \frac{5}{y-1} = 3 \text{ and } \frac{6}{x+2} + \frac{8}{y-1} = 5$$

[Hint: Let $\frac{1}{x+2} = a$, $\frac{1}{y-1} = b$ then $4a + 5b = 3$, $6a + 8b = 5$]

29. Solve the equations using Cramer's rule

$$\frac{1}{x} + \frac{3}{y} = 5 \text{ and } \frac{3}{x} - \frac{4}{y} = 2$$

30. Prove that $\begin{vmatrix} (b+c)^2 & a^2 & a^2 \\ b^2 & (c+a)^2 & b^2 \\ c^2 & c^2 & (a+b)^2 \end{vmatrix} = 2abc(a+b+c)^3$

31. Prove that $\begin{vmatrix} a^2 & bc & c^2+ac \\ a^2+ab & b^2 & ac \\ ab & b^2+bc & c^2 \end{vmatrix} = 4a^2b^2c^2$

32. Prove that $\begin{vmatrix} b+c & c+a & a+b \\ q+r & r+p & p+q \\ y+z & z+x & x+y \end{vmatrix} = 2\begin{vmatrix} a & b & c \\ p & q & r \\ x & y & z \end{vmatrix}$

[Hint: by $c_1 \to c_1 + c_2 + c_3$ and then $c_2 \to c_2 - c_1$ and $c_3 \to c_3 - c_1$]

33. If $a \neq p$, $b \neq q$, $c \neq r$, and $\begin{vmatrix} p & b & c \\ a & q & c \\ a & b & r \end{vmatrix} = 0$

evaluate $\frac{p}{p-a} + \frac{q}{q-b} + \frac{r}{r-c}$

[Hint: by $R_1 \to R_1 - R_2$ and $R_2 \to R_2 - R_3$ and expanding after divide by

$(p-q)(q-b)(r-c)$ we get $\frac{p}{p-a} + \frac{q}{q-b} + \frac{r}{r-c} = 2$]

34. If x, y and z are distinct and

$$\Delta = \begin{vmatrix} x & x^2 & 1+x^3 \\ y & y^2 & 1+y^3 \\ z & z^2 & 1+z^3 \end{vmatrix} = 0, \text{ then prove that } xyz = -1.$$

35. Prove that $\begin{vmatrix} a & b & c \\ b & c & a \\ c & a & b \end{vmatrix} = 3abc - a^3 - b^3 - c^3$

Answers

1. 24

2. (i) $-(1+x^2)$ (ii) 1 (iii) 1

3. 0

4. –8

6. 0

8. $c_{11} = -12$, $c_{12} = -2$, $c_{13} = 23$, $\Delta = 40$

9. 1 : –1

15. $x = 2, x = 7, x = -9$

16. 5/3, 1, 2

17. (i) $-1, \frac{7}{2}, \frac{7}{2}$ (ii) 0, –1, –12

18. $x = 5$

19. 5

21. $x = y = z = 1$

22. $x = 1, y = -1, z = -1$

23. No

25. (i) No (ii) same or equal

26. (i) $x = 7, y = -2$ (ii) $x = -6, y = -5$

27. $x = 3, y = 2, z = 1$

28. $x = -4, y = 2$

29. $x = \frac{1}{2}, y = 1$

CHAPTER 4

MATRICES

The theory of matrices came into existence in the nineteenth century as a result of the works of mathematician Arthen Cayley.

To define the matrix, we define the following terms.

(i) *Row*: A horizontal line is called a row.

for example a, b, c, $\longrightarrow$

(ii) *Column*: A vertical line is called a column.

$$\downarrow \begin{matrix} a_1 \\ a_2 \\ a_3 \end{matrix}$$

(iii) *Array*: A collection of numbers arranged in rows and columns forms an array.

Matrix

A collection of mn numbers arranged in the form of an ordered set of m rows and n columns in a rectangular array is called an m × n matrix. Generally we denote the matrix by rectangular [ ] or square brackets ( ), but another way to represent it by || ||

Example: **(i)** Matrix of order m × n.

$$A = \begin{bmatrix} a_{11} & a_{12} & a_{13} & -- & a_{1n} \\ a_{21} & a_{22} & a_{23} & -- & a_{2n} \\ -- & -- & -- & -- & -- \\ a_{m1} & a_{m2} & -- & -- & a_{mn} \end{bmatrix}$$

(ii) Matrix of order 2 × 3

$$B = \begin{bmatrix} 1 & 2 & 3 \\ 4 & 5 & 6 \end{bmatrix}$$

(iii) Matrix of order 3 × 2

$$C = \begin{bmatrix} 1 & 2 \\ 3 & 4 \\ 5 & 6 \end{bmatrix}$$

It is clear that order of matrix can be written by m × n, where m is number of rows and n is number of columns.

Types of Matrices

1. **Row Matrix:** It contains one row

   Example: $A = [1\ 2\ 3]_{1\times3}$ $\quad B = [1, 2, 3, - - - n]_{1\times n}$

2. **Column Matrix:** It contains one column

   Example: $A = \begin{bmatrix} 1 \\ 2 \\ 3 \end{bmatrix}_{3\times1}$, $B = \begin{bmatrix} 1 \\ 2 \\ 3 \\ : \\ n \end{bmatrix}_{n\times1}$

3. **Zero Matrix or Null Matrix:** A matrix whose all elements are zero is called zero matrix. The order of zero matrix can be made as we can.

   Example: $A = \begin{bmatrix} 0 \\ 0 \\ 0 \end{bmatrix}_{3\times1}$ $\quad B = \begin{bmatrix} 0 & 0 \\ 0 & 0 \end{bmatrix}_{2\times2}$

4. **Square Matrix:** A matrix is called square matrix if it has number of rows are equal to number of columns.

   Example: $A = \begin{bmatrix} 1 & 2 \\ 2 & 4 \end{bmatrix}_{2\times2}$ $\quad B = \begin{bmatrix} a_1 & a_2 & a_3 \\ b_1 & b_2 & b_3 \\ c_1 & c_2 & c_3 \end{bmatrix}_{3\times3}$

Generally we write square matrix of order n, at the place of n × n square matrix.

***Example*:** square matrix of order 4 is

$$A = \begin{bmatrix} 1 & 2 & 3 & 4 \\ 2 & 3 & 4 & 5 \\ 5 & 6 & 7 & 8 \\ 0 & 0 & 1 & 2 \end{bmatrix}$$

***Note*:** In every square matrix there are two diagonals. The diagonal which starts from the left hand top corner to end at the right hand bottom corner is called the principal diagonal or 'leading diagonal' i.e.

$$A = \begin{bmatrix} a_{11} & a_{12} & a_3 \\ a_{21} & a_{22} & a_{23} \\ a_{31} & a_{32} & a_{33} \end{bmatrix}$$

In short elements where i = j are formed principal diagonal of square matrix.

5. **Diagonal Matrix:** If all the elements of a square matrix are zero except those in the principal diagonal is called a diagonal matrix.

Example
$$\begin{bmatrix} 1 & 0 \\ 0 & 2 \end{bmatrix}, \begin{bmatrix} 2 & 0 & 0 \\ 0 & 1 & 0 \\ 0 & 0 & 5 \end{bmatrix}$$

are diagonal matrices. They can also denoted as diagonal (1, 2), diagonal (2, 1, 5)

6. **Scalar Matrix:** A diagonal matrix whose diagonal elements are all equal, is called a scalar matrix.

***Example*:**
$$A = \begin{bmatrix} 2 & 0 & 0 \\ 0 & 2 & 0 \\ 0 & 0 & 2 \end{bmatrix}_{3\times3} \qquad B = \begin{bmatrix} 5 & 0 \\ 0 & 5 \end{bmatrix}_{2\times2}$$

7. **Unit Matrix:** A diagonal matrix is called unit matrix if its principal diagonal elements are 1. It is denoted by I_n, where n is order of square matrix.

(a) $I_1 = [1]_{1\times1}$ (b) $I_2 = \begin{bmatrix} 1 & 0 \\ 0 & 1 \end{bmatrix}_{2\times2}$

(c) $I_3 = \begin{bmatrix} 1 & 0 & 0 \\ 0 & 1 & 0 \\ 0 & 0 & 1 \end{bmatrix}_{3\times3}$ (d) $I_4 = \begin{bmatrix} 1 & 0 & 0 & 0 \\ 0 & 1 & 0 & 0 \\ 0 & 0 & 1 & 0 \\ 0 & 0 & 0 & 1 \end{bmatrix}_{4\times4}$

Note: For unit matrix in general we can define

$$A = [a_{ij}] = 1 \text{ if } i = j$$
$$0 \text{ if } i \neq j$$

8. Triangular Matrix: If in a square matrix all the elements are above or bellow the principal diagonal are zero, is called a triangular matrix. There are two types of triangular matrix;

(a) upper triangular (b) Lower triangular

(a) *Upper triangular*: A square matrix having all elements below principal diagonal are zero is called upper triangular matrix.

$$A = \begin{bmatrix} a_{11} & a_{12} & a_{13} & \dots & a_{1n} \\ 0 & a_{22} & a_{23} & \dots & a_{2n} \\ 0 & 0 & a_{33} & \dots & a_{3n} \\ 0 & 0 & 0 & \dots & a_{nn} \end{bmatrix}$$

it is clear that $A = [a_{ij}] = 0$ for $i > j$

then A is upper triangular

(b) *Lower triangular*: A square matrix having all the elements above the principal diagonal is zero is called lower triangular matrix.

$$A = \begin{bmatrix} a_{11} & 0 & 0 & \dots & 0 \\ a_{21} & a_{22} & 0 & \dots & 0 \\ a_{31} & a_{32} & a_{33} & \dots & 0 \\ a_{n1} & a_{n2} & \dots & \dots & a_{nn} \end{bmatrix}$$

$A = [a_{ij}] = 0$ for $i < j$ is lower triangular

Example of upper and lower triangular matrices.

$$A = \begin{bmatrix} 1 & 2 & 3 & 4 \\ 0 & 3 & 4 & 5 \\ 0 & 0 & 2 & 3 \\ 0 & 0 & 0 & 5 \end{bmatrix} \qquad B = \begin{bmatrix} 1 & 0 & 0 & 0 \\ 2 & 3 & 0 & 0 \\ 4 & 5 & 6 & 0 \\ 7 & 8 & 9 & 10 \end{bmatrix}$$

Upper triangular Lower triangular

9. **Equal Matrix:** Two matrices A and B are said to be equal matrix if

(i) Their order are exactly same.

(ii) Their corresponding elements are equal.

Example: $A = \begin{bmatrix} 1 & 2 \\ 3 & 4 \end{bmatrix}$, $B = \begin{bmatrix} a & b \\ c & d \end{bmatrix}$

$$A = B \text{ if } a = 1,\ b = 2,\ c = 3, d = 4$$

10. **Singular Matrix:** A square matrix is said to be singular matrix if its corresponding determinant is zero, that is $|A| = 0$

Example: (i) $A = \begin{bmatrix} 1 & 2 \\ 2 & 4 \end{bmatrix}$

(ii) $A = \begin{bmatrix} 1 & 2 & 3 \\ 4 & 5 & 6 \\ 7 & 8 & 9 \end{bmatrix}$

11. **Non-singular or Regular or Investigable Matrix:** A square matrix is said to be non-singular if its corresponding determinant is not zero i.e. $|A| \neq 0$

Example: $A = \begin{bmatrix} 1 & 2 \\ 3 & 4 \end{bmatrix}$ $B = \begin{bmatrix} 0 & 0 & 1 \\ 2 & 3 & 4 \\ 1 & 5 & \end{bmatrix}$

Addition and Subtraction of Matrices

Two matrices A and B can be added or subtracted if their order are same. If order of matrix A is m × n, order of matrix B is m × n, then order of $A \pm B$ is also m × n.

Example: $A = \begin{bmatrix} a_1 & b_1 \\ a_2 & b_2 \end{bmatrix}_{2\times2}$ $B = \begin{bmatrix} c_1 & d_1 \\ c_2 & d_2 \end{bmatrix}_{2\times2}$ then

$$A + B = \begin{bmatrix} a_1 + c_1 & b_1 + d_1 \\ a_2 + c_2 & b_2 + d_2 \end{bmatrix}$$

$$A - B = \begin{bmatrix} a_1 - c_1 & b_1 - d_1 \\ a_2 - c_2 & b_2 - d_2 \end{bmatrix}$$

Properties of Addition of Matrices

If, three matrices A, B, C are given of same order then

(i) $A + B = B + A$ (commutative law)

(ii) $A + (B + C) = (A + B) + C$ (Associative law)

(iii) Cancellation law

(a) $A + B = A + C \Rightarrow B = C$ (left cancellation)

(b) $B + A = C + A \Rightarrow B = C$ (Right cancellation)

(iv) *Additive inverse:* For every matrix A, there exists a matrix – A such that

$$A + (-A) = 0$$

where $A = [a_{ij}]_{m\times n}$, and $-A = [-a_{ij}]_{m\times n}$ then –A is called additive inverse of A.

(v) Distributive law: If K is any number then $K(A + B) = KA + KB$

Multiplication of Two Matrices

Two matrices A and B can be multiplied if number of columns in A are equal to number of rows in B. i.e.

$$[A]_{m\times n}\ [B]_{n\times p} = [AB]_{m\times p}$$

Properties of Multiplication

(i) $AB \neq BA$ (generally)

(ii) $A(BC) = (AB)C$ (Associative law)

(iii) It follows distributive laws

$A.(B + C) = AB + AC$

Where B and C are same order matrices and AB and AC both possible

(iv) Generally it not follows cancellation law i.e. $AB = AC \Rightarrow B \neq C$

(v) Multiplication of two non square matrix may be a square matrix

Example: $A = [A]_{m\times n}$, $B = [B]_{n\times m} \Rightarrow AB = [AB]_{m\times m}$

(vi) Multiplication of two non zero matrices can be zero matrix

Example: $A = \begin{bmatrix} 1 & 0 \\ 0 & 0 \end{bmatrix}$, $B = \begin{bmatrix} 0 & 0 \\ 0 & 1 \end{bmatrix} \Rightarrow AB = \begin{bmatrix} 0 & 0 \\ 0 & 0 \end{bmatrix}$

Transpose of a Matrix

The matrix obtained by interchanging rows and columns of a given matrix A is called transpose of the matrix A and is denoted by A' or A^T

$$A = \begin{bmatrix} a_{11} & a_{12} & a_{13} \\ a_{21} & a_{22} & a_{23} \end{bmatrix} \text{ then } A' = \begin{bmatrix} a_{11} & a_{21} \\ a_{12} & a_{22} \\ a_{13} & a_{23} \end{bmatrix}$$

***Note*:** (i) If $A = [a_{ij}]_{m\times n}$ then $A' = [a_{ij}]_{n\times m}$

(ii) Transpose of row matrix is column matrix

(iii) Transpose of column matrix is row matrix

(iv) $(A+B)' = A' + B'$

(v) $(AB)' = B'A'$

(vi) $(KA)' = KA'$ where K is constant

(vii) $(A')' = A$

Examples based on the operation of matrices

1. If $A = \begin{bmatrix} 2 & 3 \\ 4 & 5 \end{bmatrix}$, then find the value of X when X + A = 0

***Solution*:** $X + A = 0$

$X = -A$

$$X = -\begin{bmatrix} 2 & 3 \\ 4 & 5 \end{bmatrix}$$

$$X = \begin{bmatrix} -2 & -3 \\ -4 & -5 \end{bmatrix}$$

2. Find (i) 3A – B (ii) 3A + 2B, where

$$A = \begin{bmatrix} 0 & 2 & 3 \\ 2 & 1 & 4 \end{bmatrix}, \quad B = \begin{bmatrix} 1 & 2 & 3 \\ 2 & 4 & 5 \end{bmatrix}$$

$$\text{(i) } 3A - B = 3\begin{bmatrix} 0 & 2 & 3 \\ 2 & 1 & 4 \end{bmatrix} - \begin{bmatrix} 1 & 2 & 3 \\ 2 & 4 & 5 \end{bmatrix}$$

$$= \begin{bmatrix} 0 & 6 & 9 \\ 6 & 3 & 12 \end{bmatrix} - \begin{bmatrix} 1 & 2 & 3 \\ 2 & 5 & 5 \end{bmatrix}$$

$$= \begin{bmatrix} -1 & 4 & 6 \\ 4 & -2 & 7 \end{bmatrix}$$

$$\text{(ii) } 3A + 2B = 3\begin{bmatrix} 0 & 2 & 3 \\ 2 & 1 & 4 \end{bmatrix} + 2\begin{bmatrix} 1 & 2 & 3 \\ 2 & 4 & 5 \end{bmatrix}$$

$$= \begin{bmatrix} 0 & 6 & 9 \\ 6 & 3 & 12 \end{bmatrix} + \begin{bmatrix} 2 & 4 & 6 \\ 4 & 8 & 10 \end{bmatrix}$$

$$= \begin{bmatrix} 2 & 10 & 15 \\ 10 & 11 & 22 \end{bmatrix}$$

***Example* 1:** If $A = \begin{bmatrix} 1 & 2 & 3 \\ 4 & 5 & 6 \end{bmatrix}$ and $B = \begin{bmatrix} 7 & 8 \\ 9 & 10 \\ 11 & 12 \end{bmatrix}$, find AB.

Solution:

$$AB = \begin{bmatrix} 1\times 7 + 2\times 9 + 3\times 11 & 1\times 8 + 2\times 10 + 3\times 12 \\ 4\times 7 + 5\times 9 + 6\times 11 & 4\times 8 + 5\times 10 + 6\times 12 \end{bmatrix}$$

$$= \begin{bmatrix} 7+18+33 & 8+20+36 \\ 28+45+66 & 32+50+72 \end{bmatrix}$$

$$= \begin{bmatrix} 58 & 64 \\ 139 & 154 \end{bmatrix}$$

***Example* 2:** If $A = \begin{bmatrix} 1 & -2 & 3 \\ -4 & 2 & 5 \end{bmatrix}$ and $B = \begin{bmatrix} 2 & 3 \\ 4 & 5 \\ 2 & 1 \end{bmatrix}$, find AB and show that $AB \neq BA$

Solution:

$$AB = \begin{bmatrix} 1 & -2 & 3 \\ -4 & 2 & 5 \end{bmatrix}\begin{bmatrix} 2 & 3 \\ 4 & 5 \\ 2 & 1 \end{bmatrix}$$

$$= \begin{bmatrix} 1\times 2 - 2\times 4 + 3\times 2 & 1\times 3 - 2\times 5 + 3\times 1 \\ -4\times 2 + 2\times 4 + 5\times 2 & -4\times 3 + 2\times 5 + 5\times 1 \end{bmatrix}$$

$$AB = \begin{bmatrix} 0 & -4 \\ 10 & 3 \end{bmatrix}$$

$$BA = \begin{bmatrix} 2 & 3 \\ 4 & 5 \\ 2 & 1 \end{bmatrix} \begin{bmatrix} 1 & -2 & 3 \\ -4 & 2 & 5 \end{bmatrix}$$

$$= \begin{bmatrix} 2\times1-3\times4 & -2\times2+3\times2 & 2\times3+3\times5 \\ 4\times1-5\times4 & -4\times2+5\times2 & 4\times3+5\times5 \\ 2\times1-1\times4 & -2\times2+1\times2 & 2\times3+1\times5 \end{bmatrix}$$

$$= \begin{bmatrix} -10 & 2 & 21 \\ -16 & 2 & 37 \\ -2 & -2 & 11 \end{bmatrix} \qquad \therefore\ AB \neq BA$$

***Example* 3:** Evaluate A(BC) and (AB)C where

$$A = [1\ 2\ 3],\quad B = \begin{bmatrix} 2 & 0 & -1 \\ -1 & 0 & 2 \\ -1 & 2 & 0 \end{bmatrix} \text{ and } C = \begin{bmatrix} 0 \\ -1 \\ 2 \end{bmatrix}$$

Also show that the result is associative under multiplication

***Solution*:** Here

$$BC = \begin{bmatrix} 2 & 0 & -1 \\ -1 & 0 & 2 \\ -1 & 2 & 0 \end{bmatrix} \begin{bmatrix} 0 \\ -1 \\ 2 \end{bmatrix}$$

$$= \begin{bmatrix} 2\times0-0\times1-1\times2 \\ -1\times0-0\times1+2\times2 \\ -1\times0-2\times1+0\times2 \end{bmatrix} = \begin{bmatrix} -2 \\ 4 \\ -2 \end{bmatrix}$$

Thus, $A(BC) = [1\ 2\ 3] \begin{bmatrix} -2 \\ 4 \\ -2 \end{bmatrix}$

$$= [-1 \times 2 + 2 \times 4 - 3 \times 2]$$

$$= [-2 + 8 - 6] = [0]$$

Also, $AB = [1\ 2\ 3]\begin{bmatrix} 2 & 0 & -1 \\ -1 & 0 & 2 \\ -1 & 2 & 0 \end{bmatrix}$

$= \begin{bmatrix} 1\times2-2\times1-3\times1 & 1\times0+2\times0+3\times2 & -1\times1+2\times2+3\times0 \end{bmatrix}$

$= [-3\ 6\ 3]$

Thus, $(AB)C = [-3\ 6\ 3]\begin{bmatrix} 0 \\ -1 \\ 2 \end{bmatrix}$

$= [-3\times0-6\times1+3\times2] = [0]$

Hence, A(BC) = (AB)C

Hence, the result is associative under multiplication.

***Example* 4:** If $A = \begin{bmatrix} 1 & 2 \\ 3 & 4 \end{bmatrix}$, $B = \begin{bmatrix} 2 & 1 \\ 4 & 2 \end{bmatrix}$ and $C = \begin{bmatrix} 5 & 1 \\ 7 & 4 \end{bmatrix}$

show that A(B + C) = AB + AC

***Solution*:** Here $B + C = \begin{bmatrix} 2 & 1 \\ 4 & 2 \end{bmatrix} + \begin{bmatrix} 5 & 1 \\ 7 & 4 \end{bmatrix}$

$= \begin{bmatrix} 2+5 & 1+1 \\ 4+7 & 2+4 \end{bmatrix} = \begin{bmatrix} 7 & 2 \\ 11 & 6 \end{bmatrix}$

$\therefore \quad A(B+C) = \begin{bmatrix} 1 & 2 \\ 3 & 4 \end{bmatrix}\begin{bmatrix} 7 & 2 \\ 11 & 6 \end{bmatrix}$

$= \begin{bmatrix} 1\times7+2\times11 & 1\times2+2\times6 \\ 3\times7+4\times11 & 3\times2+4\times6 \end{bmatrix}$

$= \begin{bmatrix} 7+22 & 2+12 \\ 21+44 & 6+24 \end{bmatrix} = \begin{bmatrix} 29 & 14 \\ 65 & 30 \end{bmatrix}$

$$AB = \begin{bmatrix} 1 & 2 \\ 3 & 4 \end{bmatrix} \begin{bmatrix} 2 & 1 \\ 4 & 2 \end{bmatrix}$$

$$= \begin{bmatrix} 1\times 2+2\times 4 & 1\times 1+2\times 2 \\ 3\times 2+4\times 4 & 3\times 1+4\times 2 \end{bmatrix}$$

$$= \begin{bmatrix} 2+8 & 1+4 \\ 6+16 & 3+8 \end{bmatrix} = \begin{bmatrix} 10 & 5 \\ 22 & 11 \end{bmatrix}$$

$$AC = \begin{bmatrix} 1 & 2 \\ 3 & 4 \end{bmatrix} \begin{bmatrix} 5 & 1 \\ 7 & 4 \end{bmatrix}$$

$$= \begin{bmatrix} 1\times 5+2\times 7 & 1\times 1+2\times 4 \\ 3\times 5+4\times 7 & 3\times 1+4\times 4 \end{bmatrix}$$

$$= \begin{bmatrix} 5+14 & 1+8 \\ 15+28 & 3+16 \end{bmatrix}$$

$$= \begin{bmatrix} 19 & 9 \\ 43 & 19 \end{bmatrix}$$

$$AB + AC = \begin{bmatrix} 10 & 5 \\ 22 & 11 \end{bmatrix} + \begin{bmatrix} 19 & 9 \\ 43 & 19 \end{bmatrix}$$

$$= \begin{bmatrix} 10+19 & 5+9 \\ 22+43 & 11+19 \end{bmatrix}$$

$$= \begin{bmatrix} 29 & 14 \\ 65 & 30 \end{bmatrix}$$

$\therefore$ $A(B + C) = AB + AC$

***Example* 5:** If $A = \begin{bmatrix} 1 & 2 & 2 \\ 2 & 1 & 2 \\ 2 & 2 & 1 \end{bmatrix}$ and $I = \begin{bmatrix} 1 & 0 & 0 \\ 0 & 1 & 0 \\ 0 & 0 & 1 \end{bmatrix}$

Evaluate $A^2 - 4A - 5I$

***Solution*:** $A^2 = A.A = \begin{bmatrix} 1 & 2 & 2 \\ 2 & 1 & 2 \\ 2 & 2 & 1 \end{bmatrix} \begin{bmatrix} 1 & 2 & 2 \\ 2 & 1 & 2 \\ 2 & 2 & 1 \end{bmatrix}$

$$= \begin{bmatrix} 1\times1+2\times2+2\times2 & 1\times2+2\times1+2\times2 & 1\times2+2\times2+2\times1 \\ 2\times1+1\times2+2\times2 & 2\times2+1\times1+2\times2 & 2\times2+1\times2+2\times1 \\ 2\times1+2\times2+1\times2 & 2\times2+2\times1+1\times2 & 2\times2+2\times2+1\times1 \end{bmatrix}$$

$$= \begin{bmatrix} 1+4+4 & 2+2+4 & 2+4+2 \\ 2+2+4 & 4+1+4 & 4+2+2 \\ 2+4+2 & 4+2+2 & 4+4+1 \end{bmatrix}$$

$$= \begin{bmatrix} 9 & 8 & 8 \\ 8 & 9 & 8 \\ 8 & 8 & 9 \end{bmatrix}$$

$$\therefore A^2 - 4A - 5I = \begin{bmatrix} 9 & 8 & 8 \\ 8 & 9 & 8 \\ 8 & 8 & 9 \end{bmatrix} - 4\begin{bmatrix} 1 & 2 & 2 \\ 2 & 1 & 2 \\ 2 & 2 & 1 \end{bmatrix} - 5\begin{bmatrix} 1 & 0 & 0 \\ 0 & 1 & 0 \\ 0 & 0 & 1 \end{bmatrix}$$

$$= \begin{bmatrix} 9-4-5 & 8-8-0 & 8-8-0 \\ 8-8-0 & 9-4-5 & 8-8-0 \\ 8-8-0 & 8-8-0 & 9-4-5 \end{bmatrix}$$

$$= \begin{bmatrix} 0 & 0 & 0 \\ 0 & 0 & 0 \\ 0 & 0 & 0 \end{bmatrix}$$

$$\therefore \mathbf{A^2 - 4A - 5I = 0}$$

***Example* 6:** If $A = \begin{bmatrix} 1 & 2 \\ 3 & 4 \end{bmatrix}$; find A^2

***Solution*:**

$$A^2 = A.A = \begin{bmatrix} 1 & 2 \\ 3 & 4 \end{bmatrix}\begin{bmatrix} 1 & 2 \\ 3 & 4 \end{bmatrix}$$

$$= \begin{bmatrix} 1\times1+2\times3 & 1\times2+2\times4 \\ 3\times1+4\times3 & 3\times2+4\times4 \end{bmatrix}$$

$$= \begin{bmatrix} 1+6 & 2+8 \\ 3+12 & 6+16 \end{bmatrix}$$

$$= \begin{bmatrix} 7 & 10 \\ 15 & 22 \end{bmatrix}$$

***Example* 7:** If $A = \begin{bmatrix} 1 & 2 & -1 \\ 3 & 0 & 2 \\ 4 & 5 & 0 \end{bmatrix}$ and $B = \begin{bmatrix} 1 & 0 & 0 \\ 2 & 1 & 0 \\ 0 & 1 & 3 \end{bmatrix}$

verify that $(AB)' = B'A'$, where A' and B' are transpose of A and B.

***Solution*:** We have

$$AB = \begin{bmatrix} 1 & 2 & -1 \\ 3 & 0 & 2 \\ 4 & 5 & 0 \end{bmatrix} \begin{bmatrix} 1 & 0 & 0 \\ 2 & 1 & 0 \\ 0 & 1 & 3 \end{bmatrix}$$

$$= \begin{bmatrix} 1\times1+2\times2+0 & 1\times0+2\times1-1\times1 & 0+0-1\times3 \\ 3\times1+0+0 & 0+0+2\times1 & 0+0+2\times3 \\ 4\times1+5\times2+0 & 0+5\times1+0 & 0+0+0 \end{bmatrix}$$

$$= \begin{bmatrix} 5 & 1 & -3 \\ 3 & 2 & 6 \\ 14 & 5 & 0 \end{bmatrix}$$

$$\therefore \quad (AB)' = \begin{bmatrix} 5 & 3 & 14 \\ 1 & 2 & 5 \\ -3 & 6 & 0 \end{bmatrix}$$

Now,

$$A' = \begin{bmatrix} 1 & 3 & 4 \\ 2 & 0 & 5 \\ -1 & 2 & 0 \end{bmatrix}, \quad B' = \begin{bmatrix} 1 & 2 & 0 \\ 0 & 1 & 1 \\ 0 & 0 & 3 \end{bmatrix}$$

$$B'A' = \begin{bmatrix} 1\times1+2\times2+0 & 1\times3+0+0 & 1\times4+2\times5+0 \\ 0+1\times2-1\times1 & 0+0+1\times2 & 0+1\times5+0 \\ 0+0-3\times1 & 0+0+3\times2 & 0+0+0 \end{bmatrix}$$

$$= \begin{bmatrix} 5 & 3 & 14 \\ 1 & 2 & 5 \\ -3 & 6 & 0 \end{bmatrix}$$

$$\therefore \quad (AB)' = B'A'$$

***Example* 8:** Let $A = \begin{bmatrix} 1 & -1 & 0 \\ 2 & 1 & 3 \\ 4 & 1 & 8 \end{bmatrix}$, $B = \begin{bmatrix} 4 & 1 & 0 \\ 2 & -3 & 1 \\ 1 & 1 & -1 \end{bmatrix}$ Verify that $(A+B)' = A' + B'$

***Solution*:** Here

$$A + B = \begin{bmatrix} 1+4 & -1+1 & 0+0 \\ 2+2 & 1-3 & 3+1 \\ 4+1 & 1+1 & 8-1 \end{bmatrix}$$

$$= \begin{bmatrix} 5 & 0 & 0 \\ 4 & -2 & 4 \\ 5 & 2 & 7 \end{bmatrix}$$

$$(A+B)' = \begin{bmatrix} 5 & 4 & 5 \\ 0 & -2 & 2 \\ 0 & 4 & 7 \end{bmatrix}$$

Now,

$$A' = \begin{bmatrix} 1 & 2 & 4 \\ -1 & 1 & 1 \\ 0 & 3 & 8 \end{bmatrix}, \quad B' = \begin{bmatrix} 4 & 2 & 1 \\ 1 & -3 & 1 \\ 0 & 1 & -1 \end{bmatrix}$$

$$A' + B' = \begin{bmatrix} 1+4 & 2+2 & 4+1 \\ -1+1 & 1-3 & 1+1 \\ 0+0 & 3+1 & 8-1 \end{bmatrix}$$

$$= \begin{bmatrix} 5 & 4 & 5 \\ 0 & -2 & 2 \\ 0 & 4 & 7 \end{bmatrix}$$

$\therefore$ $(A+B)' = A' + B'$

Singular and Non-Singular Matrix: A square matrix A is said to be singular if $|A| = 0$ and if $|A| \neq 0$ then it is called non-singular matrix.

***Example* 1:** If $A = \begin{bmatrix} 4 & -5 & -2 \\ 5 & -4 & 2 \\ 2 & 2 & 8 \end{bmatrix}$, show that A is singular

***Solution*:** Here, $|A| = 4(-32-4) + 5(40-4) - 2(10+8)$

$$= -144 + 180 - 36$$

$$= 0$$

Since $|A| = 0$, then matrix A is singular.

Orthogonal Matrix: A square matrix A is called orthogonal matrix if $AA' = I = A'A$.

***Example* 1:** Show that $\frac{1}{3}\begin{bmatrix} 1 & 2 & 2 \\ 2 & 1 & -2 \\ -2 & 2 & -1 \end{bmatrix}$ is orthogonal

***Solution*:** Let $A = \frac{1}{3}\begin{bmatrix} 1 & 2 & 2 \\ 2 & 1 & -2 \\ -2 & 2 & -1 \end{bmatrix}$

$$A' = \frac{1}{3}\begin{bmatrix} 1 & 2 & -2 \\ 2 & 1 & 2 \\ 2 & -2 & -1 \end{bmatrix}$$

$$\therefore AA' = \frac{1}{9}\begin{bmatrix} 1 & 2 & 2 \\ 2 & 1 & -2 \\ -2 & 2 & -1 \end{bmatrix}\begin{bmatrix} 1 & 2 & -2 \\ 2 & 1 & 2 \\ 2 & -2 & -1 \end{bmatrix}$$

$$= \frac{1}{9}\begin{bmatrix} 1\times1\times+2\times2+2\times2 & 1\times2+2\times1-2\times2 & -1\times2+2\times2-2\times1 \\ 2\times1+1\times2-2\times2 & 2\times2+1\times1+2\times2 & -2\times2+1\times2+2\times1 \\ -2\times1+2\times2+1\times2 & -2\times2+2\times1+1\times2 & 2\times2+2\times2+1\times1 \end{bmatrix}$$

$$= \frac{1}{9}\begin{bmatrix} 9 & 0 & 0 \\ 0 & 9 & 0 \\ 0 & 0 & 9 \end{bmatrix} = \begin{bmatrix} 1 & 0 & 0 \\ 0 & 1 & 0 \\ 0 & 0 & 1 \end{bmatrix} = I$$

$\therefore$ The given matrix is orthogonal.

Adjoint of a Matrix: Let $A = [a_{ij}]$ be a square matrix of order n × n. Then the transpose of the matrix formed by the cofactor of a_{ij} in the determinant $|A|$ is called the adjoint of A.

If c_{ij} be the cofactor of the element a_{ij} in the determinant $|a_{ij}|$ of $n \times n$ matrix then

$$a_{dj}\,A = \begin{bmatrix} c_{11} & c_{21} & \cdots & c_{n1} \\ c_{12} & c_{22} & \cdots & c_{n2} \\ \vdots & \vdots & & \vdots \\ c_{1n} & c_{2n} & \cdots & c_{nn} \end{bmatrix}$$

i.e. Adjoint A = transpose of C (cofactor)

***Example* 1:** Find the adjoint of $A = \begin{bmatrix} 1 & 1 & 3 \\ 0 & 1 & -1 \\ 2 & 0 & 4 \end{bmatrix}$

***Solution*:** Cofactor of different elements are

$$C_{11} = (-1)^{1+1}\begin{vmatrix} 1 & -1 \\ 0 & 4 \end{vmatrix} = 4 \qquad C_{21} = (-1)^{2+1}\begin{vmatrix} 1 & 3 \\ 0 & 4 \end{vmatrix} = -4$$

$$C_{12} = (-1)^{1+2}\begin{vmatrix} 0 & -1 \\ 2 & 4 \end{vmatrix} = -2 \qquad C_{22} = (-1)^{2+2}\begin{vmatrix} 1 & 3 \\ 2 & 4 \end{vmatrix} = -2$$

$$C_{13} = (-1)^{1+3}\begin{vmatrix} 0 & 1 \\ 2 & 0 \end{vmatrix} = -2 \qquad C_{23} = (-1)^{2+3}\begin{vmatrix} 1 & 1 \\ 2 & 0 \end{vmatrix} = 2$$

$$C_{31} = (-1)^{3+1}\begin{vmatrix} 1 & 3 \\ 1 & -1 \end{vmatrix} = -4 \qquad C_{32} = (-1)^{3+2}\begin{vmatrix} 1 & 3 \\ 0 & -1 \end{vmatrix} = 1$$

$$C_{33} = (-1)^{3+3}\begin{vmatrix} 1 & 1 \\ 0 & 1 \end{vmatrix} = 1$$

$$\therefore \quad C = \begin{bmatrix} 4 & -2 & -2 \\ -4 & -2 & 2 \\ -4 & 1 & 1 \end{bmatrix}$$

$$\therefore \quad \text{Adj } A = C' = \begin{bmatrix} 4 & -4 & -4 \\ -2 & -2 & 1 \\ -2 & 2 & 1 \end{bmatrix}$$

Inverse of a Matrix: If A is a non-singular square matrix of order n × n, then inverse of a matrix is given by

$$A^{-1} = \frac{\text{Adj}\,A}{|A|}$$

***Example* 1:** Find the inverse of $\begin{bmatrix} 1 & 1 & 3 \\ 1 & 3 & -3 \\ -2 & -4 & -4 \end{bmatrix}$

***Solution*:** Here $|A| = \begin{vmatrix} 1 & 1 & 3 \\ 1 & 3 & -3 \\ -2 & -4 & -4 \end{vmatrix}$

$$= 1\,(-12 - 12) - 1\,(-4 - 6) + 3\,(-4 + 6)$$

$$= -24 + 10 + 6 = -8$$

Cofactors of different elements are

$C_{11} = -12 - 12 = -24$ $\quad C_{21} = -(-4 + 12) = -8$ $\quad C_{31} = -3 - 9 = -12$

$C_{12} = -(-4 - 6) = 10$ $\quad C_{22} = -4 + 6 = 2$ $\quad C_{32} = -(-3 - 3) = 6$

$C_{13} = -4 + 6 = 2$ $\quad C_{23} = -(-4 + 2) = 2$ $\quad C_{33} = 3 - 1 = 2$

$$\therefore \quad C = \begin{bmatrix} -24 & 10 & 2 \\ -8 & 2 & 2 \\ -12 & 6 & 2 \end{bmatrix}$$

Now, Adjoint of A = C′

$$\text{Adj}A = \begin{bmatrix} -24 & -8 & -12 \\ 10 & 2 & 6 \\ 2 & 2 & 2 \end{bmatrix}$$

Hence, inverse of A i.e. $A^{-1} = \dfrac{\text{Adj A}}{|A|}$

$$= \frac{-1}{8}\begin{bmatrix} -24 & -8 & -12 \\ 10 & 2 & 6 \\ 2 & 2 & 2 \end{bmatrix}$$

$$= \begin{bmatrix} 3 & 1 & 3/2 \\ -5/4 & -1/4 & -3/4 \\ -1/4 & -1/4 & -1/4 \end{bmatrix}$$

***Example* 2:** If $A = \begin{bmatrix} 9 & 7 & 3 \\ 5 & -1 & 4 \\ 6 & 8 & 2 \end{bmatrix}$ find adjoint A and A^{-1}

***Solution*:** Here $|A| = \begin{vmatrix} 9 & 7 & 3 \\ 5 & -1 & 4 \\ 6 & 8 & 2 \end{vmatrix}$

$$= 9\,(-2-32) - 7(10-24) + 3\,(40+6)$$

$$= 9 \times (-34) - 7 \times (-14) + 3 \times 46$$

$$= -306 + 98 + 138$$

$$= -70$$

The cofactors of matrix are

$C_{11} = (-2-32) = -34$ $\quad C_{21} = -(14-24) = 10$ $\quad C_{31} = 28 + 3 = 31$

$C_{12} = -(10-24) = 14$ $\quad C_{22} = 18 - 18 = 0$ $\quad C_{32} = -(36-15) = -21$

$C_{13} = 40 + 6 = 46$ $\quad C_{23} = -(72-42) = -30$ $\quad C_{33} = (-9-35) = -44$

$$\therefore \quad C = \begin{bmatrix} -34 & 14 & 46 \\ 10 & 0 & -30 \\ 31 & -21 & -44 \end{bmatrix}$$

$$\text{Adj } A = C' = \begin{bmatrix} -34 & 10 & 31 \\ 14 & 0 & -21 \\ 46 & -30 & -44 \end{bmatrix}$$

Now, $$A^{-1} = \frac{\text{Adj } A}{|A|}$$

$$= \frac{-1}{70}\begin{bmatrix} -34 & 10 & 31 \\ 14 & 0 & -21 \\ 46 & -30 & -44 \end{bmatrix} = \frac{1}{70}\begin{bmatrix} 34 & -10 & -31 \\ -14 & 0 & 21 \\ -46 & 30 & 44 \end{bmatrix}$$

***Example* 3:** If $A = \begin{bmatrix} 3 & 5 & 7 \\ 2 & -3 & 1 \\ 1 & 1 & 2 \end{bmatrix}$ find adjoint A and A^{-1}.

***Solution*:** Here $$|A| = \begin{vmatrix} 3 & 5 & 7 \\ 2 & -3 & 1 \\ 1 & 1 & 2 \end{vmatrix}$$

$$= 3(-6 - 1) - 5(4 - 1) + 7(2 + 3)$$

$$= -21 - 15 + 35$$

$$= -1$$

Cofactors of matrix are

$C_{11} = (-6 - 1) = -7$ $\quad C_{21} = -(10 - 7) = -3$ $\quad C_{31} = 5 + 21 = 26$

$C_{12} = -(4 - 1) = -3$ $\quad C_{22} = 6 - 7 = -1$ $\quad C_{32} = -(3 - 14) = 11$

$C_{13} = (2 + 3) = 5$ $\quad C_{23} = -(3 - 5) = 2$ $\quad C_{33} = -9 - 10 = -19$

$$\therefore \quad C = \begin{bmatrix} -7 & -3 & 5 \\ -3 & -1 & 2 \\ 26 & 11 & -19 \end{bmatrix}$$

$$\therefore \text{ Adj } A = C' = \begin{bmatrix} -7 & -3 & 26 \\ -3 & -1 & 11 \\ 5 & 2 & -19 \end{bmatrix}$$

Now, $A^{-1} = \dfrac{\text{Adj A}}{|A|}$

$$= \frac{-1}{1}\begin{bmatrix} -7 & -3 & 26 \\ -3 & -1 & 11 \\ 5 & 2 & -19 \end{bmatrix}$$

$$= \begin{bmatrix} 7 & 3 & -26 \\ 3 & 1 & -11 \\ -5 & -2 & 19 \end{bmatrix}$$

***Example* 4:** Find the inverse of the matrix $\begin{bmatrix} 1 & 2 & 3 \\ 3 & 4 & 5 \\ 2 & 7 & 11 \end{bmatrix}$

***Solution*:** Here $|A| = \begin{vmatrix} 1 & 2 & 3 \\ 3 & 4 & 5 \\ 2 & 7 & 11 \end{vmatrix}$

$$= 1\,(44 - 35) - 2\,(33 - 10) + 3\,(21 - 8)$$
$$= 9 - 46 + 39$$
$$= 2$$

Cofactors of matrix are

$C_{11} = 44 - 35 = 9$ $C_{21} = -(22 - 21) = -1$ $C_{31} = 10 - 12 = -2$

$C_{12} = -(33 - 10) = -23$ $C_{22} = 11 - 6 = 5$ $C_{32} = -(5 - 9) = 4$

$C_{13} = 21 - 8 = 13$ $C_{23} = -(7 - 4) = -3$ $C_{33} = 4 - 6 = -2$

$$\therefore \quad C = \begin{bmatrix} 9 & -23 & 13 \\ -1 & 5 & -3 \\ -2 & 4 & -2 \end{bmatrix}$$

$$\therefore \quad \text{Adj A} = C' = \begin{bmatrix} 9 & -1 & -2 \\ -23 & 5 & 4 \\ 13 & -3 & -2 \end{bmatrix}$$

$$\therefore \qquad A^{-1} = \frac{\text{Adj}A}{|A|} = \frac{1}{2}\begin{bmatrix} 9 & -1 & -2 \\ -23 & 5 & 4 \\ 13 & -3 & -2 \end{bmatrix}$$

***Example* 5:** Find the inverse of $\begin{bmatrix} 2 & 1 & 2 \\ 2 & 2 & 1 \\ 1 & 2 & 2 \end{bmatrix}$

***Solution*:** Here $|A| = \begin{vmatrix} 2 & 1 & 2 \\ 2 & 2 & 1 \\ 1 & 2 & 2 \end{vmatrix}$

$$= 2(4-2) - 1(4-1) + 2\,(4-2)$$

$$= 4 - 3 + 4 = 5$$

Cofactor of matrix are

$C_{11} = 4 - 2 = 2$ $\qquad C_{12} = -\,(4-1) = -3$ $\qquad C_{13} = (4-2) = 2$

$C_{21} = -\,(2-4) = 2$ $\qquad C_{22} = 4 - 2 = 2$ $\qquad C_{23} = -\,(4-1) = -3$

$C_{31} = 1 - 4 = -3$ $\qquad C_{32} = -\,(2-4) = 2$ $\qquad C_{33} = 4 - 2 = 2$

$$\therefore \qquad C = \begin{bmatrix} 2 & -3 & 2 \\ 2 & 2 & -3 \\ -3 & 2 & 2 \end{bmatrix}$$

$$\text{Adj } A = \begin{bmatrix} 2 & 2 & -3 \\ -3 & 2 & 2 \\ 2 & -3 & 2 \end{bmatrix}$$

$$\therefore \text{ Inverse of A, } A^{-1} = \frac{\text{Adj } A}{|A|}$$

$$= \frac{1}{5}\begin{bmatrix} 2 & 2 & -3 \\ -3 & 2 & 2 \\ 2 & -3 & 2 \end{bmatrix}$$

***Example* 6:** Find the inverse of the matrix $\begin{bmatrix} 2 & 7 & 1 \\ 3 & 1 & 2 \\ 4 & 5 & 6 \end{bmatrix}$

Solution: Here $|A| = \begin{vmatrix} 2 & 7 & 1 \\ 3 & 1 & 2 \\ 4 & 5 & 6 \end{vmatrix}$

$$= 2\,(6 - 10) - 7\,(18 - 8) + 1\,(15 - 4)$$

$$= -8 - 70 + 11$$

$$= -67$$

Cofactor of matrix are

$C_{11} = 6 - 10 = -4$ $\quad C_{21} = -(42 - 5) = -37$ $\quad C_{31} = 14 - 1 = 13$

$C_{12} = -(18 - 8) = -10$ $\quad C_{22} = 12 - 4 = 8$ $\quad C_{32} = -(4 - 3) = -1$

$C_{13} = 15 - 4 = 11$ $\quad C_{23} = -(10 - 28) = 18$ $\quad C_{33} = 2 - 21 = -19$

$$\therefore \quad C = \begin{bmatrix} -4 & -10 & 11 \\ -37 & 8 & 18 \\ 13 & -1 & -19 \end{bmatrix}$$

$$\text{Adj } A = \begin{bmatrix} -4 & -37 & 13 \\ -10 & 8 & -1 \\ 11 & 18 & -19 \end{bmatrix}$$

$$\therefore \text{ Inverse of A, } A^{-1} = \frac{\text{Adj A}}{|A|}$$

$$= \frac{-1}{67}\begin{bmatrix} -4 & -37 & 13 \\ -10 & 8 & -1 \\ 11 & 18 & -19 \end{bmatrix}$$

Solutions of Linear Equations: Let AX = B be a system of n linear equations with n unknown. If A is non singular then A^{-1} exists and then $X = A^{-1}B$ is a solution of given system of equations.

Criterion of Consistency: Suppose AX = B be a system of n-linear equations in n unknowns.

(i) if $|A| \neq 0$, then the system is consistent and has a unique solution given by $X = A^{-1}B$.

(ii) If $|A| = 0$ and (Adj A) B = 0, then the system is consistent and has infinitely many solutions.

(iii) If $|A| = 0$ and (Adj A) $B \neq 0$, then the system is inconsistent and it has no solutions.

***Example* 1:** Using matrices solve the equations

$$2x - y + 3z = 9$$
$$x + y + z = 6$$
$$x - y + z = 2$$

***Solution*:** Given system is equivalent to Ax = B, where

$$A = \begin{bmatrix} 2 & -1 & 3 \\ 1 & 1 & 1 \\ 1 & -1 & 1 \end{bmatrix}, \quad X = \begin{bmatrix} x \\ y \\ z \end{bmatrix}, \quad B = \begin{bmatrix} 9 \\ 6 \\ 2 \end{bmatrix}$$

$\therefore$ $X = A^{-1}B$

Now
$$|A| = \begin{vmatrix} 2 & -1 & 3 \\ 1 & 1 & 1 \\ 1 & -1 & 1 \end{vmatrix}$$

$$= 2(1 + 1) + 1(1 - 1) + 3(-1 - 1)$$
$$= 4 - 6$$
$$= -2$$

Cofactor of matrix A is

$C_{11} = (1 + 1) = 2$ $\quad C_{21} = -(-1 + 3) = -2$ $\quad C_{31} = -1 - 3 = -4$

$C_{12} = -(1 - 1) = 0$ $\quad C_{22} = 2 - 3 = -1$ $\quad C_{32} = -(2 - 3) = 1$

$C_{13} = (-1 - 1) = -2$ $\quad C_{23} = -(-2 + 1) = 1$ $\quad C_{33} = 2 + 1 = 3$

$$\therefore \quad C = \begin{bmatrix} 2 & 0 & -2 \\ -2 & -1 & 1 \\ -4 & 1 & 3 \end{bmatrix}$$

$$\therefore \text{Adj } A = C' = \begin{bmatrix} 2 & -2 & -4 \\ 0 & -1 & 1 \\ -2 & 1 & 3 \end{bmatrix}$$

$$A^{-1} = \frac{\text{Adj} A}{|A|} = \frac{-1}{2}\begin{bmatrix} 2 & -2 & -4 \\ 0 & -1 & 1 \\ -2 & 1 & 3 \end{bmatrix}$$

Then $X = A^{-1}B$

$$= \begin{bmatrix} x \\ y \\ z \end{bmatrix} = \frac{-1}{2}\begin{bmatrix} 2 & -2 & -4 \\ 0 & -1 & 1 \\ -2 & 1 & 3 \end{bmatrix}\begin{bmatrix} 9 \\ 6 \\ 2 \end{bmatrix}$$

$$= \frac{-1}{2}\begin{bmatrix} 2\times 9 - 2\times 6 - 4\times 2 \\ 0 - 1\times 6 + 1\times 2 \\ -2\times 9 + 1\times 6 + 3\times 2 \end{bmatrix}$$

$$= \frac{-1}{2}\begin{bmatrix} -2 \\ -4 \\ -6 \end{bmatrix}$$

$$\begin{bmatrix} x \\ y \\ z \end{bmatrix} = \begin{bmatrix} 1 \\ 2 \\ 3 \end{bmatrix}$$

$\Rightarrow$ $x = 1, \quad y = 2, \quad z = 3$

***Example* 2:** Solve the system of equation by using matrix method

$$x + y + 3z = 4$$
$$x + 2z = 5$$
$$2y + 3x = 7$$

***Solution*:** Given system is equivalent to AX = B, where

$$A = \begin{bmatrix} 1 & 1 & 3 \\ 1 & 0 & 2 \\ 3 & 2 & 0 \end{bmatrix}, \quad X = \begin{bmatrix} x \\ y \\ z \end{bmatrix}, \quad B = \begin{bmatrix} 4 \\ 5 \\ 7 \end{bmatrix}$$

Now $|A| = \begin{vmatrix} 1 & 1 & 3 \\ 1 & 0 & 2 \\ 3 & 2 & 0 \end{vmatrix}$

$= 1\,(0-4) - 1(0-6) + 3\,(2-0)$

$= -4 + 6 + 6$

$= 8$

Cofactor of matrix A is

$C_{11} = 0 - 4 = -4$ $\quad C_{21} = -(0-6) = 6$ $\quad C_{31} = 2 - 0 = 2$

$C_{12} = -(0-6) = 6$ $\quad C_{22} = 0 - 9 = -9$ $\quad C_{32} = -(2-3) = 1$

$C_{13} = (2-0) = 2$ $\quad C_{23} = -(2-3) = 1$ $\quad C_{33} = 0 - 1 = -1$

$$\therefore \quad C = \begin{bmatrix} -4 & 6 & 2 \\ 6 & -9 & 1 \\ 2 & 1 & -1 \end{bmatrix}$$

$$\text{Adj } A = \begin{bmatrix} -4 & 6 & 2 \\ 6 & -9 & 1 \\ 2 & 1 & -1 \end{bmatrix}$$

$$\therefore \quad A^{-1} = \frac{\text{Adj } A}{|A|} = \frac{1}{8}\begin{bmatrix} -4 & 6 & 2 \\ 6 & -9 & 1 \\ 2 & 1 & -1 \end{bmatrix}$$

Then $X = A^{-1}B$

$$\begin{bmatrix} x \\ y \\ z \end{bmatrix} = \frac{1}{8}\begin{bmatrix} -4 & 6 & 2 \\ 6 & -9 & 1 \\ 2 & 1 & -1 \end{bmatrix}\begin{bmatrix} 4 \\ 5 \\ 7 \end{bmatrix}$$

$$= \frac{1}{8}\begin{bmatrix} -4\times 4 + 6\times 5 + 2\times 7 \\ 6\times 4 - 9\times 5 + 1\times 7 \\ 2\times 4 + 1\times 5 - 1\times 7 \end{bmatrix}$$

$$= \frac{1}{8}\begin{bmatrix} 28 \\ -14 \\ 6 \end{bmatrix}$$

$$\begin{bmatrix} x \\ y \\ z \end{bmatrix} = \begin{bmatrix} 7/2 \\ -7/4 \\ 3/4 \end{bmatrix}$$

$$\Rightarrow \quad x = \frac{7}{2}, \quad y = -\frac{7}{4}, \quad z = \frac{3}{4}$$

***Example* 3:** Solve

$$2x + 3y + z = 9$$
$$x + 2y + 3z = 6$$
$$3x + y + 2z = 8$$

by finding inverse of coefficient matrix.

***Solution*:** Given system is equivalent to AX = B, where

$$A = \begin{bmatrix} 2 & 3 & 1 \\ 1 & 2 & 3 \\ 3 & 1 & 2 \end{bmatrix}, \qquad X = \begin{bmatrix} x \\ y \\ z \end{bmatrix}, \qquad B = \begin{bmatrix} 9 \\ 6 \\ 8 \end{bmatrix}$$

Now,
$$|A| = \begin{vmatrix} 2 & 3 & 1 \\ 1 & 2 & 3 \\ 3 & 1 & 2 \end{vmatrix}$$
$$= 2(4-3) - 3(2-9) + 1(1-6)$$
$$= 2 + 21 - 5$$
$$= 18$$

Cofactors are

$C_{11} = 4 - 3 = 1$ $\quad C_{21} = -(6-1) = -5$ $\quad C_{31} = 9 - 2 = 7$

$C_{12} = -(2-9) = 7$ $\quad C_{22} = 4 - 3 = 1$ $\quad C_{32} = -(6-1) = -5$

$C_{13} = (1-6) = -5$ $\quad C_{23} = -(2-9) = 7$ $\quad C_{33} = 4 - 3 = 1$

$$\therefore \quad C = \begin{bmatrix} 1 & 7 & -5 \\ -5 & 1 & 7 \\ 7 & -5 & 1 \end{bmatrix}$$

$$\text{Adj } A = \begin{bmatrix} 1 & -5 & 7 \\ 7 & 1 & -5 \\ -5 & 7 & 1 \end{bmatrix}$$

$$A^{-1} = \frac{\text{Adj}\,A}{|A|} = \frac{1}{18}\begin{bmatrix} 1 & -5 & 7 \\ 7 & 1 & -5 \\ -5 & 7 & 1 \end{bmatrix}$$

Then $X = A^{-1}B$

$$\begin{bmatrix} x \\ y \\ z \end{bmatrix} = \frac{1}{18}\begin{bmatrix} 1 & -5 & 7 \\ 7 & 1 & -5 \\ -5 & 7 & 1 \end{bmatrix}\begin{bmatrix} 9 \\ 6 \\ 8 \end{bmatrix}$$

$$= \frac{1}{18}\begin{bmatrix} 9-30+56 \\ 63+6-40 \\ -45+42+8 \end{bmatrix}$$

$$= \frac{1}{18}\begin{bmatrix} 35 \\ 29 \\ 5 \end{bmatrix} \quad \Rightarrow \; x = \frac{35}{18},\; y = \frac{29}{18},\; z = \frac{5}{18}$$

***Example* 4:** Solve the following system of equations by matrix method.

$$x + 2y + z = 4$$
$$2x + y = 3$$
$$x + z = 2$$

***Solution*:** The given system is equivalent to AX = B, where

$$A = \begin{bmatrix} 1 & 2 & 1 \\ 2 & 1 & 0 \\ 1 & 0 & 1 \end{bmatrix}, \quad X = \begin{bmatrix} x \\ y \\ z \end{bmatrix}, \quad B = \begin{bmatrix} 4 \\ 3 \\ 2 \end{bmatrix}$$

Now, $$|A| = \begin{vmatrix} 1 & 2 & 1 \\ 2 & 1 & 0 \\ 1 & 0 & 1 \end{vmatrix}$$

$$= 1\,(1-0) - 2\,(2-0) + 1\,(0-1)$$
$$= 1 - 4 - 1$$
$$= -4$$

Cofactors are

$C_{11} = 1 - 0 = 1$ $\quad C_{21} = -(2-0) = -2$ $\quad C_{31} = 0 - 1 = -1$

$C_{12} = -(2-0) = -2$ $\quad C_{22} = 1 - 1 = 0$ $\quad C_{32} = -(0-2) = 2$

$C_{13} = (0-1) = -1$ $\quad C_{23} = -(0-2) = 2$ $\quad C_{33} = 1 - 4 = -3$

$$\therefore \qquad C = \begin{bmatrix} 1 & -2 & -1 \\ -2 & 0 & 2 \\ -1 & 2 & -3 \end{bmatrix}$$

$$\text{Adj A} = \begin{bmatrix} 1 & -2 & -1 \\ -2 & 0 & 2 \\ -1 & 2 & -3 \end{bmatrix}$$

Then $A^{-1} = \dfrac{\text{Adj A}}{|A|} = \dfrac{-1}{4}\begin{bmatrix} 1 & -2 & -1 \\ -2 & 0 & 2 \\ -1 & 2 & -3 \end{bmatrix}$

Now, $x = A^{-1}B$

$$\Rightarrow \qquad \begin{bmatrix} x \\ y \\ z \end{bmatrix} = \frac{-1}{4}\begin{bmatrix} 1 & -2 & -1 \\ -2 & 0 & 2 \\ -1 & 2 & -3 \end{bmatrix}\begin{bmatrix} 4 \\ 3 \\ 2 \end{bmatrix}$$

$$= \frac{-1}{4}\begin{bmatrix} 4-6-2 \\ -8+0+4 \\ -4+6-6 \end{bmatrix}$$

$$= \frac{-1}{4}\begin{bmatrix} -4 \\ -4 \\ -4 \end{bmatrix}$$

$$\Rightarrow \qquad x = 1, \quad y = 1, \quad z = 1$$

***Example* 5:** Solve by matrix method, the equation.

$$x + y = 0$$

$$y + z = 1$$

$$z + x = 3$$

***Solution*:** The given system is equivalent to AX = B, where

$$A = \begin{bmatrix} 1 & 1 & 0 \\ 0 & 1 & 1 \\ 1 & 0 & 1 \end{bmatrix}, \quad X = \begin{bmatrix} x \\ y \\ z \end{bmatrix}, \quad B = \begin{bmatrix} 0 \\ 1 \\ 3 \end{bmatrix}$$

Now, $|A| = \begin{vmatrix} 1 & 1 & 0 \\ 0 & 1 & 1 \\ 1 & 0 & 1 \end{vmatrix}$

$$= 1(1-0) - 1(0-1) + 0(0-1)$$
$$= 1 + 1$$
$$= 2$$

Cofactors are

$C_{11} = 1$ $\quad C_{21} = -1$ $\quad C_{31} = 1$

$C_{12} = -(-1) = 1$ $\quad C_{22} = 1$ $\quad C_{32} = -1$

$C_{13} = -1$ $\quad C_{23} = -(-1) = 1$ $\quad C_{33} = 1$

$\therefore$ $C = \begin{bmatrix} 1 & 1 & -1 \\ -1 & 1 & 1 \\ 1 & -1 & 1 \end{bmatrix}$

$\therefore$ $\text{Adj } A = \begin{bmatrix} 1 & -1 & 1 \\ 1 & 1 & -1 \\ -1 & 1 & 1 \end{bmatrix}$

$$A^{-1} = \frac{\text{Adj} A}{|A|} = \frac{1}{2}\begin{bmatrix} 1 & -1 & 1 \\ 1 & 1 & -1 \\ -1 & 1 & 1 \end{bmatrix}$$

Then $x = A^{-1}B$

$$\begin{bmatrix} x \\ y \\ z \end{bmatrix} = \frac{1}{2}\begin{bmatrix} 1 & -1 & 1 \\ 1 & 1 & -1 \\ -1 & 1 & 1 \end{bmatrix}\begin{bmatrix} 0 \\ 1 \\ 3 \end{bmatrix}$$

$$= \frac{1}{2}\begin{bmatrix} 0 - 1 + 3 \\ 0 + 1 - 3 \\ 0 + 1 + 3 \end{bmatrix}$$

$$\begin{bmatrix} x \\ y \\ z \end{bmatrix} = \frac{1}{2}\begin{bmatrix} 2 \\ -2 \\ 4 \end{bmatrix}$$

$$\Rightarrow \quad x = 1, \quad y = -1, \quad z = 2$$

Example Based on Pharmaceutical Application of matrices

***Example* 1:** M/s Gaya pharmacy employs 20 unskilled workers and 10 skilled workers for manufacturing a product. If each unskilled worker is paid at the rate of Rs. 70 per day whereas skilled worker is paid at the rate of Rs. 105 per day. Find the total payment made to the workers per day.

***Solution*:** Let A and B are the workers and rate of payment. Each of the matrix A and B can be written either as a row or as a column matrix

$$A = \begin{matrix} \text{unskilled} & \text{skilled} \\ [20 & 10] \end{matrix}, \qquad B = \begin{bmatrix} 70 \\ 105 \end{bmatrix} \begin{matrix} \text{rate per unskilled} \\ \text{rate per skilled} \end{matrix}$$

The total payment per day, we can find by multiplication matrix.

$$AB = [20 \quad 10]\begin{bmatrix} 70 \\ 105 \end{bmatrix}$$

$$= [20 \times 70 + 10 \times 105]$$

$$= [1400 + 1050]$$

$$= [2450]$$

***Example* 2:** There are two families A and B. There are 2 men, 3 women and 1 child in family A and 1 man, 1 woman, 2 children in family B. The recommended daily allowance for calories is; men : 2400, women : 1990, children : 1800 and for proteins is; men : 55 gm, women : 45 gm, children : 33 gm. Represent the above information by matrix using matrix multiplication. Calculate the total requirement of calories and protein for each of the family.

***Solution*:** Let A, B, C represent men, women, and children. Then from the given data the family matrix is

$$x = \begin{matrix} & A & B & C & \\ & \begin{bmatrix} 2 & 3 & 1 \\ 1 & 1 & 2 \end{bmatrix} & & & \begin{matrix} I^{st} \text{ family} \\ II^{nd} \text{ family} \end{matrix} \end{matrix}$$

Also, the required matrix y is given by

$$y = \begin{bmatrix} 2400 & 55 \\ 1990 & 45 \\ 1800 & 33 \end{bmatrix} \begin{matrix} A \\ B \\ C \end{matrix}$$

(columns: calories, proteins)

So, the product matrix is

$$xy = \begin{bmatrix} 2 & 3 & 1 \\ 1 & 1 & 2 \end{bmatrix} \begin{bmatrix} 2400 & 55 \\ 1990 & 45 \\ 1800 & 33 \end{bmatrix}$$

(columns: calories, proteins)

$$= \begin{bmatrix} 2 \times 2400 + 3 \times 1900 + 1 \times 1800 & 2 \times 55 + 3 \times 45 + 1 \times 33 \\ 2400 + 1900 + 2 \times 1800 & 55 + 45 + 2 \times 33 \end{bmatrix} \begin{matrix} I^{st} \\ II^{nd} \end{matrix}$$

$$= \begin{bmatrix} 12300 & 278 \\ 7900 & 166 \end{bmatrix} \begin{matrix} I^{st} \text{ family} \\ II^{nd} \text{ family} \end{matrix}$$

(columns: calories, proteins)

Hence, the first family requires 12, 300 calories and 278 gm proteins and IInd family requires 7900 calories and 166 gm proteins.

***Example* 3:** A manufacturer produces three products x, y, z which are sold in Bhopal and Rajasthan. The annual sales of these products are given below.

	x	y	z
Bhopal	5000	7500	15000
Rajastan	9000	12000	8700

If the sale price of the products x, y, z per unit by Rs. 2, Rs. 3 and Rs. 4 respectively. Calculate the total revenue in each centre by using matrices.

***Solution*:** The sale matrix S and price matrix P are given by

$$S = \begin{matrix} \text{Bhopal} \\ \text{Rajas tan} \end{matrix} \begin{bmatrix} 5000 & 7500 & 15000 \\ 9000 & 12000 & 8700 \end{bmatrix}$$

$$P = \begin{bmatrix} 2 \\ 3 \\ 4 \end{bmatrix}$$

$$\therefore \quad \text{Revenue matrix is S.P} = \begin{bmatrix} 5000 & 7500 & 15000 \\ 9000 & 12000 & 8700 \end{bmatrix} \begin{bmatrix} 2 \\ 3 \\ 4 \end{bmatrix}$$

$$= \begin{bmatrix} 5000 \times 2 + 7500 \times 3 + 15000 \times 4 \\ 9000 \times 2 + 12000 \times 3 + 8700 \times 4 \end{bmatrix}$$

$$= \begin{bmatrix} 92500 \\ 88800 \end{bmatrix}$$

Hence, the total revenue in Bhopal is Rs. 92500 and in Rajasthan is Rs. 88800.

***Example* 4:** The daily cost of operating a hospital C is a linear function of the number of in patients I, and out patients O, plus a cost a, i.e.,

$$C = a + bO + dI$$

Given the following data from three days, find the values of a, b and d by setting up a linear system of equations and using the matrix inverse.

Days	Cost in Rs.	No. of in Patients (I)	No. of out Patients (O)
1	6950	40	10
2	6725	35	9
3	7100	40	12

***Solution*:** The above data can be expressed as a linear equations.

$$a + 10b + 40d = 6950$$

$$a + 9b + 35d = 6725$$

$$a + 12b + 40d = 7100$$

We have

$$A = \begin{bmatrix} 1 & 10 & 40 \\ 1 & 9 & 35 \\ 1 & 12 & 40 \end{bmatrix}, \quad X = \begin{bmatrix} a \\ b \\ d \end{bmatrix}, \quad B = \begin{bmatrix} 6950 \\ 6725 \\ 7100 \end{bmatrix}$$

Now,
$$|A| = \begin{vmatrix} 1 & 10 & 40 \\ 1 & 9 & 35 \\ 1 & 12 & 40 \end{vmatrix}$$

$$= 1\,(9 \times 40 - 12 \times 35) - 10\,(40 - 35) + 40\,(12 - 9)$$

$$= -60 - 50 + 120$$

$$= 10$$

Cofactors are

$C_{11} = 360 - 420 = -60$ $\quad C_{21} = -(400 - 480) = 80$ $\quad C_{31} = 350 - 360 = -10$

$C_{12} = -(40 - 35) = -5$ $\quad C_{22} = 40 - 40 = 0$ $\quad C_{32} = -(35 - 40) = 5$

$C_{13} = 12 - 9 = 3$ $\quad C_{23} = -(12 - 10) = -2$ $\quad C_{33} = 9 - 10 = -1$

$$\therefore \qquad C = \begin{bmatrix} -60 & -5 & 3 \\ 80 & 0 & -2 \\ -10 & 5 & -1 \end{bmatrix}$$

$$\text{Adj } A = \begin{bmatrix} -60 & 80 & -10 \\ -5 & 0 & 5 \\ 3 & -2 & -1 \end{bmatrix}$$

$$A^{-1} = \frac{\text{Adj} A}{|A|} = \frac{1}{10}\begin{bmatrix} -60 & 80 & -10 \\ -5 & 0 & 5 \\ 3 & -2 & -1 \end{bmatrix}$$

Then $\qquad X = A^{-1}B$

$$\begin{bmatrix} a \\ b \\ c \end{bmatrix} = \frac{-1}{10}\begin{bmatrix} 60 & -80 & 10 \\ 5 & 0 & -5 \\ -3 & 2 & 1 \end{bmatrix}\begin{bmatrix} 6950 \\ 6725 \\ 7100 \end{bmatrix}$$

$$= \frac{-1}{10}\begin{bmatrix} 60 \times 6950 - 80 \times 6725 + 10 \times 7100 \\ 5 \times 6950 + 0 - 5 \times 7100 \\ -3 \times 6950 + 2 \times 6725 + 1 \times 7100 \end{bmatrix}$$

$$= \frac{-1}{10}\begin{bmatrix} -50000 \\ -750 \\ -300 \end{bmatrix}$$

$$\begin{bmatrix} a \\ b \\ c \end{bmatrix} = \begin{bmatrix} 5000 \\ 75 \\ 30 \end{bmatrix}$$

$\Rightarrow \qquad a = 5000, \quad b = 75, \quad c = 30$

Exercises

1. Find x, y, z satisfying the equations

$$\begin{bmatrix} x-3 & 7 \\ 0 & -5 \end{bmatrix} = \begin{bmatrix} 2x+5 & 7 \\ 0 & 3y-7 \end{bmatrix}$$

2. If $A = \begin{bmatrix} 2 & 3 & 1 \\ 0 & -1 & 5 \end{bmatrix}$ and $B = \begin{bmatrix} 1 & 2 & -6 \\ 0 & -1 & 0 \end{bmatrix}$, evaluate 3A – 4B

3. $A = \begin{bmatrix} 1 & 2 \\ 3 & 4 \\ 5 & 6 \end{bmatrix}$, $B = \begin{bmatrix} -3 & -2 \\ 1 & -5 \\ 4 & 3 \end{bmatrix}$ find C such that A + B – C = 0

4. If $x + y = \begin{bmatrix} 1 & -2 \\ 3 & 4 \end{bmatrix}$, $x - y = \begin{bmatrix} 3 & 2 \\ -1 & 0 \end{bmatrix}$ find x and y

5. If $A = \begin{bmatrix} -1 & 2 & -1 \\ 0 & -3 & -1 \\ 1 & 4 & 0 \end{bmatrix}$, $B = \begin{bmatrix} 1 & 3 & 5 \\ 2 & 4 & 6 \\ 3 & 5 & 7 \end{bmatrix}$, $C = \begin{bmatrix} 1 & 2 & 1 \\ 0 & 3 & 5 \\ 1 & 1 & 0 \end{bmatrix}$

   then evaluate 3A – B + 5C

6. If $A = \begin{bmatrix} 0 & -\tan\frac{\alpha}{2} \\ \tan\frac{\alpha}{2} & 0 \end{bmatrix}$ and $I = \begin{bmatrix} 1 & 0 \\ 0 & 1 \end{bmatrix}$ show that

$$I + A = (I - A)\begin{bmatrix} \cos\alpha & -\sin\alpha \\ \sin\alpha & \cos\alpha \end{bmatrix}$$

7. If $A = \begin{bmatrix} 1 \\ 2 \\ 3 \end{bmatrix}$ and B = [3 1 2] find AB and BA

8. If $A = \begin{bmatrix} 2 & 3 & 4 \\ 1 & 2 & 3 \\ -1 & 1 & 2 \end{bmatrix}$ and $B = \begin{bmatrix} 1 & 3 & 0 \\ -1 & 2 & 1 \\ 0 & 0 & 2 \end{bmatrix}$ find AB and BA

9. If $A = \begin{bmatrix} 1 & 1 & 3 \\ 2 & 2 & 6 \\ -1 & -1 & -3 \end{bmatrix}$ find A^2

10. If $A = \begin{bmatrix} \cos\alpha & -\sin\alpha \\ \sin\alpha & \cos\alpha \end{bmatrix}$, $B = \begin{bmatrix} \cos\beta & -\sin\beta \\ \sin\beta & \cos\beta \end{bmatrix}$ show that AB = BA

11. If $A = \begin{bmatrix} 0 & 1 \\ 1 & 1 \end{bmatrix}$ and $B = \begin{bmatrix} 0 & -1 \\ 1 & 0 \end{bmatrix}$ show that $(A + B)(A - B) \neq A^2 - B^2$

12. If $A = \begin{bmatrix} \cos\theta & -\sin\theta \\ \sin\theta & \cos\theta \end{bmatrix}$ show that $A^n = \begin{bmatrix} \cos n\theta & -\sin n\theta \\ \sin n\theta & \cos n\theta \end{bmatrix}$

13. If $A = \begin{bmatrix} i & 0 \\ 0 & i \end{bmatrix}$, $B = \begin{bmatrix} 0 & -1 \\ 1 & 0 \end{bmatrix}$, $C = \begin{bmatrix} 0 & i \\ i & 0 \end{bmatrix}$ and $I = \begin{bmatrix} 1 & 0 \\ 0 & 1 \end{bmatrix}$ show that

(i) $A^2 = B^2 = C^2 = -1$ (ii) $AB = -C = -BA$

14. Evaluate $([1\ 2] + 5\,[3\ 4]) \begin{bmatrix} 1 & 4 & 8 \\ 2 & 6 & 10 \end{bmatrix}$

15. If $A = \begin{bmatrix} 2 & 0 & 1 \\ 2 & 1 & -3 \\ 1 & -1 & 0 \end{bmatrix}$, $f(x) = x^2 - 5x + 6$, find f(A)

16. If $A = \begin{bmatrix} 4 & 2 \\ -1 & 1 \end{bmatrix}$, show that $(A - 2I)(A - 3I) = 0$

17. If $A = \begin{bmatrix} 0 & 0 & 1 \\ 0 & 1 & 0 \\ 1 & 0 & 0 \end{bmatrix}$, find A^2 and A^6

18. Find the transpose of A. If $A = \begin{bmatrix} 1 & -1 & 2 \\ 3 & 1 & 4 \\ -2 & 7 & 8 \end{bmatrix}$

19. If $A = \begin{bmatrix} 1 & 1 & 2 \\ 2 & 1 & 0 \end{bmatrix}$ and $B = \begin{bmatrix} 1 & 3 \\ 2 & 0 \\ 2 & 4 \end{bmatrix}$ show that $(AB)' = B'A'$

20. If $A = \begin{bmatrix} 1 & 1 & 1 \\ 2 & 2 & 3 \\ 2 & 4 & 9 \end{bmatrix}$ and $B = \begin{bmatrix} 0 & 1 & -1 \\ -3 & 2 & 4 \\ 1 & 1 & 0 \end{bmatrix}$ show that $(AB)' = B'A'$

21. Find the adjoint A if $A = \begin{bmatrix} 1 & 2 & 1 \\ 3 & 2 & 3 \\ 1 & 1 & 2 \end{bmatrix}$

22. Find the adjoint of A if $A = \begin{bmatrix} 0 & 1 & 1 \\ 1 & 2 & 0 \\ 3 & -1 & 4 \end{bmatrix}$

23. Find the inverse of $A = \begin{bmatrix} 1 & 2 & 3 \\ 2 & 4 & 5 \\ 3 & 5 & 6 \end{bmatrix}$

24. Find the inverse of $A = \begin{bmatrix} 1 & 1 & 0 \\ 0 & 1 & 1 \\ 0 & 0 & 1 \end{bmatrix}$

25. Solve the following system of equations

(i) $4x - 3y = 5$

$3x - 5y = 1$

(ii) $x - y + z = 4$

$2x + y - 3z = 0$

$x + y + z = 2$

(iii) $3x - 2y + 3z = 8$

$2x + y - z = 1$

$4x - 3y + 2z = 4$

(iv) $2x - y + 3z = 9$

$x + y + z = 6$

$x - y + z = 2$

26. If $A = \begin{bmatrix} 4 & -7 \\ -8 & 3 \end{bmatrix}$, $B = \begin{bmatrix} 11 & -23 \\ -22 & 8 \end{bmatrix}$ show that $A + 2A^1 - B = I$

27. If $A = \begin{bmatrix} 1 & -1 & 1 \\ 2 & 3 & 0 \\ 18 & 2 & 10 \end{bmatrix}$, prove that A(Adj A) = 0

28. If $A = \begin{bmatrix} 4 & 1 \\ 6 & 5 \end{bmatrix}$, find A^{-1} and verify $AA^{-1} = A^{-1}A = I$

29. If $A = \begin{bmatrix} 1 & 1 & 2 \\ 1 & 9 & 3 \\ 1 & 4 & 2 \end{bmatrix}$, $B = \begin{bmatrix} 1 & 2 & 0 \\ 2 & 3 & -1 \\ 1 & -1 & 3 \end{bmatrix}$ verify that $(AB)^{-1} = B^{-1}A^{-1}$

30. Show that the following system of equations is consistent. $2x + 3y - z = 5$, $3x + 2y - z = 7$ and $4x + 5y - 5z = 9$. Also find the solution.

31. A mixture is to be made of three foods A, B, C. The three foods contains nutrients P, Q, R as shown below:

Food ounces per pound of nutrient

	P	Q	R
A	1	2	5
B	3	1	1
C	4	2	1

How to form a mixture which will have 8 ounces of P, 5 ounces of Q and 7 ounces of R.

32. The prices of Isabgoal, honey and liquorice for a pharmaceutical companies A, B and C are x, y and z per tonne respectively. A purchases 4 tonnes of liquorice and sells 3 tonnes of Isabgoal and 5 tonnes of honey. B purchases 3 tonnes of honey and sells 2 tonnes of Isabgoal and 1 tonnes of liquorice. C purchases 1 tonne of Isabgoal and sells 4 tonnes of honey and 6 tonnes of liquorice. In this process A, B, C earns Rs. 6000, Rs. 5000, Rs. 13000 respectively. Find the price per tonne of three commodities.

[Hint: linear equations are $-3x - 5y + 4z = 6000$, $-2x + 3y - z = 5000$, $x - 4y - 6z = 13000$].

33. Three raw materials X, Y and Z have prices x, y, z respectively. A purchases 4 units of Z and sells 3 units of X and 5 units of Y, B purchases 3 units of Y and sells 2 units of X and 1 unit of Z. C purchases 1 unit of X and sells 4 units of Y and 6 units of Z. In the process A, B, C earn Rs. 6000, 5000 and 13000 respectively. Using matrices, find the price per unit of the three materials. (Note that selling the units is positive earnings and buying the units is negative earnings).

[Hint: linear equations are $3x + 5y - 4z = 6000$, $2x - 3y + z = 5000$, $-x + 4y + 6z = 13000$].

Answers

1. $x = -8, y = \frac{2}{3}$

2. $\begin{bmatrix} 2 & 1 & 27 \\ 0 & 1 & 3 \end{bmatrix}$

3. $\begin{bmatrix} -2 & 0 \\ 4 & -1 \\ 9 & 9 \end{bmatrix}$

4. $x = \begin{bmatrix} 2 & 0 \\ 1 & 1 \end{bmatrix}, y = \begin{bmatrix} -1 & -2 \\ 2 & 2 \end{bmatrix}$

5. $\begin{bmatrix} 1 & 13 & -3 \\ -2 & 2 & 16 \\ 5 & 12 & -7 \end{bmatrix}$

7. $AB = BA = \begin{bmatrix} 3 \\ 2 \\ 0 \end{bmatrix}$

8. $AB = \begin{bmatrix} -1 & 12 & 11 \\ -1 & 7 & 8 \\ -2 & -1 & 5 \end{bmatrix}, BA = \begin{bmatrix} 5 & 9 & 13 \\ -1 & 2 & 4 \\ -2 & 2 & 4 \end{bmatrix}$

9. $\begin{bmatrix} 0 & 0 & 0 \\ 0 & 0 & 0 \\ 0 & 0 & 0 \end{bmatrix}$

14. [60, 196, 348]

15. $\begin{bmatrix} 1 & -1 & -3 \\ -7 & 5 & 14 \\ -5 & 4 & 10 \end{bmatrix}$

17. $\begin{bmatrix} 1 & 0 & 0 \\ 0 & 1 & 0 \\ 0 & 0 & 1 \end{bmatrix}$

18. $\begin{bmatrix} 1 & 3 & -2 \\ -1 & 1 & 7 \\ 2 & 4 & 8 \end{bmatrix}$

21. $\begin{bmatrix} 1 & -3 & 4 \\ -3 & -1 & 0 \\ 1 & 1 & -4 \end{bmatrix}$

22. $\begin{bmatrix} 8 & -5 & 2 \\ -4 & -3 & 1 \\ -7 & 3 & -1 \end{bmatrix}$

23. $\begin{bmatrix} 7/2 & -3 & 1/2 \\ -1/2 & 0 & 1/2 \\ -1/3 & 1 & -1/2 \end{bmatrix}$

24. $\begin{bmatrix} 1 & -1 & 0 \\ 0 & 1 & -1 \\ 0 & 0 & 1 \end{bmatrix}$

25. (i) x = 2, y = 1

(ii) x = 2, y = –1, z = 1

(iii) x = 1, y = 2, z = 3

(iv) x = 1, y = 2, z = 3

31. x = 1, y = 1, z = 1

32. x = 1, y = 2, z = 4

33. x = 3000, y = 1000, z = 2000

CHAPTER 5

ANALYTICAL PLANE GEOMETRY

Coordinate Geometry

It is a branch of geometry which set up a definite correspondence between the position of a point in a plane and a pair of algebraic numbers called co-ordinates.

Cartesian Co-ordinates (Rectangular Co-ordinates)

In Cartesian co-ordinates the position of a point P is determined by knowing the distance from a fixed point parallel to two perpendicular lines passing through the fixed point.

Let O be the fixed point called *origin*.

XOX′ and YOY′ are called *Cartesian* or *rectangular co-ordinates axis*.

OM is called the x co-ordinate or abscissa

ON is called the y *co-ordinate or ordinate*.

P(x,y) is called the *Cartesian co-ordinates*.

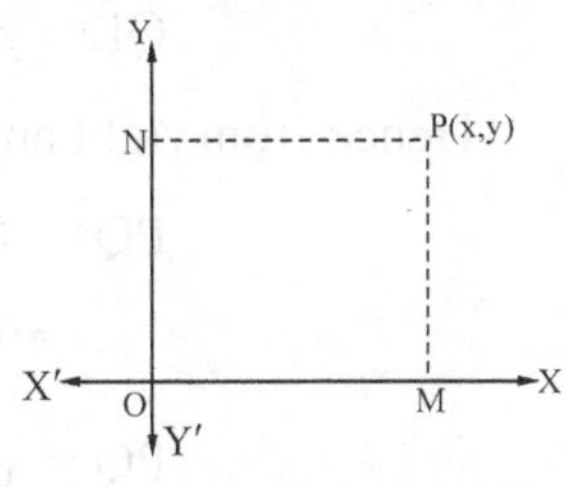

Quadrants

A quadrant is $\frac{1}{4}$ parts of a plane divided by co-ordinate axes.

(i) In the first quadrant, both co-ordinates i.e., x and y of a point are positive.

(ii) In the second quadrant, x is negative any y is positive.

(iii) In the third quadrant, both co-ordinates x and y are negative.

(iv) In the fourth quadrant, x is positive and y is negative.

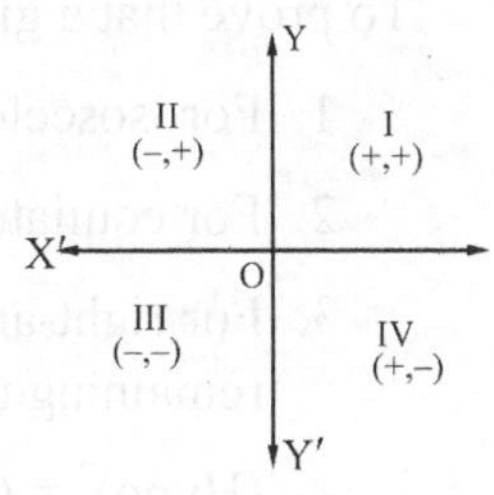

***Example* 1 :** In which quadrants do the following points lie?

(i) (3, –4) (ii) (–5, –7)

(iii) (–1, 8) (iv) (10, 15)

Solution:

(i) (3, –4) lies in the fourth quadrant.

(ii) (–5, –7) lies in the third quadrant

(iii) (–1, 8) lies in the second quadrant

(iv) (10, 15) lies in the first quadrant.

Distance between two points (Distance Formula)

Let $P(x_1, y_1)$ and $Q(x_2, y_2)$ be the two given points in the plane. From P and Q draw the perpendicular PM and QN on the x-axis. From P draw PL perpendicular to QN. Join PQ. Then

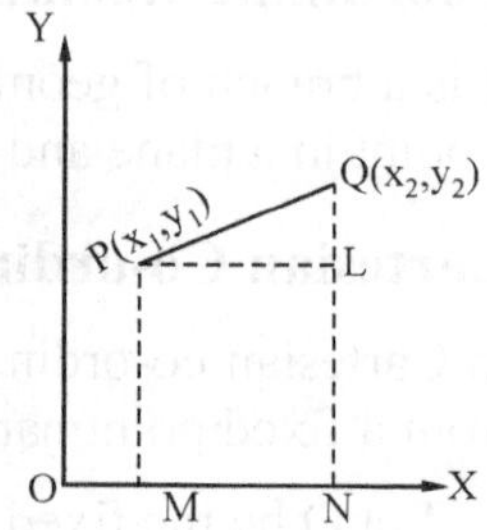

$$OM = x_1,\ MP = y_1,\ ON = x_2,\ NQ = y_2$$

$$PL = MN = ON - OM$$

$$PL = x_2 - x_1$$

and $$QL = QN - LN$$

$$= QN - PM$$

$$QL = y_2 - y_1$$

Hence from right angled triangle PL Q,

$$PQ^2 = PL^2 + QL^2$$

$$= (x_2 - x_1)^2 + (y_2 - y_1)^2$$

$$PQ = \sqrt{(x_2 - x_1)^2 + (y_2 - y_1)^2}$$

Note:

To prove that a given triangle is

1. For isosceles – Two sides are equal
2. For equilateral – All sides are equal
3. For right-angled triangle – the sum of squares of two sides is equal to the square of remaining third side. i.e.,

   $(Hypo)^2 = (Perp)^2 + (base)^2$

To prove that a four sides figures is -

1. For parallelogram-	(a)	Opposite sides are equal
	(b)	diagonals are not equal
2. For rhombus -	(a)	Four sides are equal
	(b)	Diagonals are not equal
3. For square -	(a)	Four sides are equal
	(b)	Diagonals are equal
4. For rectangle -	(a)	Opposite sides are equal
	(b)	Diagonals are equal

***Example* 2 :** Find the distance between the points –

(i) (3, 4) and (7, 7)

(ii) $(a \cos \theta_1, a \sin \theta_1)$ and $(a \cos \theta_2, a \sin \theta_2)$

Solution :

(i) Required distance

$$= \sqrt{(x_2 - x_1)^2 + (y_2 - y_1)^2}$$

$$= \sqrt{(7-3)^2 + (7-4)^2}$$

$$= \sqrt{(4)^2 + (3)^2}$$

$$= \sqrt{16+9}$$

$$= \sqrt{25}$$

$$= \mathbf{5}$$

(ii) Required distance

$$= \sqrt{(a \cos \theta_2 - a \cos \theta_1)^2 + (a \sin \theta_2 - a \sin \theta_1)^2}$$

$$= a\sqrt{(\cos \theta_2 - \cos \theta_1)^2 + (\sin \theta_2 - \sin \theta_1)^2}$$

$$= a\sqrt{(\cos^2 \theta_2 + \cos^2 \theta_1 - 2 \cos \theta_2 \cos \theta_1 + \sin^2 \theta_2 + \sin^2 \theta_1 - 2 \sin \theta_2 \sin \theta_1)}$$

$$= a\sqrt{1+1-2(\cos \theta_1 \cos \theta_2 + \sin \theta_1 \sin \theta_2)} \qquad (\because \sin^2 \theta + \cos^2 \theta = 1)$$

$$= a\sqrt{2-2\cos(\theta_1-\theta_2)}$$

$$= a\sqrt{2-2\left(1-2\sin^2\frac{\theta_1-\theta_2}{2}\right)} \qquad (\because \cos\theta = 1-2\sin^2\theta/2)$$

$$= a\sqrt{2-2+4\sin^2\frac{\theta_1-\theta_2}{2}}$$

$$= \mathbf{2a\sin\frac{\theta_1-\theta_2}{2}}$$

***Example* 3:** Prove that the points (a, a) (–a, –a) and $(-a\sqrt{3}, a\sqrt{3})$ are the vertices of an equilateral triangle.

Solution : Let the coordinates of vertices A, B, C of ΔABC are A(a, a), B(–a, –a) and $C(-a\sqrt{3}, a\sqrt{3})$. By using distance formula.

$$AB = \sqrt{(a+a)^2+(a+a)^2}$$

$$= \sqrt{(2a)^2+(2a)^2}$$

$$= \sqrt{4a^2+4a^2} = \sqrt{8a^2} = 2a\sqrt{2}$$

$$BC = \sqrt{\left(-a+a\sqrt{3}\right)^2+\left(-a-a\sqrt{3}\right)^2}$$

$$= \sqrt{a^2+3a^2-2a^2\sqrt{3}+a^2+3a^2+2a^2\sqrt{3}} = \sqrt{8a^2} = 2a\sqrt{2}$$

$$CA = \sqrt{\left(a+a\sqrt{3}\right)^2+\left(a-a\sqrt{3}\right)^2}$$

$$= \sqrt{a^2+3a^2+2\sqrt{3}a^2+a^2+3a^2-2\sqrt{3}a^2} = \sqrt{8a^2} = 2a\sqrt{2}$$

Thus, AB = BC = CA

∴ A, B, C are the vertices of an equilateral triangle

***Example* 4:** Find a point P which is equidistance from the three points A(0, 1), B(1, 0) and C (4, 3).

Solution : Let P(x, y) be the required point. Then PA = PB = PC (equidistance)

Taking $PA^2 = PB^2$

$$(x-0)^2+(y-1)^2 = (x-1)^2+(y-0)^2$$

$$x^2+y^2-2y+1 = x^2-2x+1+y^2$$

$$\Rightarrow \quad x = y$$

again taking $PA^2 = PC^2$

$$(x-0)^2 + (y-1)^2 = (x-4)^2 + (y-3)^2$$

$$x^2 + y^2 - 2y + 1 = x^2 - 8x + 16 + y^2 - 6y + 9$$

$$8x + 4y - 24 = 0$$

putting $y = x$, we get $12x = 24$

$$x = 2$$

and $y = 2$

Hence, the required point is (2, 2)

***Example* 5:** Show that the points A(1, –1), B (5, 2) and C(9, 5) are collinear.

Solution : $AB = \sqrt{(5-1)^2 + (2+1)^2}$

$$= \sqrt{(4)^2 + (3)^2} = \sqrt{16+9} = \sqrt{25} = 5$$

$$BC = \sqrt{(9-5)^2 + (5-2)^2} = \sqrt{16+9} = \sqrt{25} = 5$$

$$CA = \sqrt{(1-9)^2 + (-1-5)^2} = \sqrt{64+36} = \sqrt{100} = 10$$

Since AB + BC = CA, ∴ the points A, B, and C are collinear.

Section Formula

For Internal Division

A point C divides the straight line joining two points $A(x_1, y_1)$ and $B(x_2, y_2)$ internally in the ratio m : n. To find the co-ordinate of C.

Let Co-ordinate of C be (x, y). Draw perpendiculars AL, BM, CN on the x-axis from A, B, and C. Draw AP and CQ perpendicular from A and C on CN and BM respectively.

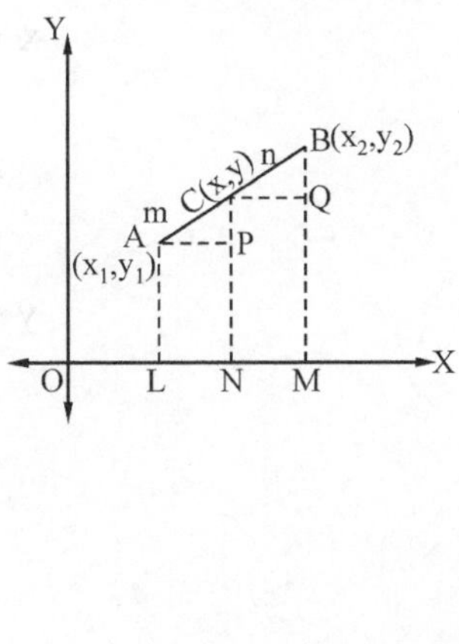

Now $AP = LN = ON - OL$

$$= x - x_1$$

$$CQ = NM = OM - ON$$

$$= x_2 - x$$

$$CP = CN - PN = y - y_1$$

$$BQ = MB - MQ = y_2 - y$$

Then from the similar triangles ACP and CBQ, we have,

$$\frac{AP}{CQ} = \frac{AC}{CB} \Rightarrow \frac{x - x_1}{x_2 - x} = \frac{m}{n}$$

$$\Rightarrow n(x - x_1) = m(x_2 - x)$$

$$\Rightarrow nx + mx = mx_2 + nx_1$$

$$\Rightarrow x(n + m) = mx_2 + nx_1 \qquad \Rightarrow x = \frac{mx_2 + nx_1}{m + n}$$

Similarly, $y = \dfrac{my_2 + ny_1}{m + n}$

Co-ordinates of the mid-point

If C be the mid-point of AB, then it will divide AB in the ratio 1:1, then the coordinate of C are

$$x = \frac{x_1 + x_2}{2}, y = \frac{y_1 + y_2}{2}$$

Co-ordinate for external division

The co-ordinates of a point which divided the straight line joining two points (x_1, y_1) and (x_2, y_2) externally in the ratio m : n are

$$x = \frac{mx_2 - nx_1}{m - n}, \qquad y = \frac{my_2 - ny_1}{m - n}$$

***Example* 6:** Find the co-ordinates of the point which divides (i) internally and (ii) externally, the line joining (–1, 2) to (4, –5) in the ratio 2:3.

Solution : (i) for internally – $x_1 = -1, x_2 = 4, y_1 = 2, y_2 = -5, m = 2, n = 3$

$$x = \frac{mx_2 + nx_1}{m + n} = \frac{2 \times 4 + 3 \times (-1)}{2 + 3}$$

$$= \frac{8 - 3}{5} = 1$$

$$y = \frac{my_2 + ny_1}{m + n} = \frac{2 \times (-5) + 3 \times 2}{2 + 3}$$

$$= \frac{-10 + 6}{5} = \frac{-4}{5}$$

(ii) For externally

$$x = \frac{mx_2 - nx_1}{m-n} = \frac{8+3}{-1} = -11$$

$$y = \frac{my_2 - ny_1}{m-n} = \frac{-10-6}{-1} = 16$$

∴ co-ordinates are $\left(1, -\frac{4}{5}\right)$ and (–11, 16)

***Example* 7:** In what ratio is the line joining the points A(4, 4) and B(7, 7) divided by P(–1, –1)?

Solution : Let P divide AB in the ratio of $\lambda : 1$

Here $x_1 = 4$, $x_2 = 7$, $x = -1$, $m = \lambda$, $n = 1$

$$\therefore \quad x = \frac{mx_2 + nx_1}{m+n}$$

$$-1 = \frac{\lambda \times 7 + 1 \times 4}{\lambda + 1} \Rightarrow -(\lambda + 1) = 7\lambda + 4$$

$$\Rightarrow -1 - 4 = 7\lambda + \lambda$$

$$\Rightarrow \quad -5 = 8\lambda \quad \Rightarrow \lambda = \frac{-5}{8}$$

Hence $\lambda : 1 = -5 : 8$

∴ P divides AB externally in the ration 5 : 8 ('–' sign shows externally)

Area of triangle

Let $A(x_1, y_1)$, $B(x_2, y_2)$, $C(x_3, y_3)$ be the vertices of the ΔABC.

Draw AL, BM, CN perpendiculars from A, B, C to x-axis

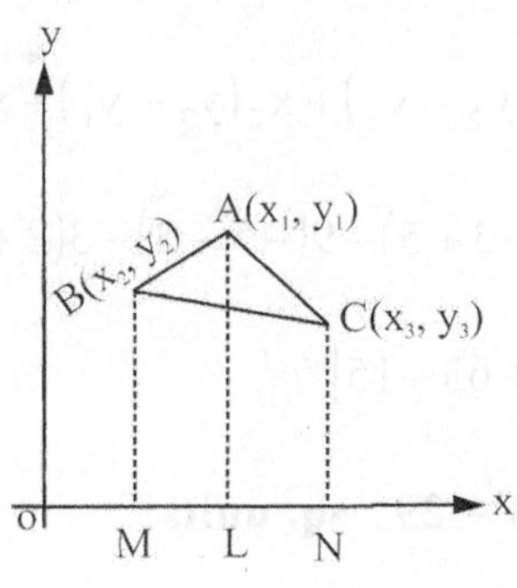

Then $ML = OL - OM = x_1 - x_2$
$LN = ON - OL = x_3 - x_1$
$MN = ON - OM = x_3 - x_2$

Now, Area of Δ ABC = area of trapezium ABML + area of trapezium ALNC – area of trapezium BMNC

$$= \frac{1}{2}(MB + LA)ML + \frac{1}{2}(LA + NC)LN - \frac{1}{2}(MB + NC)MN$$

$$= \frac{1}{2}(y_2 + y_1)(x_1 - x_2) + \frac{1}{2}(y_1 + y_3)(x_3 - x_1) - \frac{1}{2}(y_2 + y_3)(x_3 - x_2)$$

$$\Delta\, ABC = \frac{1}{2}[y_1(x_3 - x_2) + y_2(x_1 - x_3) + y_3(x_2 - x_1)]$$

or

$$\Delta\, ABC = \frac{1}{2}[x_1(y_2 - y_3) + x_2(y_3 - y_1) + x_3(y_1 - y_2)]$$

Area of ΔABC in determinant form is

$$\Delta ABC = \frac{1}{2}\begin{vmatrix} x_1 & y_1 & 1 \\ x_2 & y_2 & 1 \\ x_3 & y_3 & 1 \end{vmatrix}$$

Note : If three points are collinear, then area of triangle formed by them is zero.

i.e., $$\begin{vmatrix} x_1 & y_1 & 1 \\ x_2 & y_2 & 1 \\ x_3 & y_3 & 1 \end{vmatrix} = 0$$

Area of quadrilateral $= \frac{1}{2}[(x_1y_2 - x_2y_1) + (x_2y_3 - x_3y_2) + (x_3y_4 - x_4y_3) + (x_4y_1 - x_1y_4)]$

Example 8 : Find the area of the triangle, the co-ordinates of vertices are (5, 2), (–9, –3) and (–3, –5).

Solution : Area of $\Delta = \frac{1}{2}[x_1(y_2 - y_3) + x_2(y_3 - y_1) + x_3(y_1 - y_2)]$

$$= \frac{1}{2}[5(-3+5) - 9(-5-2) - 3(2+3)]$$

$$= \frac{1}{2}[10 + 63 - 15]$$

$$= \frac{1}{2} \times 58 = 29 \text{ sq. units}$$

***Example* 9 :** Show that the points (a, b + c), (b, c + a) and (c, a + b) are collinear

Solution : Area of $\Delta = \frac{1}{2}[a(c+a-a-b)+b(a+b-b-c)+c(b+c-c-a)]$

$$= \frac{1}{2}[ac - ab + ab - bc + bc - ac]$$

$$= 0$$

Since area of ΔABC is zero. Then the points are collinear.

Centroid of triangle

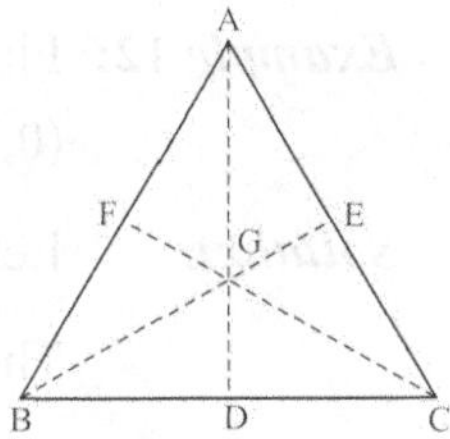

G, the point where the three medians of a triangle meet is called centroid of ΔABC

$$G = \left(\frac{x_1 + x_2 + x_3}{3}, \frac{y_1 + y_2 + y_3}{3}\right)$$

***Example* 10 :** The vertices of a triangle are (2, 1), (5, 2) and (3, 4). Find the co-ordinates of the centroid.

Solution : Here $x = \frac{x_1 + x_2 + x_3}{3}$, $y = \frac{y_1 + y_2 + y_3}{3}$

$$= \frac{2+5+3}{3} \quad , \quad = \frac{1+2+4}{3}$$

$$x = \frac{10}{3} \quad , \quad y = \frac{7}{3}$$

∴ **centroid is** $\left(\frac{\mathbf{10}}{\mathbf{3}}, \frac{\mathbf{7}}{\mathbf{3}}\right)$

***Example* 11 :** Two vertices of a triangle are (1, 2), (3, 5) and its centroid is at the origin. Find the co-ordinates of the third vertex.

Solution : Let the co-ordinates of third vertex is (x, y) and centroid is at origin i.e., (0, 0) then.

$$x = \frac{x_1 + x_2 + x_3}{3} \qquad y = \frac{y_1 + y_2 + y_3}{3}$$

$$0 = \frac{1+3+x}{3} \qquad 0 = \frac{2+5+y}{3}$$

$$0 = 4 + x \qquad 0 = 7 + y$$

$$x = -4 \qquad y = -7$$

∴ **co-ordinates of third vertex are (–4, –7)**

Incentre of triangle

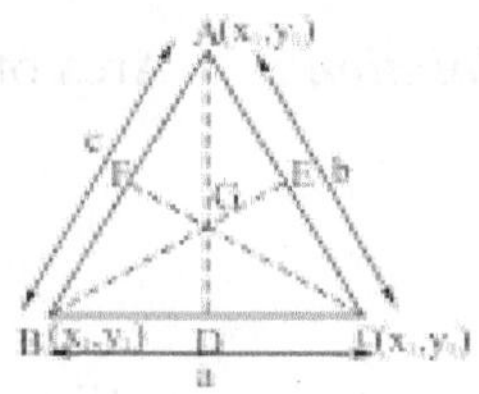

The point of concurrence of the bisectors of the angles of a triangle is called the incentre of a triangle.

Its coordinates are $\left(\dfrac{ax_1 + bx_2 + cx_3}{a+b+c}, \dfrac{ay_1 + by_2 + cy_3}{a+b+c}\right)$

***Example* 12:** Find the incentre of the triangle given by the points (–36, 7), (20, 7) and (0, –8)

Solution: Let A(–36, 7), B(20,7) and C(0, –8) be the vertices

Then $a = BC = \sqrt{(20-0)^2 + (7+8)^2}$

$$= \sqrt{400+225} = \sqrt{625} = 25$$

$$b = CA = \sqrt{(-36-0)^2 + (7+8)^2}$$

$$= \sqrt{1296+225} = \sqrt{1521} = 39$$

$$c = AB = \sqrt{(-36-20)^2 + (7-7)^2}$$

$$= \sqrt{(56)^2} = 56$$

∴ Co-ordinates of incentre,

$$x = \frac{ax_1 + bx_2 + cx_3}{a+b+c}, \quad y = \frac{ay_1 + by_2 + cy_3}{a+b+c}$$

$$= \frac{25\times(-36)+39\times 20+56\times 0}{25+39+56}, \quad = \frac{25\times 7+39\times 7+56\times(-8)}{25+39+56}$$

$$= \frac{-120}{120} = -1, \quad = \frac{0}{120} = 0$$

∴ **Incentre is (–1, 0)**

Locus and its equation

When a point moves under a certain condition, the path it traces out is called its locus under these conditions.

The equation of a locus is the equation of x and y that is satisfied by the co-ordinates of every point on the locus.

Method of finding the locus

1. Let the point whose locus is to be calculated to be (h, k).
2. Set up a relation between h and k with the help of given conditions.
3. Replace h and k by x and y.

   This is the required equation of locus.

***Example* 13:** Find the equation to the locus of a point equidistance from the points. A(1, 3) and B (–2, 1)

Solution : Let p(h. k) by any point on the locus. Then PA = PB (Given)

$\Rightarrow \quad PA^2 = PB^2$

$\Rightarrow \quad (h-1)^2 + (k-3)^2 = (h+2)^2 + (k-1)^2$

$\Rightarrow \quad h^2 - 2h + 1 + k^2 - 6k + 9 = h^2 + 4h + 4 + k^2 - 2k + 1$

$\Rightarrow \quad 6h + 4k = 5$

Hence, locus of (h. k) is 6x + 4y = 5

***Example* 14:** Find the equation of the locus of points twice as far from (3, 2) as from (1, 1)

Solution: Let point P(h, k) be any point on locus. Then PA = 2PB (Given)

$\Rightarrow \quad PA^2 = 4\,PB^2$

$\Rightarrow \quad (h-3)^2 + (k-2)^2 = 4\,[(h-1)^2 + (k-1)^2]$

$\Rightarrow \quad h^2 - 6h + 9 + k^2 - 4k + 4 = 4[h^2 - 2h + 1 + k^2 - 2k + 1]$

$\Rightarrow \quad 3h^2 + 3k^2 - 2h - 4k - 5 = 0$

Hence, locus of (h, k) is $3x^2 + 3y^2 - 2x - 4y - 5 = 0$

***Example* 15:** Find the equation of locus of a point whose sum of squares of distances from two fixed points (a, 0) and (–a, 0) is $2c^2$.

Solution: Let A(a, 0) and B(–a, 0) are the co-ordinates and p(x, y) be a point on locus.

Then $\quad PA^2 + PB^2 = 2c^2 \quad$ (Given)

$\Rightarrow \quad (x-a)^2 + (y-0)^2 + (x+a)^2 + (y-0)^2 = 2c^2$

$\Rightarrow \quad x^2 + a^2 - 2ax + y^2 + x^2 + 2ax + a^2 + y^2 + = 2c^2$

$\Rightarrow \quad 2x^2 + 2y^2 + 2a^2 = 2c^2$

$\Rightarrow \quad \mathbf{x^2 + y^2 + a^2 = c^2}$ which is the required locus

The Straight Line

Equation of Straight line Parallel to Axis of Co-ordinates

(i) ***Lines parallel to x-axis :*** The equation of straight line parallel to x-axis at a distance b from it is $y = b$ and equation of x-axis is $y = 0$.

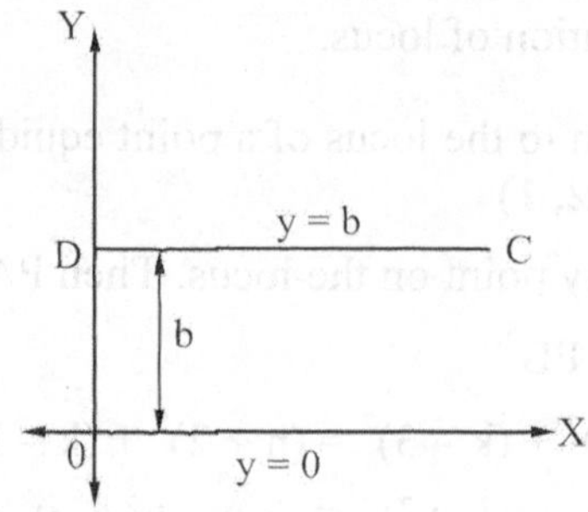

(ii) ***Lines parallel to y-axis :*** The equation of straight line parallel to y-axis at a distance a from it is $x = a$ and equation of y-axis is $x = 0$.

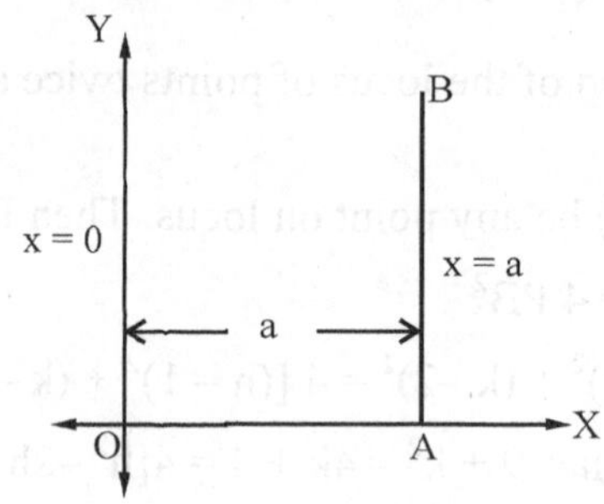

Slope of a line :

(i) The slope of a line is the tangent of the angle which the part of the above x-axis makes with the positive direction of x-axis. The slope of line is denoted by 'm'.

i.e., $m = \text{Tan}\,\theta$ It is also called gradient

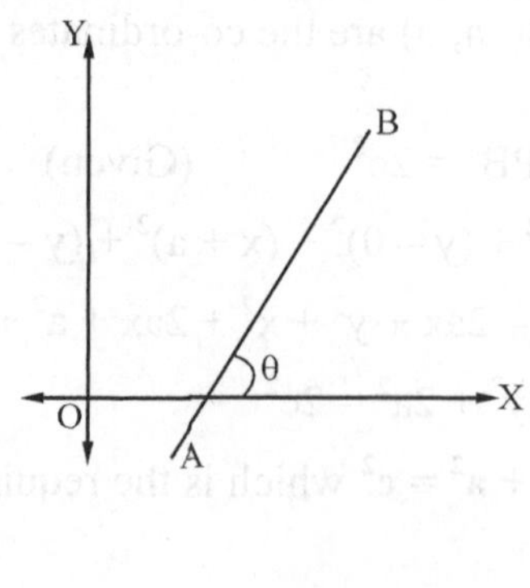

(ii) Slope of a line passing through two points (x_1, y_1) and (x_2, y_2) is

$$m = \frac{y_2 - y_1}{x_2 - x_1}$$

(iii) Let the slope of two lines be m_1 and m_2.

(a) These lines are said to be parallel to each other if $m_1 = m_2$

(b) These lines are said to be perpendicular to each other if $m_1 m_2 = -1$

Equations of Straight Lines

(a) ***Slope form :*** The equation of straight line which cuts off given intercept c on the y-axis and is inclined at a given angle θ to the x-axis is

$$y = mx + c, \text{ where } m = \tan\theta$$

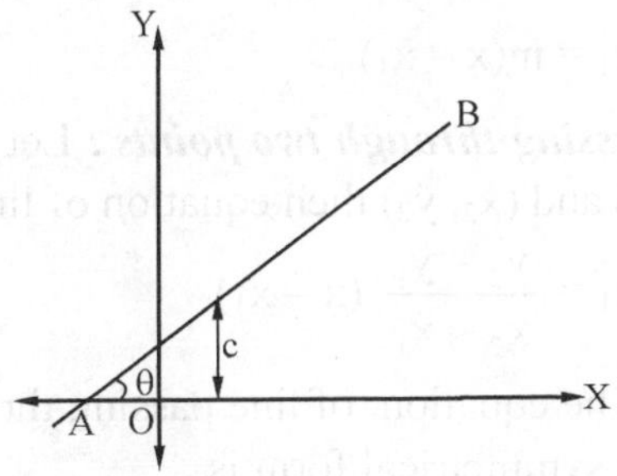

(b) ***Intercept form :*** If the line cuts off intercepts a and b of x and y-axis respectively, then the equation of straight line is

$$\frac{x}{a} + \frac{y}{b} = 1$$

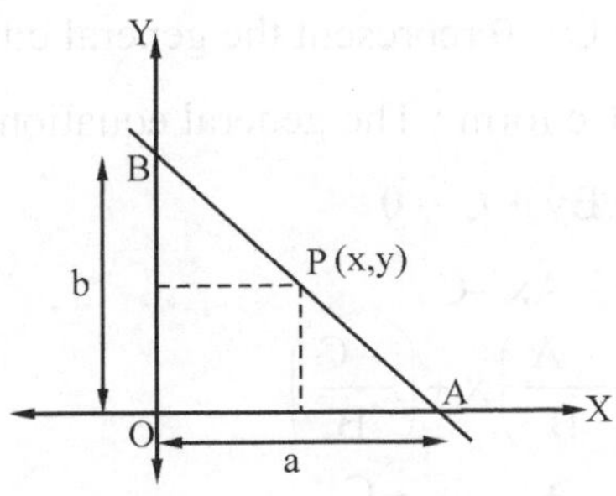

(c) ***Perpendicular (Normal) form :*** If p be the length of perpendicular from the origin on the line and α be the angle with the perpendicular makes with the x-axis, then the equation of the line is

$$x \cos \alpha + y \sin \alpha = p$$

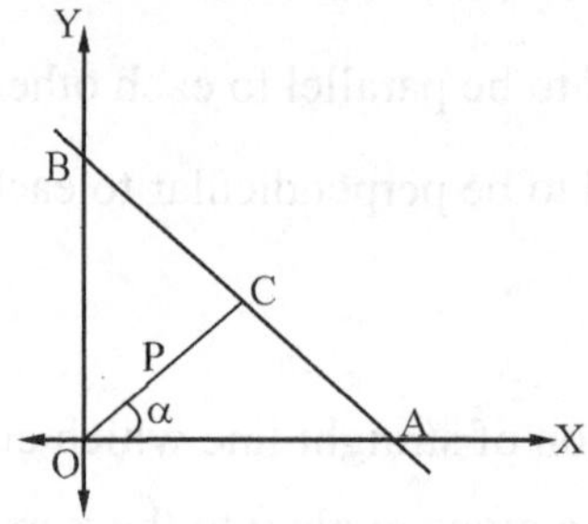

(d) ***Lines of given slope through a point :*** The equation of a line which passes through a point (x_1, y_1) and whose slope is m is given by

$$y - y_1 = m(x - x_1)$$

(e) ***Equation of a line passing through two points :*** Let the line be passing through the two points (x_1, y_1) and (x_2, y_2) then equation of line is

$$y - y_1 = \frac{y_2 - y_1}{x_2 - x_1}(x - x_1)$$

(f) ***Symmetrical form :*** The equation of line passing through (x_1, y_1) and making an angle θ with x-axis in symmetrical form is

$$\frac{x - x_1}{\cos \theta} = \frac{y - y_1}{\sin \theta} = r$$

where r is the distance of point of intersection

General Equation of Straight Line

The liner equation $Ax + By + C = 0$ represent the general equation of a straight line.

(i) Reduction in $y = mx + c$ form : The general equation of straight line is

$$Ax + By + C = 0$$

$$By = -Ax - C$$

$$y = \left(\frac{-A}{B}\right)x + \left(\frac{-C}{B}\right)$$

where $$m = \frac{-A}{B}, \ c = \frac{-C}{B}$$

(ii) Reduction in $\frac{x}{a}+\frac{y}{b}=1$ form:

$$Ax + By + C = 0$$

$$Ax + By = -C$$

$$\left(-\frac{A}{C}\right)x + \left(-\frac{B}{C}\right)y = 1$$

$$\frac{x}{-C/A} + \frac{y}{-C/B} = 1$$

where $a = -C/A$, $b = -C/B$

(iii) Reduction in $x \cos \alpha + y \sin \alpha = p$ form:

$$Ax + By + C = 0$$

$$Ax + By = -C$$

divide by $\sqrt{A^2 + B^2}$, we get

$$\frac{A}{\sqrt{A^2+B^2}}x + \frac{B}{\sqrt{A^2+B^2}}y = \frac{-C}{\sqrt{A^2+B^2}}$$

where $\cos \alpha = \frac{A}{\sqrt{A^2+B^2}}$, $\sin \alpha = \frac{B}{\sqrt{A^2+B^2}}$, $P = \frac{-C}{\sqrt{A^2+B^2}}$

***Example* 16 :** Find the equation of straight line which is parallel to x-axis and passes through the point (5, –3)

Solution : Any straight line parallel to x-axis is $y = b$

Since it passes through the point (5, –3) $\therefore b = -3$

$\therefore$ required equation of line is $y = -3$

$$\mathbf{y + 3 = 0}$$

***Example* 17:** Find the equation to a line which cut an intercept of $\sqrt{2}$ on the negative direction of y-axis and make an angle $\tan^{-1}\left(\frac{1}{\sqrt{2}}\right)$ with the x-axis.

Solution: Here $c = \sqrt{2}$, $\theta = \tan^{-1}\frac{1}{\sqrt{2}}$

$$\Rightarrow \tan\theta = \frac{1}{\sqrt{2}} \qquad \Rightarrow m = \frac{1}{\sqrt{2}}$$

Required equation of line is $y = mx + c$

$$y = \frac{1}{\sqrt{2}}x + \sqrt{2}$$

$$\mathbf{x - \sqrt{2}\,y + 2 = 0}$$

***Example* 18:** Find the equation to line which passes through (2, 3) and whose m is 3.

Solution : Equation of line passing through one point is

$$y - y_1 = m(x - x_1)$$

Here $m = 3, x_1 = 2, y_1 = 3$

$\therefore$ required equation of line is $(y - 3) = 3(x - 2)$

$$y - 3 = 3x - 6$$

$$\mathbf{3x - y - 3 = 0}$$

***Example* 19:** Find the equation of line passing through the point (3, –2) and perpendicular to the line $x - 3y + 5 = 0$

Solution : Slope of given line $x - 3y + 5 = 0$

$$3y = x + 5$$

$$y = \frac{1}{3}\,x + 5$$

$$m_1 = \frac{1}{3}$$

as the line perpendicular, then $m_1m_2 = -1 \Rightarrow \frac{1}{3}\,m_2 = -1 \Rightarrow m_2 = -3$

$\therefore$ required equation of line passing through (3, –2) is

$$y + 2 = -3\,(x - 3)$$

$$y + 2 = -3x + 9$$

$$\mathbf{3x + y - 7 = 0}$$

***Example* 20:** Find the equation of line passing through the points (–3, 4) and (1, –3).

Solution : Equation of straight line passing though (x_1, y_1) and (x_2, y_2) is

$$y - y_1 = \frac{y_2 - y_1}{x_2 - x_1}\,(x - x_1)$$

here $x_1 = -3, x_2 = 1, y_1 = 4, y_2 = -3$

$\therefore$ required equation of line is $y - 4 = \frac{-3-4}{1+3}(x+3)$

$$4(y-4) = -7(x+3)$$

$$4y - 16 = -7x - 21$$

$$\mathbf{7x + 4y + 5 = 0}$$

***Example* 21 :** Find the slope and the equation of line joining the points (2, –5) and (4, 1)

Solution : Slope of a line passing through (x_1, y_1) and (x_2, y_2) is

$$m = \frac{y_2 - y_1}{x_2 - x_1}$$

here $x_1 = 2, x_2 = 4, y_1 = -5, y_2 = 1$

$\therefore$ required slope is $m = \frac{1+5}{4-2}$

$$m = 3$$

$\therefore$ required equation of line is $y - y_1 = m(x - x_1)$

$$y + 5 = 3(x - 2)$$

$$y + 5 = 3x - 6$$

$$\mathbf{3x - y - 11 = 0}$$

***Example* 22:** A straight line, drawn through the point A(2, 1), make an angle $\frac{\pi}{4}$ with positive x-axis and intersect another line $x + 2y + 1 = 0$ at point B. Find the length AB.

Solution : The equation of line in symmetrical form is $\frac{x - x_1}{\cos\theta} = \frac{y - y_1}{\sin\theta} = r$

Here $x_1 = 2, y_1 = 1, \theta = \frac{\pi}{4}$ $\Rightarrow$ $\frac{x-2}{\cos\frac{\pi}{4}} = \frac{y-1}{\sin\frac{\pi}{4}} = r$

$$\Rightarrow \frac{x-2}{\frac{1}{\sqrt{2}}} = \frac{y-1}{\frac{1}{\sqrt{2}}} = r$$

$$\Rightarrow x - 2 = \frac{r}{\sqrt{2}}, y - 1 = \frac{r}{\sqrt{2}}$$

$$\Rightarrow x = \frac{r}{\sqrt{2}} + 2, y = \frac{r}{\sqrt{2}} + 1$$

Thus co-ordinates of B are $\left(2+\frac{r}{\sqrt{2}}, 1+\frac{r}{\sqrt{2}}\right)$

Since B lies on $x + 2y + 1 = 0$

$$\Rightarrow \quad 2+\frac{r}{\sqrt{2}}+2\left(1+\frac{r}{\sqrt{2}}\right)+1=0$$

$$\Rightarrow \quad 2+\frac{r}{\sqrt{2}}+2+\frac{2r}{\sqrt{2}}+1=0$$

$$\Rightarrow \quad \frac{3r}{\sqrt{2}}=-5 \quad \Rightarrow \quad r=\frac{-5\sqrt{2}}{3}$$

∴ length of AB is $\frac{5\sqrt{2}}{3}$, (neglecting '–' sign)

***Example* 23:** Find the equation of the line through (2, 3), so that the segment of the line intercepted between the axes is bisected at this point.

Solution: Let the equation of the line $\frac{x}{a}+\frac{y}{b}=1$

It meets axes of x and y at the point (a, 0) and (0, b) respectively

A(2, 3) divides the joining of B(a, 0)and C(0, b) in the ratio 1:1.

∴ x co-ordinate of $\quad 2=\frac{1\times 0+1\times a}{1+1}$

$$4 = a$$

y co-ordinate of $\quad 3=\frac{1\times b+1\times 0}{2}$

$$6 = b$$

∴ required equation of line is $\quad \frac{x}{4}+\frac{y}{6}=1$

$$\frac{3x+2y}{12}=1$$

$$\mathbf{3x + 2y = 12}$$

***Example* 24 :** Find the length of intercept cut by the line $4x + 6y – 15 = 0$ from both the axes.

Solution : Given line is $\quad 4x + 6y – 15 = 0$

$$4x + 6y = 15$$

$$\frac{4}{15}x+\frac{6}{15}y=1$$

$$\frac{x}{\frac{15}{4}}+\frac{y}{\frac{15}{6}}=1$$

On comparing $\frac{x}{a}+\frac{y}{b}=1$

∴ Intercepts are a = $\frac{15}{4}$, b = $\frac{15}{6}$

***Example* 25:** Find the equation of line which passes through the point (3, 4) and whose intercept on y-axis is twice the intercept on x-axis.

Solution: Let a be the intercept on x-axis and 2a is intercept on y-axis

∴ equation of line be $\frac{x}{a}+\frac{y}{2a}=1$

Since the line passes through (3, 4), then

$$\frac{3}{a}+\frac{4}{2a}=1$$

$$\frac{6+4}{2a}=1$$

$$10 = 2a \qquad \therefore a = 5$$

Hence, the equation of line is $\frac{x}{5}+\frac{y}{10}=1$

$$\mathbf{2x + y = 10}$$

***Example* 26:** A line is at a unit distance from the origin and the perpendicular from the origin makes an angle $\tan^{-1}\frac{3}{4}$ with x-axis. Find the equation of line.

Solution: Here $\alpha = \tan^{-1}\frac{3}{4} \Rightarrow \tan\alpha = \frac{3}{4}$

Then $\sin\alpha = \frac{3}{5}$, $\cos\alpha = \frac{4}{5}$, P = 1 (unit distance)

∴ required equation of line be $x\cos\alpha + y\sin\alpha = P$

$$x \times \frac{4}{5} + y \times \frac{3}{5} = 1$$

$$\mathbf{4x + 3y = 5}$$

***Example* 27:** Reduce $4x + 3y - 9 = 0$ to the normal form and find the distance from origin.

Solution: Given line is $4x + 3y = 9$

divided by $\sqrt{(4)^2 + (3)^2} = 5$

We get $\frac{4}{5}x + \frac{3}{5}y = \frac{9}{5}$

On comparing $x \cos \alpha + y \sin \alpha = P$

$\therefore \quad \cos \alpha = \frac{4}{5}, \sin \alpha = \frac{3}{5}, P = \frac{9}{5}$

$\therefore$ **Normal distance from origin is** $\mathbf{P = \frac{9}{5}}$

Point of Intersection of Two Straight Lines

Let the equations of two given lines be

$$a_1x + b_1y + c_1 = 0 \text{ and } a_2x + b_2y + c_2 = 0$$

by solving these two lines for x and y, we have

$$\frac{x}{b_1c_2 - b_2c_1} = \frac{y}{c_1a_2 - c_2a_1} = \frac{1}{a_1b_2 - a_2b_1}$$

$$\therefore \quad x = \frac{b_1c_2 - b_2c_1}{a_1b_2 - a_2b_1}, \qquad y = \frac{c_1a_2 - c_2a_1}{a_1b_2 - a_2b_1}$$

$\therefore$ The point of intersection is $\left(\frac{b_1c_2 - b_2c_1}{a_1b_2 - a_2b_1}, \frac{c_1a_2 - c_2a_1}{a_1b_2 - a_2b_1}\right)$

Conditions

(i) If the two lines are perpendicular then $a_1a_2 + b_1b_2 = 0$

(ii) If the two lines are parallel then $\frac{a_1}{b_1} = \frac{a_2}{b_2}$

Angle between Two Lines

Let the equation of two lines are $y = m_1x + c_1$ and $y = m_2x + c_2$, inclined at angle θ_1, and θ_2 on x-axis.

Then $m_1 = \tan \theta_1, m_2 = \tan \theta_2$

If θ be the angle between the given lines, then

$$\theta = \theta_1 - \theta_2$$

$$\tan \theta = \tan (\theta_1 - \theta_2)$$

$$= \frac{\tan \theta_1 - \tan \theta_2}{1 + \tan \theta_1 \tan \theta_2}$$

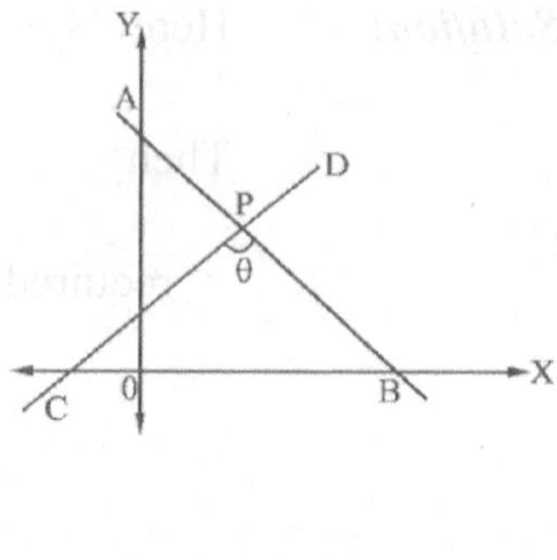

$$\tan\theta = \frac{m_1 - m_2}{1 + m_1 m_2}$$

$$\theta = \tan^{-1}\left(\frac{m_1 - m_2}{1 + m_1 + m_2}\right)$$

Note : $\tan\theta$ will be positive or negative accordingly as θ is acute or obtuse.

Any line through the point of intersection of two given lines

Let the equation of two lines are

$$a_1x + b_1y + c_1 = 0 \quad .....(9.1)$$

and

$$a_2x + b_2y + c_2 = 0 \quad .....(9.2)$$

Consider the equation $(a_1x + b_1y + c_1) + \lambda\,(a_2x + b_2y + c_2) = 0$ .....(9.3)

The co-ordinate of the points which satisfy both the equations 9.1 and 9.2 so also satisfy 9.3. Therefore 9.3 is the equation of any straight line through the point of intersection of the lines 9.1 and 9.2.

***Example* 28:** Find the point of intersection of the pair of lines

$$3x - y = 19 \text{ and } 2x - 5y - 30 = 0$$

Solution: By solving simultaneously the equations

$3x - y = 19$ and

$2x - 5y = 30,$ we get

$$\frac{x}{30 - 95} = \frac{y}{-38 + 90} = \frac{1}{-15 + 2}$$

$$\frac{x}{-65} = \frac{y}{52} = \frac{1}{-13}$$

$$\therefore \quad x = \frac{-65}{-13} = 5 \text{ and } y = \frac{52}{-13} = -4$$

∴ required point of intersection is (5, –4)

***Example* 29:** Find the equation of the line passing through the intersection of the lines $x - 2y + 3 = 0$, $2x - 3y + 4 = 0$ and also through the point (1, 1).

Solution: The equation of any line passes through the point of intersection of the given lines is

$$(x - 2y + 3) + \lambda(2x - 3y + 4) = 0 \quad .....(i)$$

If it passes through the point (1, 1), we get

$$(1 - 2 + 3) + \lambda\,(2 - 3 + 4) = 0$$

$$2 + 3\lambda = 0 \qquad \Rightarrow \qquad \lambda = \frac{-2}{3}$$

Substituting $\lambda = \frac{-2}{3}$ in equation (i), we have

$$(x - 2y + 3) - \frac{2}{3}(2x - 3y + 4) = 0$$

$$3x - 6y + 9 - 4x + 6y - 8 = 0$$

$$-x + 1 = 0$$

$$\mathbf{x - 1 = 0}$$

Which is required equation of line.

***Example* 30 :** Find the angle between the two straight lines

$$2x + 3y + 4 = 0 \text{ and } x - 2y + 3 = 0$$

Solution: Let the lines be $2x + 3y + 4 = 0$ .....(i)

and $x - 2y + 3 = 0$ .....(ii)

Slope of line (i) is $m_1 = \frac{-2}{3}$

Slope of line (ii) is $m_2 = \frac{1}{2}$

Now,
$$\tan\theta = \frac{m_1 - m_2}{1 + m_1 m_2}$$

$$= \frac{\frac{-2}{3} - \frac{1}{2}}{1 + \left(\frac{-2}{3}\right)\left(\frac{1}{2}\right)}$$

$$= \frac{-7}{6} \times \frac{3}{2} = \frac{-7}{4}$$

$$\boldsymbol{\theta = \tan^{-1}\left(\frac{-7}{4}\right)} \text{ \textbf{i.e., obtuse angle}}$$

Equation of a straight Line parallel to a given line

The equation of a line parallel to a given line ax + by + c = 0 by changing the value of c by k.

i.e., $ax + by + k = 0$

Equation of a line perpendicular to a given line

The equation of a line perpendicular to a given line $ax + by + c = 0$ is

$$bx - ay + k = 0$$

***Example* 31:** Find the equation of the line which is parallel to the line $4x + 7y + 8 = 0$ and passes through the point (5, 6)

Solution: The equation of any line parallel to the line $4x + 7y + 8 = 0$ is

$$4x + 7y + k = 0 \quad .....(i)$$

If it passes through the point (5, 6), we have from eq. (i)

$$4 \times 5 + 7 \times 6 + k = 0$$

$$20 + 42 + k = 0 \quad \Rightarrow \quad k = -62$$

Substituting the value of k in eq. (i), we get

$$4x + 7y - 62 = 0$$

which is the required line.

***Example* 32:** Find the equation of line which is perpendicular to the line $3x - 4y + 7 = 0$ and passes through the point (–3, 2).

Solution: The equation of any line perpendicular to the line $3x - 4y + 7 = 0$ is

$$4x + 3y + k = 0 \quad .....(i)$$

If it passes through the point (–3, 2), we have, from eq. (i)

$$4 \times (-3) + 3 \times 2 + k = 0 \quad \Rightarrow \quad k = 6$$

Substituting the value of k in eq. (i), we get $4x + 3y + 6 = 0$

Which is the required line.

Condition of concurrency of three lines

Let the three lines are

$$a_1x + b_1y + c_1 = 0 \quad .....(1)$$

$$a_2x + b_2y + c_2 = 0 \quad .....(2)$$

$$a_3x + b_3y + c_3 = 0 \quad .....(3)$$

Let the point of intersection of (1) and (2) are (x_1, y_1), which satisfy (3) then three lines are concurrent.

The required condition for concurrent is $\begin{vmatrix} a_1 & b_1 & c_1 \\ a_2 & b_2 & c_2 \\ a_3 & b_3 & c_3 \end{vmatrix} = 0$

***Example* 33:** Show that the lines 3x + 2y – 5 = 0, 4x + 3y + 7 = 0 and 21x + 13y – 76 = 0 are concurrent.

Solution : By solving the equation 3x + 2y – 5 = 0

and 21x + 13y – 76 = 0 simultaneously

we get,
$$\frac{x}{14+15} = \frac{y}{-20-21} = \frac{1}{9-8}$$

$$\frac{x}{29} = \frac{y}{-41} = 1$$

$\Rightarrow$ x = 29,

and y = – 41

Hence, point of intersection of first two lines is (29, –41)

Substituting these co-ordinates in equation 21x + 13y – 76 = 0, we get

$$= 21 \times 29 + 13 \times (-41) - 76$$

$$= 609 - 533 - 76 = 0$$

Since the co-ordinates satisfy the third equation, then three lines are concurrent.

Length of perpendicular

The length of perpendicular p from the point (x_1, y_1) to the line ax + by + c = 0 is given by

$$p = \frac{ax_1 + by_1 + c}{\sqrt{a^2 + b^2}}$$

It is also called perpendicular distance of a point.

***Example* 34:** Find the length of the perpendicular from the point (3, –5) on the line

3x + 4y + 7 = 0

Solution: Apply the formula, we have the perpendicular distance

$$p = \frac{3 \times 3 + 4 \times (-5) + 7}{\sqrt{(3)^2 + (4)^2}}$$

$$= \frac{9 - 20 + 7}{\sqrt{9+16}}$$

$$= \frac{-4}{5}$$

***Example* 35:** Find the distance between the parallel lines.

$$2x + 3y - 25 = 0 \qquad \text{and} \qquad 4x + 6y + 11 = 0$$

Solution : The given lines are $2x + 3y - 25 = 0$ .....(i)

and $4x + 6y + 11 = 0$ .....(ii)

Length of perpendicular from origin to (i) $= \dfrac{0+0-25}{\sqrt{4+9}} = \dfrac{-25}{\sqrt{13}}$

Length of perpendicular from origin to (ii) $= \dfrac{0+0+11}{\sqrt{16+36}} = \dfrac{11}{\sqrt{52}}$

Distance between (i) and (ii) be $= \dfrac{11}{\sqrt{52}} - \left(\dfrac{-25}{\sqrt{13}}\right)$

$$= \frac{11+50}{\sqrt{52}} = \frac{61}{\sqrt{52}}$$

***Example* 36:** Find the distance between the parallel lines

$$3x + 4y = 12 \text{ and } 3x + 4y = 3$$

Solution : Put $y = 0$ in $3x + 4y = 12$, we get $x = 4$

Thus $(4, 0)$ is a point on the line $3x + 4y = 12$

Now, length of perpendicular from the point $(4, 0)$ to the line

$$3x + 4y = 3$$

$$\therefore \quad p = \frac{3 \times 4 + 4 \times 0 - 3}{\sqrt{9+16}}$$

$$= \frac{12-3}{5} = \frac{9}{5}$$

Practice Questions

***Example* 1:** Prove the following points are collinear; (–3, 0) (0, –9) and (–2, –3)

Solution: Area of triangle $= \dfrac{1}{2}[x_1(y_2 - y_3) + x_2(y_3 - y_1) + x_3(y_1 - y_2)]$

$$= \frac{1}{2}[-3(-9+3) + 0(-3-0) - 2(0+9)]$$

$$= \frac{1}{2}[18-18]$$

$$= 0$$

Since area of a triangle is zero, hence given points are collinear.

***Example* 2:** Show that the point (1, 1) (2, 2) (3, 3) are collinear

Solution : Area of a triangle $= \frac{1}{2}[x_1(y_2 - y_3) + x_2(y_3 - y_1) + x_3(y_1 - y_2)]$

$$= \frac{1}{2}[1(2-3) + 2(3-1) + 3(1-2)]$$

$$= \frac{1}{2}[-1+4-3]$$

$$= \frac{1}{2}[-4+4] = 0$$

Since area of triangle is zero, hence given points are collinear.

***Example* 3:** Find the area of triangle whose vertices are given as

(1, 2), (–2, 4) and (2, 6)

Solution : Area of triangle $= \frac{1}{2}[1(4-6) - 2(6-2) + 2(2-4)]$

$$= \frac{1}{2}[-2-8-4]$$

$$= \frac{1}{2}[-14] = -7 \text{ sq. units}$$

***Example* 4:** The co-ordinates of vertices B and C of a triangle are (1, –2), (2, 4) and A lies on the line 2x + y – 2 = 0. The area of the triangle is 8 units. Find the vertex co-ordinates of A.

Solution: Since A lies on the line $2x + y - 2 = 0$

$\therefore \quad y = 2 - 2x$

$\therefore \quad A(x_1, 2-2x), B(1, -2), C(2, 4)$

$\therefore$ Area of triangle $= \frac{1}{2}[x_1(y_2 - y_3) + x_2(y_3 - y_1) + x_3(y_1 - y_2)]$

$$8 = \frac{1}{2}[x(-2-4) + 1(4-2+2x) + 2(2-2x+2)]$$

$$8 = \frac{1}{2}[-6x + 2 + 2x + 8 - 4x]$$

$$16 = -8x + 10$$

$$16 - 10 = -8x$$

$$6 = -8x$$

$$\Rightarrow \quad x = \frac{-6}{8}$$

$$\Rightarrow \quad x = \frac{-3}{4}$$

Now, $\quad y = 2 - 2x$

$$= 2 + 2 \times \frac{3}{4} \quad \Rightarrow \quad y = 2 + \frac{3}{2} = \frac{7}{2}$$

$\therefore$ Co-ordinates of A are $\left(\frac{-3}{4}, \frac{7}{2}\right)$

***Example* 5:** Find the co-ordinates of point which divides the line joining points (1, 2) and (–3, 4) in the ratio 2 : 3.

Solution: Let A(1, 2) and B(–3, 4), p divides in the ratio 2 : 3

$\therefore$ co-ordinate of p are $\left(\frac{mx_2 + nx_1}{m+n}, \frac{my_2 + ny_1}{m+n}\right)$

$$= \left(\frac{2\times(-3)+3\times1}{2+3}, \frac{2\times4+3\times2}{2+3}\right)$$

$$= \left(\frac{-6+3}{5}, \frac{8+6}{5}\right)$$

$$= \left(\frac{-3}{5}, \frac{14}{5}\right)$$

***Example* 6:** Find the equation of the straight line which cuts an intercept 3 from y-axis and make angle of 135° from x-axis.

Solution: Given $\quad \theta = 135^\circ, c = 3$

$$m = \tan\theta$$

$$= \tan 135^\circ$$

$$m = -1$$

Now, required equation of straight line

$$y = mx + c$$

$$y = (-1)x + 3$$

$$y = -x + 3$$

$$x + y - 3 = 0$$

***Example* 7:** Find the equation to the straight line cutting off an intercept –3 from the axis of y and inclined at an angle $\tan^{-1}\left(\frac{3}{5}\right)$ to the axis of x.

Solution : Given $c = -3, \quad \theta = \tan^{-1}\left(\frac{3}{5}\right)$

$$\Rightarrow \quad \tan\theta = \frac{3}{5}$$

$$\Rightarrow \quad m = \frac{3}{5}$$

Required equation of straight line is

$$y = mx + c$$

$$y = \frac{3}{5}x - 3$$

$$5y = 3x - 15$$

$$3x - 5y - 15 = 0$$

***Example* 8:** Find the equation of a line passing through the points (2, 3) and (2, –5)

Solution: Equation of straight line is

$$y - y_1 = \frac{y_2 - y_1}{x_2 - x_1}(x - x_1)$$

$$y - 3 = \frac{-5-3}{2-2}(x-2)$$

$$y - 3 = \frac{-8}{0}(x-2)$$

$$0 = -8x + 16$$

$$8x - 16 = 0$$

***Example* 9:** Find the equation of straight line whose perpendicular distance from the origin is 4 and the angle this perpendicular makes with x-axis is 30°.

Solution: Given $\theta = 30^\circ, p = 4$

Required equation of line is

$$x\cos\theta + y\sin\theta = p$$

$$x\cos 30^\circ + y\sin 30^\circ = 4$$

$$x\frac{\sqrt{3}}{2} + y \times \frac{1}{2} = 4$$

$$x\sqrt{3} + y = 8$$

***Example* 10:** Find the intercepts cut by the straight line 2x – 3y + 18 = 0 from the axes.

Solution : Given line is

$$2x - 3y + 18 = 0$$

$$2x - 3y = -18$$

$$\frac{2x}{-18} - \frac{3y}{-18} = 1$$

$$\frac{x}{-9} + \frac{y}{6} = 1$$

On comparing $\frac{x}{a} + \frac{y}{b} = 1$

$\therefore$ intercepts are a = –9, b = 6

***Example* 11:** Find the equation to the straight line cutting off intercepts – 5 and 6 from the axes.

Solution : Equation of line is $\frac{x}{a} + \frac{y}{b} = 1$

Here a = –5, b = 6

$\therefore$ required line is $\frac{x}{-5} + \frac{y}{6} = 1$

$$\frac{-6x + 5y}{30} = 1$$

$$-6x + 5y = 30$$

$$6x - 5x + 30 = 0$$

***Example* 12:** Find the equation to the straight line which passes through the point (4, –5) and is perpendicular to the line 3x + 4y + 5 = 0.

Solution: The equation of a line perpendicular to 3x + 4y + 5 = 0 is

$$4x - 3y + k = 0 \qquad \text{.....(i)}$$

The line (i) passes through (4, –5), we have

$$4 \times 4 - 3 \times (-5) + k = 0$$

$$16 + 15 + k = 0$$

$$k = -31$$

Putting k value in eq. (i), we get

$$4x - 3y - 31 = 0$$

***Example* 13:** Find the equation of straight line passing through the point (–6, 10) and parallel to the line $7x + 8y = 5$

Solution : The equation of line parallel to $7x + 8y - 5 = 0$ is

$$7x + 8y + k = 0 \qquad \text{.....(i)}$$

Which passes through (–6, 10), we have

$$7 \times (-6) + 8 \times 10 + k = 0$$

$$-42 + 80 + k = 0$$

$$38 + k = 0$$

$$k = -38$$

Putting $k = -38$ in eq. (i), we get $7x + 8y - 38 = 0$

***Example* 14:** Which of the lines $2x - y + 3 = 0$ and $x - 4y - 7 = 0$ is farther from the origin.

Solution: The length of perpendicular from (x_1, y_1) on $ax + by + c = 0$ is

$$= \frac{ax_1 + by_1 + c}{\sqrt{a^2 + b^2}}$$

$\therefore$ length of perpendicular from (0, 0) on $2x - y + 3 = 0$ is

$$p_1 = \frac{2 \times 0 - 0 + 3}{\sqrt{4+1}} = \frac{3}{\sqrt{5}}$$

and length of perpendicular from (0, 0) on $x - 4y - 7 = 0$ is

$$p_2 = \frac{0 - 4 \times 0 - 7}{\sqrt{1+16}} = \frac{-7}{\sqrt{17}}$$

here $p_1 > p_2$

$\therefore$ $2x - y + 3$ is farther from the origin.

***Example* 15:** Find the equation to the straight line which passes through the intersection of the lines $2x - 3y + 4 = 0$, $3x + 4y + 5 = 0$ and is perpendicular to the straight line $6x - 7y + 8 = 0$.

Solution: Given straight lines are (intersection form)

$$(2x - 3y + 4) + k\,(3x + 4y + 5) = 0$$

$$(2 + 3k)\,x + (-3 + 4k)\,y + (4 + 5k) = 0 \qquad \text{.....(i)}$$

Now, slope of (i) is $m_1 = \frac{-(2+3k)}{4k-3}$

Slope of the line $6x - 7y + 8 = 0$ is .....(ii)

$$m_2 = \frac{6}{7}$$

If line (i) is perpendicular to line (ii), then

$$m_1 . m_2 = \frac{-(2+3k)}{4k-3} \times \frac{6}{7} = -1$$

$$\Rightarrow \quad k = \frac{33}{10}$$

Putting $k = \frac{33}{10}$ value in eq. (i), we have $119x + 102y + 205 = 0$

***Example* 16:** Find the angle made by the line $x \cos 30^\circ + y \sin 30^\circ + \sin 120^\circ = 0$ with the positive direction of x-axis.

Solution: Let $x \cos 30^\circ + y \sin 30^\circ + \sin 120^\circ = 0$

$$x \times \frac{\sqrt{3}}{2} + y \times \frac{1}{2} + \frac{\sqrt{3}}{2} = 0$$

$$\frac{\sqrt{3}}{2} x + \frac{1}{2} y + \frac{\sqrt{3}}{2} = 0$$

$$\sqrt{3}x + y + \sqrt{3} = 0$$

$$y = -\left(\sqrt{3}\,x + \sqrt{3}\right)$$

on comparing, $y = mx + c$

here $m = -\sqrt{3}$

$$\tan\theta = -\sqrt{3}$$

$$\tan\theta = \tan 120^\circ$$

$$\theta = 120^\circ$$

Exercises

1. Find the distance between the following pair of points
   (i) $(am_1^2, 2am_1)$, $(am_2^2, 2am_2)$
   (ii) (3, 4) and (7, 7)
   (iii) (–3, 2) and (4, 3)
2. If A(x, y) is equidistant from (–4, 3) and (3, –4). Show that x = y.
3. If the points (x, y) is equidistant from the point (a + b, b – a) and (a – b, a + b) prove that bx = ay.
4. Prove that points are collinear if
   (i) (6, 9), (0, 1) and (–6, –7)
   (ii) (5, 1), (1, –1) and (3, 0)
   (iii) (2, –1), (5, 3), (–4, –9)
5. Prove that the points A(0, 1), B(1, 4), C(4, 3), D(3, 0) are the vertices of a square.
6. Prove that the points A(0, 5), B(–2, –2), C(5, 0), D(7, 7) are the vertices of rhombus.
7. Prove that the points (a, a), (–a, –a) and $(-a\sqrt{3}, a\sqrt{3})$ are vertices of an equilateral triangle.
8. Find the co-ordinates of the point which divides (i) internally and (ii) externally, the line joining (–1, 2) to (4, –5) in the ratio 2:3.
9. Determine the ratio in which 2x + 3y + 7 = 0, divides joint of (3, 4) and (7, 8) and at what point.
10. The line joining (5, –6), (–7, 5) is trisected. Find the coordinates of the point of trisection.
11. Find the co-ordinates of the mid-point of the line joining the points (4, 7) and (6, 9).
12. Prove that the straight line y – x + 2 = 0 cuts the line joining (3, –1) and (8, 9) in the ratio 2:3.
13. The mid-point of the sides of a triangle are (1, 2), (0, –1) and (2, –1). Find the co-ordinates of the vertices of the triangle.
14. Find the centroid of the triangle whose vertices are
    (i) (–4, 6), (2, –2) and (2, 5)
    (ii) (1, –4), (3, –2) and (–3, 16)

15. Find the centroid of the triangle formed by the straight lines
$x - 2y = 0$, $x + y = 3$ and $2x + y + 4 = 0$

16. Find the co-ordinates of incentre of a triangle whose vertices are
(–36, 7), (20, 7) and (0, 8)

17. Find the area of the triangle formed by the following points:
   (i) (3, 4), (2, –1), (4, –6)
   (ii) (5, 2), (–9, –3), (–3, –5)
   (iii) (a, c + a), (a, c) and (–a, c – a)

18. Find the area of the quadrilaterals whose points are:
   (i) (1, 2), (6, 2), (5, 3) and (3, 4)
   (ii) (1, 1), (3, 4), (5, –2) and (4, –7)

19. Find the area of triangle formed by straight lines
$y = x$, $y = 2x$ and $y = 3x + 4$

20. Find the equation of the line
   (i) Which pass through the origin and makes an angle of $\frac{3}{4}\pi$ with x-axis.
   (ii) Which cuts an intercept –5 on y-axis and is inclined at 60° with x-axis.
   (iii) Which makes intercepts 3 and –5 on the axes.
   (iv) Which makes equal intercepts on the axes and passes through the point (6, –5)

21. The length of the perpendicular from the origin on a straight line is 3 and the perpendicular is inclined to the x-axis at 120°. Find the equation.

22. If A(a, 0) and B(–a, 0) are two fixed points. Find the locus of a point p which moves so that 3 |PA| = 2 |PB|.

23. Find the equation of the locus of the point p if the sum of squares of its distances from (1, 2) and (3, 4) is 25.

24. Find the slope of the line through the points.
   (i) (–2, 6) and (4, 8)
   (ii) (2, –2) and (–6, 4)

25. Find y if the slope of the line joining (–8, 11) and (2, y) is $\frac{-4}{3}$.

26. Find the equation of the line which passes through the intersection of the line $x - 2y = a$ and $x + 3y = 2a$, and is parallel to the line $3x + 4y = 0$

27. Find the equation of the straight line which passes through the intersection of the straight lines $2x - 3y + 4 = 0$, $3x + 4y + 5 = 0$ and is perpendicular to the line $6x - 7y + 8 = 0$.

28. Find the equation of line which passes through the intersection of the lines $3x - 4y + 1 = 0$, $5x + y = 1$ and cut off equal intercepts from the axes.

29. Show that the lines $5x - y = 6$, $2x + y - 1 = 0$ and $4x + y - 3 = 0$ are concurrent.

30. Find the equation of the line through the point (4, –3) and (0, 1).

31. Find the angle made by the line $x + \sqrt{3}y - 6 = 0$ with the positive direction of x-axis.

32. Find the equation of straight lines whose slope is –3 and which passes through (–2, 3).

33. Find the equation of the line passing through
    (i) (3, –2) and (–5, –7)
    (ii) (a, b) and (–a, b)

34. The vertices of a triangle are (0, 0), (2, 4) and (6, 4). Find the equation of its sides.

35. Find the equation of the line that is parallel to $2x + 5y = 7$ and passes through the mid point of the line joining (2, 7) and (–4, 1).

36. Find the equation of a line passing through (5, 1) and is parallel to the line $7x - 2y + 5 = 0$.

37. Find the equation of a line passing through (0, 3) and perpendicular to the line $x - 2y + 5 = 0$.

38. Find the equation of straight line
    (i) Through the point (–6, 10) and parallel to the line $7x + 8y = 5$
    (ii) Through the point (–6, 10) and perpendicular to the line $7x + 8y = 5$
    (iii) Joining of points (2, 8) and (7, –9)
    (iv) Through the point (8, –3) and making an angle of 135° with x-axis.

39. Show that the points (3, 7) and (–3, –1) are on the same side of the line $3x - 8y = 7$. [***Hint :*** Find length of perpendicular from both the points on the line. If signs are same then it is on same side.]

40. Find the distance between the parallel lines
$4x - 3y + 2 = 0$ and $4x - 3y + 9 = 0$

41. Which of the two lines $2x - y + 3 = 0$ and $x - 4y - 7 = 0$ is farther from the origin.

42. Find the equation of a line which passes through (1, –2) and cuts of equal intercepts on the two axes.

43. Find the equation of a line which passes through (3, –5) and cuts off intercept on the axes which are equal in magnitude but opposite in sign.

44. Find the intercepts cut off by the line $3x - 4y = 12$ on the axes.

45. Find the equation of line for which
(i) $p = 5, \alpha = 30^0$
(ii) $p = 2, \alpha = 30^0$

46. Reduce the equation $4x + 5y = 20$ to the form
(i) $y = mx + c$
(ii) $\frac{x}{a} + \frac{y}{b} = 1$

47. Reduce the equation $3x - 4y + 1 = 0$ to
(i) Slope intercept form and find slop and the y-intercept
(ii) Intercept form and find intercepts on the axes
(iii) The normal form and find p and α.

Answers

1. (i) $a(m_1 - m_2)\sqrt{(m_1 + m_2)^2 + 4}$ (ii) 5 (iii) $5\sqrt{2}$

8. (i) $\left(1, \frac{-4}{5}\right)$ (ii) (–11, 16)

9. 5 : 9, (–2, –1)

10. $\left(1, \frac{-7}{3}\right), \left(-3, \frac{4}{3}\right)$

11. (5, 8)

13. (3, 2), (–1, 2), (1, –4)

14. (i) (0, 3) (ii) $\left(\frac{1}{3}, \frac{10}{3}\right)$

15. $\left(-\frac{11}{5}, \frac{17}{5}\right)$

17. (i) $\frac{15}{2}$ (ii) 29 (iii) a^2

18. (i) $5\frac{1}{2}$ (ii) $20\frac{1}{2}$ 19. 4

20. (i) $y + x = 0$ (ii) $y = \sqrt{3}\,x - 5$ (iii) $5x - 3y = 15$ (iv) $x + y = 1$

21. $x - \sqrt{3}\,y + 6 = 0$ 22. $5(x^2 + y^2) + 5a^2 - 26ax = 0$

23. $2x^2 + 2y^2 - 8x - 12y + 5 = 0$ 24. (i) $\frac{1}{3}$ (ii) $\frac{-3}{4}$

25. $\frac{-7}{3}$ 26. $3x + 4y = 5a$

27. $119x + 102y + 250 = 0$ 28. $23x + 23y = 11$

30. $y + x - 1 = 0$ 31. $\theta = 150^o$

32. $3x + y + 3 = 0$ 33. (i) $5x - 8y = 31$ (ii) $y = b$

34. $y = 4,\ 2x - 3y = 0,\ 2x - y = 0$ 35. $2x + 5y - 18 = 0$

36. $7x - 2y - 33 = 0$ 37. $2x + y - 3 = 0$

38. (i) $7x + 8y = 38$ (ii) $8x - 7x + 118 = 0$ (iii) $17x + 5y = 74$ (iv) $x + y - 5 = 0$

40. $\frac{7}{5}$ 41. $x - 4y - 7 = 0$

42. $x + y + 1 = 0$ 43. $x - y - 8 = 0$

44. 4 and –3 45. (i) $y + \sqrt{3}\,x = 10$ (ii) $x - \sqrt{3}\,y = 4$

46. (i) $y = \frac{-4}{5}x + 4$ (ii) $\frac{x}{5} + \frac{y}{4} = 1$

47. (i) Slope $= \frac{3}{4}$, intercept $= \frac{1}{4}$

(ii) x-intercept $= \frac{1}{3}$, y-intercept $= \frac{1}{4}$

(iii) $p = \frac{1}{5},\ \alpha = \pi - \tan^{-1}\left(\frac{4}{3}\right)$

CHAPTER 6

CALCULUS

Functions and Limits

***Related Quantities*:** If the value of one quantity depends upon the other, then they are called Related quantities.

e.g.: The area of circle $A = \pi r^2$.

Here A depends upon r, hence A and r are related quantities.

Variable: A quantity which may assume different values in a set of mathematical operation is called variable. Generally a variable is expressed by x, y, z, u, v …. etc.

Variables are of two types.

(i) *Independent variable :* A variable which may be assigned any arbitrary value is called independent variable.

(ii) *Dependent variable:* A variable whose value depend upon the value of independent variable.
e.g. In the equation $y = 2x^2 + 5x + 1$ when $x = 1, y = 8$; $x = -1, y = -2$
here x is independent and y is dependent variable.

Constant: A quantity which remains unchanged in the mathematical operation is called a constant. It is denoted by a, b, c … etc.

Constants are of two types.

(i) ***Absolute constant:*** A constant which remains the same value during the mathematical operation.

$$\text{e.g. } 1, 2, 3, \pi, \sqrt{7}, -\frac{3}{5} \text{ etc.}$$

(ii) ***Arbitrary constant:*** A constant which remains same during a particular operation but it changes with the change of reference.

e.g: In the equation $y = mx + c$ of a straight line. Here m and c are constants for a given line, but for a different line, the values of m and c assume different values. So m and c are arbitrary constants.

Function: A function is a relationship between two non empty sets X and Y such that there exists a rule which associates each element of set X to a unique element of set Y.

We write this function from X to Y as

$$f : X \rightarrow Y \text{ or } y = f(x), x \in X, y \in Y.$$

e.g. Let $y = x^2 - 3x + 2$

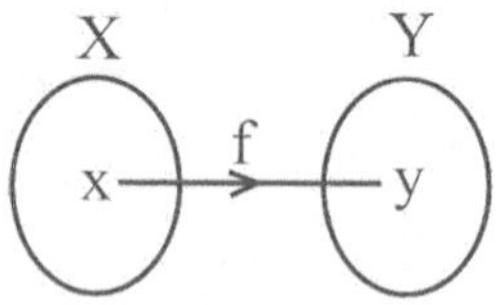

here $y = f(x)$

$\Rightarrow \quad f(x) = x^2 - 3x + 2$

If we put $x = 2$ then $f(x) = (2)^2 - 3 \times 2 + 2$

$$= 4 - 6 + 2$$

$$= 0$$

At $x = 2$, the value of function f(x) is 0.

Domain and range: If a function is defined by $f : X \rightarrow Y$, then the set X is called the domain and Y is called the co-domain of the function f. The range of the function is a subset of Y.

Intervals: The set of real numbers which lie between two specific numbers is called the interval.

(i) ***Open interval:*** If the value of the x lies between two definite numbers a and b then $\{x : a < x < b\}$ is called an open interval. It is denoted by (a, b) or]a, b[.

(ii) **Closed interval:** Intervals $\{ x : a \le x \le b\}$ in which both a and b lie, is called closed interval and is denoted by [a, b].

(iii) ***Semi closed or semi open interval:*** The interval $\{x : a \le x < b\}$ or $\{ x : a < x \le b\}$ are called semi closed or semi open interval. They are denoted by [a, b] or [a, b[and (a, b) or] a, b].

Types of Functions

Constant function: Let f: X→k be a function, where k is a fixed real number. Then $f(x) = k$ represents a constant function.

Identity function: A function f(x) = x is called Identity function as it associates to each real number x the same number x. The Domain of Identity function is R and its range is also R.

Polynomial function: A function of the type

$P(x) = a_0 x^n + a_1x^{n-1} + a_2 x^{n-2} + ... + a_n$ is known as polynomial function. Where a_0, a_1, a_2 ....a_n are real numbers and n is a whole number.

e.g. $$p(x) = 3x^4 + 2x^3 - x^2 + 5x - 7$$

The Domain and range of a polynomial are the set of real numbers.

Note : $x^{\frac{1}{3}} + 5x + 7$ is not a polynomial.

Rational function: A rational function is a quotient of two polynomial functions. i.e., $R(x) = \dfrac{P(x)}{Q(x)}$, where p(x) and Q(x) are polynomical but $Q(x) \neq 0$.

e.g. $$\frac{x^2 + 5x + 7}{x^3 + 2x^2 + 2x + 1}$$

Reciprocal function: A function $f(x) = \dfrac{1}{x}$, where x is a non zero real number ($x \neq 0$) is called a reciprocal function.

The Domain is R –[0] and range is the set of real numbers.

Exponential function: A function $f(x) = e^x$ is called exponential function.

Where $$e^x = 1 + \frac{x}{\lfloor 1} + \frac{x^2}{\lfloor 2} + ... + \frac{x^n}{\lfloor n} + ...$$

Whose domain is the set of real numbers R and range is the set of positive real numbers.

Logarithmic function: A function $f(x) = \log_a x$ or $e^y = x$, where x is a positive real number and a > 0, is called logarithmic function. Domain is the set of positive real numbers and range is the set of real numbers R.

Note: A function $f(x) = \log_e x$ is natural logarithmic function.

Trigonometric Function: sinx, cosx, tanx, cotx, sec x, cosec x are called trigonometric function. As f(x) = sin x is called sine function. Angle x is in radian.

Function	Domain	Range	Name of the function
y = sinx	R	[–1, 1]	sine function
y = cos x	R	[–1, 1]	cosine function
y = tanx	$R - \left\{x : x = \left(\frac{2n+1}{2}\right)\pi, n \in I\right\}$	R	tangent function
y = cot x	$R - \{x : x = n\pi, n \in I\}$	R	cotangent function
y = sec x	$R - \{x : x = (2n+1)\frac{\pi}{2}, n \in I\}$	R – (–1, 1)	secant function
y = cosec x	$R - \{x : x = n\pi, n \in I\}$	R –(–1, 1)	cosecant function

Periodic function: A function f(x + p) = f(x) is called a periodic function of x and the least positive value of p is called the period of the function.

e.g. Let a function y = f(x) = sin x is a periodic function because

$$\sin(x + 2\pi) = \sin(x + 4\pi) = \ldots\ldots = \sin x$$

The least value of p is 2π. So, the period of sinx is 2π.

Even function: If y = f(x) and f(–x) = f(x),

then y = f(x) is called an even function of x.

e.g:

$$\cos(-x) = \cos x$$
$$(-x)^2 = x^2$$

Odd function: If y = f(x) and f(–x) = –f(x), then y = f(x) is called an odd function of x.

e.g.

$$\sin(-x) = -\sin x$$
$$(-x)^3 = -x^3$$

Illustrative Examples

Example 1: If f(x) = 1 + sin x, then find $f\left(\frac{\pi}{3}\right)$

Solution: Given f(x) = 1 + sinx

Put $x = \frac{\pi}{3}$, $f\left(\frac{\pi}{3}\right) = 1 + \sin\frac{\pi}{3}$

$$= 1 + \sin 60^\circ$$

$$= 1 + \frac{\sqrt{3}}{2}$$

***Example* 2:** If $f(x) = x^2 + 3x - 4\sin^3 x$, then prove that f(x) is an odd function of f(x).

Solution: Given function $f(x) = x^3 + 3x - 4\sin^3 x$

Put $x = -x$, we get

$$f(-x) = (-x)^3 + 3(-x) - 4\{\sin(-x)\}^3$$
$$= -x^3 - 3x - 4 - (-\sin x)^3$$
$$= -x^3 - 3x + 4\sin^3 x$$
$$= -(x^3 + 3x - 4\sin^3 x)$$
$$f(-x) = -f(x)$$

$\therefore$ f(x) is an odd function of x.

***Example* 3:** Name the function $f(x) = \dfrac{x^3 - x^2 + 4x - 7}{3x + 11}$ and find its domain.

Solution: Here $f(x) = \dfrac{x^3 - x^2 + 4x - 7}{3x + 11}$

It is a quotient of polynomial functions $x^3 - x^2 + 4x - 7$ and $3x + 11$

Therefore, f(x) is a rational function

Now, f(x) is not defined when $3x + 11 = 0 \quad \Rightarrow \quad x = \dfrac{-11}{3}$

For any other value of x which is not equal to $\dfrac{-11}{3}$, f(x) has a unique and defined value.

$\therefore$ Domain of f is the set of real number except $\dfrac{-11}{3}$

i.e., Domain = $R - \left\{\dfrac{-11}{3}\right\}$

***Example* 4:** Find the range of function $f(x) = 11 - 7\sin x$

Solution: For all values of x, we have $-1 \le \sin x \le 1$

Multiplying by –7, the direction of inequality is reversed.

$$7 \ge -7\sin x \ge -7$$
$$\Rightarrow \quad 11 + 7 \ge 11 - 7\sin x \ge 11 - 7$$
$$\Rightarrow \quad 18 \ge 11 - 7\sin x \ge 4$$
$$\Rightarrow \quad 4 \le 11 - 7\sin x \le 18$$

$\therefore$ range = [4, 18]

***Example* 5:** If f(x) = 1 – cosx, then find $f\left(\frac{\pi}{4}\right)$

Solution: Here $f(x) = 1 - \cos x$

Put $x = \frac{\pi}{4},\ f\left(\frac{\pi}{4}\right) = 1 - \cos\left(\frac{\pi}{4}\right)$

$$= 1 - \cos 45^{\circ}$$

$$= 1 - \frac{1}{\sqrt{2}}$$

***Example* 6:** If f(x) = x² + 2x – 3, then find the values of f(0), f(–1), $f\left(\frac{1}{3}\right)$, f(sinx).

Solution: Here $f(x) = x^2 + 2x - 3$

(i) put $x = 0,\ f(0) = (0)^2 + 2(0) - 3$

$$= 0 + 0 - 3$$

$$= -3$$

(ii) Put $x = -1,\ f(-1) = (-1)^2 + 2(-1) - 3$

$$= 1 - 2 - 3$$

$$= -4$$

(iii) Put $x = \frac{1}{3},\ f\left(\frac{1}{3}\right) = \left(\frac{1}{3}\right)^2 + 2\left(\frac{1}{3}\right) - 3$

$$= \frac{1}{9} + \frac{2}{3} - 3$$

$$= \frac{1 + 6 - 27}{9}$$

$$= \frac{-20}{9}$$

(iv) Put $x = \sin x,\ f(\sin x) = (\sin x)^2 + 2(\sin x) - 3$

$$= \sin^2 x + 2\sin x - 3$$

Limits

Meaning of 'x → a': Let x be a variable and a be a constant. If x assumes values nearer and nearer to a, then we say that x tends to a and write as $x \to a$.

Limit of a function: Let f(x) be a function of x. If x assumes nearer and nearer to a, except possibly the value a, f(x) assumes values nearer and nearer to l, then f(x) tends to the limit l as x tends to a and is represented by

$$\lim_{x \to a} f(x) = l.$$

Rules for limits:

(i) $$\lim_{x \to a} [f_1(x) + f_2(x)] = \lim_{x \to a} f_1(x) + \lim_{x \to a} f_2(x)$$

(ii) $$\lim_{x \to a} [f_1(x) - f_2(x)] = \lim_{x \to a} f_1(x) - \lim_{x \to a} f_2(x)$$

(iii) $$\lim_{x \to a} [f_1(x).f_2(x)] = \lim_{x \to a} f_1(x). \lim_{x \to a} f_2(x)$$

(iv) $$\lim_{x \to a} \frac{f_1(x)}{f_2(x)} = \frac{\lim_{x \to a} f_1(x)}{\lim_{x \to a} f_2(x)}, \quad \lim_{x \to a} f_2(x) \neq 0$$

(v) $$\lim_{x \to a} c.f_1(x) = c. \lim_{x \to a} f_1(x)$$

Evaluation of limits: The limit of f(x) when $x \to a$ can be calculated by the following methods.

1. ***By factorization:*** when f(x) is of the form $\frac{f_1(x)}{f_2(x)}$, factorise $f_1(x)$ and $f_2(x)$. Cancel common factor and put the value of x.
2. ***By substitution:*** The method consists of the following steps.
   (a) put x = a + h, Hence as $x \to a$, $h \to 0$.
   (b) simplify numerator and denominator and cancel h throughout.
   (c) Put h = 0, we get the required limit.
3. ***By rationalization:*** we rationalize the factor containing the square root and simplify and we put the value of x.

Illustrative Examples

***Example* 1:** Evaluate $\lim_{x \to 0} \frac{y^3}{x^2}$, where $y^3 = ax^2 + bx^4 + cx^5$

Solution:

$$\lim_{x \to 0} \frac{y^3}{x^2} = \lim_{x \to 0} \frac{ax^2 + bx^4 + cx^5}{x^2}$$

$$= \lim_{x \to 0} \frac{x^2(a + bx^2 + cx^3)}{x^2}$$

$$= \lim_{x \to 0} (a + bx^2 + cx^3)$$

$$= a + b(0)^2 + c(0)^3$$

$$= a$$

***Example* 2:** Evaluate $= \lim_{x \to 5} \frac{x^2 + x - 30}{x - 5}$

Solution:

$$\lim_{x \to 5} \frac{x^2 + x - 30}{x - 5} = \lim_{x \to 5} \frac{(x-5)(x+6)}{(x-5)}$$

$$= \lim_{x \to 5} (x+6)$$

$$= 5 + 6$$

$$= 11$$

***Example* 3:** Evaluate $\lim_{x \to 0} \frac{x}{\sqrt{1+x} - 1}$

Solution:

$$\lim_{x \to 0} \frac{x}{\sqrt{1+x} - 1} = \lim_{x \to 0} \frac{x}{\sqrt{1+x} - 1} \cdot \frac{\sqrt{1+x} + 1}{\sqrt{1+x} + 1}$$

$$= \lim_{x \to 0} \frac{x(\sqrt{1+x} + 1)}{1 + x - 1}$$

$$= \lim_{x \to 0} (\sqrt{1+x} + 1)$$

$$= \sqrt{1+0} + 1$$

$$= \sqrt{1} + 1$$

$$= 1 + 1 = 2$$

***Example* 4:** Evaluate $\lim_{x \to 2} \frac{x^2 - 4}{x - 2}$

Solution:

$$\lim_{x \to 2} \frac{x^2 - 4}{x - 2} = \lim_{x \to 2} \frac{(x-2)(x+2)}{x-2}$$

$$= \lim_{x \to 2} (x+2)$$

$$= 2 + 2$$

$$= 4$$

***Example* 5:** Evaluate $\lim_{x \to 5} \frac{x^3 - 125}{x - 5}$

Solution:

$$\lim_{x \to 5} \frac{x^3 - 125}{x - 5} = \lim_{x \to 5} \frac{(x-5)(x^2 + 5x + 25)}{(x-5)}$$

$$= \lim_{x \to 5} (x^2 + 5x + 25)$$

$$= (5)^2 + 5(5) + 25$$

$$= 25 + 25 + 25$$

$$= 75$$

***Example* 6:** Evaluate $\lim_{x \to 3} \frac{x^2 - 7x + 12}{x^2 - 9}$

Solution:

$$\lim_{x \to 3} \frac{x^2 - 7x + 12}{x^2 - 9} = \lim_{x \to 3} \frac{(x-3)(x-4)}{(x-3)(x+3)}$$

$$= \lim_{x \to 3} \frac{x-4}{x+3}$$

$$= \frac{3-4}{3+3} = \frac{-1}{6}$$

***Example* 7:** Evaluate $\lim_{x \to 0} \frac{\sqrt{1+x}-1}{x}$

Solution:

$$\lim_{x \to 0} \frac{\sqrt{1+x}-1}{x} = \lim_{x \to 0} \frac{\sqrt{1+x}-1}{x} \cdot \frac{\sqrt{1+x}+1}{\sqrt{1+x}+1}$$

$$= \lim_{x \to 0} \frac{1+x-1}{x(\sqrt{1+x}+1)}$$

$$= \lim_{x \to 0} \frac{1}{(\sqrt{1+x}+1)}$$

$$= \frac{1}{\sqrt{1+0}+1}$$

$$= \frac{1}{\sqrt{1}+1}$$

$$= \frac{1}{1+1} = \frac{1}{2}$$

Some Important Limits

Formula :

$$\lim_{x \to a} \frac{x^n - a^n}{x-a} = na^{n-1}, \qquad (a > 0)$$

***Example* 8:** Evaluate $\lim_{x \to 3} \frac{x^3 - 27}{x-3}$

Solution:

$$\lim_{x \to 3} \frac{x^3 - 27}{x-3} = \lim_{x \to 3} \frac{x^3 - 3^3}{x-3}$$

$$= 3(3)^{3-1}$$

$$= 3.\ 3^2 = 27$$

Formula:

(i) $\lim_{\theta \to 0} \sin\theta = 0$ (ii) $\lim_{\theta \to 0} \cos\theta = 1$

(iii) $\lim_{\theta \to 0} \frac{\sin\theta}{\theta} = 1$ (iv) $\lim_{\theta \to 0} \frac{\tan\theta}{\theta} = 1$

***Example* 9:** Evaluate $\lim_{x \to 0} \frac{\sin 5x - \sin 3x}{\sin x}$

Solution:

$$\lim_{x \to 0} \frac{\sin 5x - \sin 3x}{\sin x} = \lim_{x \to 0} \frac{2\cos 4x.\sin x}{\sin x}$$

$$= \lim_{x \to 0} 2\cos 4x$$

$$= 2 \lim_{x \to 0} \cos 4x$$

$$= 2.\cos o^0$$

$$= 2.1$$

$$= \mathbf{2}$$

***Example* 10:** Evaluate $\lim_{x \to 0} \frac{\sin ax}{\sin bx}$

Solution:

$$\lim_{x \to 0} \frac{\sin ax}{\sin bx} = \lim_{x \to 0} \frac{\sin ax}{ax} \cdot \frac{bx}{\sin bx} \cdot \frac{a}{b}$$

$$= \frac{a}{b} \lim_{x \to 0} \frac{\sin ax}{ax} \cdot \lim_{x \to 0} \frac{bx}{\sin bx}$$

$$= \frac{a}{b} \frac{\lim_{x \to 0} \frac{\sin ax}{ax}}{\lim_{x \to 0} \frac{\sin bx}{bx}}$$

$$= \frac{a}{b} \cdot \frac{1}{1} = \frac{a}{b}$$

Formula:

(i) $\lim_{x \to \infty} \left(1 + \frac{1}{x}\right)^x = e$ (ii) $\lim_{x \to 0} (1 + x)^{\frac{1}{x}} = e$

(iii) $\lim_{x \to 0} \left(1 + \frac{x}{a}\right)^{\frac{a}{x}} = e$ (iv) $\lim_{x \to 0} \frac{e^x - 1}{x} = 1$

(v) $\lim_{x \to 0} \frac{\log(1 + x)}{x} = 1$ (vi) $\lim_{x \to 0} \frac{a^x - 1}{x} = \log_e a$

***Example* 11:** Evaluate $\lim_{x \to 0} \frac{e^x - e^{-x}}{\sin x}$

Solution:

$$\lim_{x \to 0} \frac{e^x - e^{-x}}{\sin x} = \lim_{x \to 0} \frac{e^{2x} - 1}{e^x \sin x}$$

$$= \lim_{x \to 0} \frac{(e^{2x} - 1)}{2x} \cdot \frac{2x}{\sin x} \cdot \frac{1}{e^x}$$

$$= \lim_{2x \to 0} \frac{e^{2x} - 1}{2x} \cdot 2 \lim_{x \to 0} \frac{x}{\sin x} \cdot \lim_{x \to 0} \frac{1}{e^x}$$

$$= 1 \cdot 2 \cdot 1 \cdot 1$$

$$= 2$$

***Example* 12:** Evaluate $= \lim_{x \to 0} \frac{e^{ax} - 1}{x}$

Solution:

$$\lim_{x \to 0} \frac{e^{ax} - 1}{x} = \lim_{x \to 0} \frac{e^{ax} - 1}{ax} \cdot a$$

$$= \lim_{ax \to 0} \frac{e^{ax} - 1}{ax} \cdot a$$

$$= 1 \cdot a = a$$

Meaning of the symbol: $\lim_{x \to \infty} f(x) = 1$

Let f(x) be a function of x. If x assumes larger and larger values f(x) assumes values nearer and nearer to 1. Then we say f(x) tends to the limit 1 as x tends to infinity.

***Example* 13:** Evaluate $\lim_{x \to \infty} \frac{(2x - 3)(3x - 4)}{(4x - 5)(5x - 6)}$

Solution:

$$= \lim_{x \to \infty} \frac{(2x - 3)(3x - 4)}{(4x - 5)(5x - 6)} = \lim_{x \to \infty} \frac{x(2 - \frac{3}{x}) \cdot x\,(3 - \frac{4}{x})}{x(4 - \frac{5}{x}) \cdot x\,(5 - \frac{6}{x})}$$

$$= \lim_{x \to \infty} \frac{(2-\frac{3}{x})(3-\frac{4}{x})}{(4-\frac{5}{x})(5-\frac{6}{x})}$$

$$= \frac{(2-0)(3-0)}{(4-0)(5-0)}$$

$$= \frac{6}{20} = \frac{3}{10}$$

***Example* 14:** Evaluate $\lim_{x \to \infty} \frac{2x^2 - 3x + 1}{x^2 - 1}$

Solution:

$$\lim_{x \to \infty} \frac{2x^2 - 3x + 1}{x^2 - 1} = \lim_{x \to \infty} \frac{x^2(2-\frac{3}{x}+\frac{1}{x^2})}{x^2(1-\frac{1}{x^2})}$$

$$= \lim_{x \to \infty} \frac{(2-\frac{3}{x}+\frac{1}{x^2})}{(1-\frac{1}{x^2})}$$

$$= \frac{(2-0+0)}{(1-0)} = 2$$

Meaning of left limit and Right limit :

(i) A function f(x) tends to a limit l_1, when x → a through values greater than a. Then l_1 of f(x) is called Right limit of f(x) and denoted by $f(a^+)$ and is written as $\lim_{x \to a^+} f(x) = l_1 = \lim_{h \to 0} f(a+h)$

(ii) A function f(x) tends to a limit l_2, when x → a through values less than a. Then l_2 of f(x) is called left limit of f(x) and denoted by $f(a^-)$ and is written as $\lim_{x \to a^-} f(x) = l_2 = \lim_{h \to 0} f(a-h)$.

(iii) If right limit and left limit both are equal. Then the limit exists.

i.e., $\lim_{x \to a^+} f(x) = \lim_{x \to a^-} f(x)$

***Example* 1:** If f(x) is defined as follows:

$$f(x) = \begin{cases} x^2, \text{ when } x < 1 \\ 2, \text{ when } x = 1 \\ x^2 + 1, \text{ when } x > 1 \end{cases}$$

Does limit exists at x = 1

Solution:

$$\text{Left limit} = \lim_{x \to 1^-} f(x) = \lim_{x \to 1^-} x^2$$

$$= \lim_{h \to 0} (1 - h)^2 \qquad \text{(put } x = 1 - h\text{)}$$

$$= 1$$

$$\text{Right limit} = \lim_{x \to 1^+} f(x) = \lim_{x \to 1^+} (x^2 + 1)$$

$$= \lim_{h \to 0} [(1 + h)^2) + 1] \qquad \text{(put } x = 1 + h\text{)}$$

$$= 1 + 1 = 2$$

here, $\lim_{x \to 1^-} f(x) \neq \lim_{x \to 1^+} (fx)$

Hence limit of f(x) does not exists at x = 1.

***Example* 2:** Let $f(x) = \frac{1-x}{1+x}$, $0 \le x \le 1$, then evaluate $\lim_{x \to \frac{1}{2}} f(x)$ if it exists.

Solution: Let $f(x) = \frac{1-x}{1+x}$

$$\text{Left limit} = \lim_{x \to \frac{1}{2}^-} f(x) = \lim_{x \to \frac{1}{2}^-} \frac{1-x}{1+x}$$

$$= \lim_{h \to 0} \frac{1 - \left(\frac{1}{2} - h\right)}{1 + \left(\frac{1}{2} - h\right)}$$

$$= \lim_{h \to 0} \frac{\frac{1}{2} + h}{\frac{3}{2} - h}$$

$$= \frac{\frac{1}{2}}{\frac{3}{2}} = \frac{1}{3}$$

Right limit $= \lim_{x \to \frac{1}{2}^+} f(x) = \lim_{x \to \frac{1}{2}^+} \frac{1-x}{1+x}$

$$= \lim_{h \to 0} \frac{1 - \left(\frac{1}{2} + h\right)}{1 + \left(\frac{1}{2} + h\right)}$$

$$= \frac{1 - \frac{1}{2}}{1 + \frac{1}{2}} = \frac{1}{3}$$

Since left limit = Right limit

Hence $\lim_{x \to \frac{1}{2}} f(x)$ exists and is equal to $\frac{1}{3}$.

Continuity of a function at a point: A function is continuous at a point a if the left limit and right limit exists and both are equal to the value of function at x = a.

i.e., $\lim_{x \to a^-} f(x) = \lim_{x \to a^+} f(x) = f(a)$

***Example* 3:** Test the continuity of the function defined as

$$f(x) = \begin{cases} \frac{x^2 - 4x + 3}{x - 1}, & x \neq 1 \\ -2, & x = 1 \end{cases}$$

Solution: Right limit $= \lim_{x \to 1^+} f(x) = \lim_{x \to 1^+} \frac{x^2 - 4x + 3}{x - 1}$

$$= \lim_{h \to 0} \frac{(1+h)^2 - 4(1+h) + 3}{(1+h) - 1}$$

$$= \lim_{h \to 0} \frac{h^2 - 2h}{h}$$

$$= \lim_{h \to 0} (h - 2) = -2$$

$$\text{Left limit} = \lim_{x \to 1^-} f(x) = \lim_{x \to 1} \frac{x^2 - 4x + 3}{x - 1}$$

$$= \lim_{h \to 0} \frac{(1-h)^2 - 4(1-h) + 3}{(1-h) - 1}$$

$$= \lim_{h \to 0} \frac{h^2 + 2h}{-h}$$

$$= \lim_{h \to 0} -(h + 2) = -2$$

Value $f(1) = -2$ (given)

Since Left limit = Right limit = value

i.e., $\lim_{x \to 1^-} f(x) = \lim_{x \to 1^+} f(x) = f(1)$

$\therefore$ f(x) is continuous at x = 1

***Example* 4:** A function f(x) is defined as

$$f(x) = \frac{1}{2} - x, \text{ when } 0 < x < \frac{1}{2}$$

$$= \frac{1}{2}, \quad \text{when } x = \frac{1}{2}$$

$$= \frac{3}{2} - x, \text{ when } \frac{1}{2} < x < 1$$

Prove that f(x) is discontinuous at $x = \frac{1}{2}$

Solution:

$$\text{left limit} = \lim_{x \to \frac{1}{2}^-} f(x) = \lim_{x \to \frac{1}{2}^-} \left(\frac{1}{2} - x\right)$$

$$= \lim_{h \to 0} \left[\frac{1}{2} - \left(\frac{1}{2} - h\right)\right]$$

$$= \lim_{h \to 0} h = 0$$

$$\text{Right limit} = \lim_{x \to \frac{1}{2}^+} f(x) = \lim_{x \to \frac{1}{2}^+} \left(\frac{3}{2} - x\right)$$

$$= \lim_{h \to 0} \left[\frac{3}{2} - \left(\frac{1}{2} + h\right)\right]$$

$$= \lim_{h \to 0} (1 - h) = 1$$

Since $\lim_{x \to \frac{1}{2}^-} f(x) \neq \lim_{x \to \frac{1}{2}^+} f(x) \neq f(\frac{1}{2})$

Hence f(x) is discontinuous at $x = \frac{1}{2}$

***Example* 5:** prove that $\lim_{x \to 1} |x|$ does not exists

Solution: $f(x) = |x| = x$ (when $x > 0$)

$= -x$ (when $x < 0$)

$$\text{left limit} = \lim_{x \to 1^-} f(x) = \lim_{x \to 1^-} (-x)$$

$$= \lim_{h \to 0} -(1 - h) = -1$$

$$\text{Right limit} = \lim_{x \to 1^+} f(x) = \lim_{x \to 1^+} (x)$$

$$= \lim_{h \to 0} (1 + h) = 1$$

since $\lim_{x \to 1^-} f(x) \neq \lim_{x \to 1^+} f(x)$

∴ The given function does not exists

***Example* 6:** At what point is the following function discontinuous?

$$f(x) = \frac{x^2 + 1}{x^2 - 3x + 2}$$

***Solution*:** Let $x^2 - 3x + 2 = x^2 - 2x - x + 2$

$= x(x - 2) - 1(x - 2)$

$= (x - 2)(x - 1)$

$x = 1, 2$

since f(x) is not defined at x = 1 and x = 2

Hence f(x) is discontinuous at x = 1 and x = 2.

***Example* 7:** Prove that the sine function is continuous.

Solution: we know that $\lim_{x \to 0} \sin x = 0$ and $\lim_{x \to 0} \cos x = 1$

Then $\lim_{x \to a} \sin x = \lim_{h \to 0} \sin(a + h)$

$$= \lim_{h \to 0} (\sin a \cos h + \cos a \sin h)$$

$$= \sin a \lim_{h \to 0} \cos h + \cos a \lim_{h \to 0} \sin h$$

$$= \sin a \times 1 + \cos a \times 0$$

$$= \sin a$$

$\therefore$ sin x is a continuous function

***Example* 8:** The exponential function is continuous i.e., $f(x) = e^x$.

Solution: we know that $\lim_{x \to 0} e^x = 1$

$$\therefore \lim_{x \to a} e^x = \lim_{h \to 0} e^{a+h}$$

$$= \lim_{h \to 0} e^a \,.\, e^h$$

$$= e^a \,.\, \lim_{h \to 0} e^h$$

$$= e^a \times 1$$

$$= e^a$$

$\therefore$ e^x is continuous function.

Exercises

1. If $f(x) = x^2 - 4x + 6$, find f(1), $f\left(\frac{1}{2}\right)$, find f(2 + h) and $\frac{f(2+h) - f(2)}{h}$

2. Given $f(x) = 3x - 1$, if $x > 3$

   $= x^2 - 2$, if $-2 \le x \le 3$

   $= 2x + 3$, if $x < -2$

   Evaluate f(2), f(4), f(–1), f(–3), f(0)

3. Determine the domain and range of the following functions.

(i) $\dfrac{x+7}{x-5}$ (ii) $\dfrac{x^2-1}{x-1}$ (iii) $\sqrt{x-3}$

4. If f(x) = log x, prove that f(u v w) = f(u) + f(v) + f(w)

5. If $f(x) = \log\left(\dfrac{1-x}{1+x}\right)$, prove that $f(a) + f(b) = f\left(\dfrac{a+b}{1+ab}\right)$

6. If f(x) = sin x and F(x) = cos x, prove that $[f(x)]^2 + [F(x)]^2 = 1$.

7. Determine which of the following are even or odd?

(i) $f(x) = (x^4 + 3)\sqrt{x^6 - 3}$ (ii) $f(x) = x^3 (3x + x^5)^3$

(iii) $f(x) = 2^x$ (iv) $f(x) = 3^x - 3^{-x}$

8. If $f(x) = a^x$, show that f(x) . f(y) = f(x + y)

9. If $f(x) = \dfrac{1-x}{1+x}$ then prove that $f(\tan\theta) = \tan\left(\dfrac{\pi}{4} - \theta\right)$

10. Evaluate the limits

(i) $\lim\limits_{x\to\frac{1}{2}} \dfrac{4x^2-1}{2x-1}$ (ii) $\lim\limits_{x\to-2} \dfrac{x^3+8}{x+2}$

(iii) $\lim\limits_{x\to 0} \dfrac{\sqrt{1+x}-\sqrt{1-x}}{x}$ (iv) $\lim\limits_{x\to 2} \dfrac{x^2-5x+6}{x^2-4}$

(v) $\lim\limits_{x\to 1} \dfrac{x^2-4x+3}{x^2+2x-3}$ (vi) $\lim\limits_{x\to 2} \left(\dfrac{1}{x-2} - \dfrac{4}{x^3-2x^2}\right)$

11. Evaluate

(i) $\lim\limits_{x\to 4} \dfrac{x^4-256}{x-4}$ (ii) $\lim\limits_{x\to a} \dfrac{x^7-a^7}{x^5-a^5}$ (iii) $\lim\limits_{x\to a} \dfrac{(x+2)^{\frac{5}{3}}-(a+2)^{\frac{5}{3}}}{x-a}$

12. Evaluate

(i) $\lim\limits_{x\to 0} \dfrac{\sin px}{\sin qx}$ (ii) $\lim\limits_{x\to 0} \dfrac{1-\cos x}{x^2}$

(iii) $\lim\limits_{x\to 0} \dfrac{\sin x - \sin a}{\sqrt{x}-\sqrt{a}}$ (iv) $\lim\limits_{x\to 0} \dfrac{\cos ax - \cos bx}{x^2}$

13. Evaluate

(i) $\lim_{x \to \frac{\pi}{2}} \frac{1+\cos 2x}{(\pi - 2x)^2}$ (ii) $\lim_{x \to \frac{\pi}{4}} \frac{1-\tan x}{x - \frac{\pi}{4}}$

14. Evaluate

(i) $\lim_{x \to 0} \frac{e^x - e^{-x}}{\sin x}$ (ii) $\lim_{x \to 0} \frac{e^{\sin x} - 1}{x}$

15. Show that $\lim_{x \to 2} \frac{|x-2|}{x-2}$ does not exist.

16. Evaluate $\lim_{x \to \infty} \frac{x-1}{x}$

17. Evaluate

(i) $\lim_{x \to \infty} \frac{4x^2 - 5x + 7}{2x - 3}$ (ii) $\lim_{x \to \infty} \sqrt{x}(\sqrt{x+3} - \sqrt{x})$

(iii) $\lim_{x \to 0} e^x \cos x$ (iv) $\lim_{x \to 0} \frac{x(e^x - 1)}{1 - \cos x}$

18. Discuss the continuity of $f(x) = \frac{x+1}{x^2+1}$ at x = 1

19. Discuss the continuity of

$$f(x) = \begin{cases} \frac{1-\cos x}{x^2} & ; \quad x \neq 0 \\ \frac{1}{2} & ; \quad x = 0 \end{cases}$$

20. Given $f(x) = \begin{cases} \frac{\sin x}{x} + \cos x & ; \quad x \neq 0 \\ 2 & ; \quad x = 0 \end{cases}$

Show that f(x) is continuous at x = 0

21. Show that function is discontinuous at x = 2.

$$f(x) = \begin{cases} 2x & ; \quad x < 2 \\ 2 & ; \quad x = 2 \\ x^2 & ; \quad x > 2 \end{cases}$$

22. If $f(x) = \begin{cases} \dfrac{1-\cos 4x}{x^2} & ; \quad x \neq 0 \\ 4 & ; \quad x = 0 \end{cases}$

Find whether function f(x) is continuous at x = 0.

23. At what points is the following function discontinuous?

$$f(x) = \frac{x+1}{x^2 - 5x + 6}$$

24. Find where the following function is discontinuous

$$f(x) = \frac{x-1}{x^2 - 4x + 3}$$

25. Prove that f(x) defined as below is discontinuous at $x = \dfrac{1}{2}$.

$$f(x) = \begin{cases} x & , \quad \text{if } 0 \leq x < \dfrac{1}{2} \\ 1 & , \quad \text{if } x = \dfrac{1}{2} \\ 1 - x & , \quad \text{if } \dfrac{1}{2} < x \leq 1 \end{cases}$$

26. Test the continuity for the function.

$$f(x) = \begin{cases} e^{\frac{1}{x}} & , \quad \text{if } x \neq 0 \\ 0 & , \quad \text{if } x = 0 \text{ at} \end{cases} \qquad x = 0.$$

Answers

1. 3, $4\frac{1}{4}$, $h^2 + 2$, h
2. 2, 11, –1, –3, –2
3. (i) D = R – { 5 }, Range = R

   (ii) D = R – { 1 }, Rang = R–{2}

   (iii) D = [3, ∞], Range = all positive real numbers.
7. (i) even (ii) even (iii) neither even nor odd (iv) odd

10. (i) 2 (ii) 12 (iii) 1 (iv) $\frac{-1}{4}$ (v) $\frac{-1}{2}$
(vi) 1

11. (i) 256 (ii) $\frac{7}{5}a^2$ (iii) $\frac{5}{3}(a+2)^{\frac{2}{3}}$

12. (i) $\frac{p}{q}$ (ii) $\frac{1}{2}$ (iii) $2\sqrt{a}\cos a$ (iv) $\frac{b^2-a^2}{2}$

13. (i) $\frac{1}{2}$ (ii) -2

14. (i) 2 (ii) 1

16. 1

17. (i) ∞ (ii) $\frac{3}{2}$ (iii) 1 (iv) 2

18. Continuous

19. Continuous

22. Discontinuous

23. $x = 2, x = 3$

24. $x = 1, x = 3$

26. Discontinuous

CHAPTER 7

DIFFERENTIATION

Introduction

If f(x) is a function of x, then $\lim\limits_{h\to 0}\dfrac{f(a+h)-f(a)}{h}$ is defined as the differential coefficient of f(x) at x = a, provided that this limit exists and is finite.

The differential coefficient of f(x) with respect to x is generally written as $\dfrac{d}{dx}f(x)$ or $f'(x)$ or Df(x).

Note: Generally, $\dfrac{d}{dx}f(x) = \lim\limits_{h\to 0}\dfrac{f(x+h)-f(x)}{h}$ is known as the first principle of differential calculus.

Differential Coefficient of a Function at a Point

The value of the derivative of f(x) obtained by putting x = a is called the differential coefficient of f(x) at x = a and is denoted by $f'(a)$ or $\left(\dfrac{dy}{dx}\right)_{x=a}$.

List of Formulae

1. $\dfrac{d}{dx}x^n = nx^{n-1}$

2. $\dfrac{d}{dx}\log_e x = \dfrac{1}{x}$

3. $\frac{d}{dx} e^x = e^x$

4. $\frac{d}{dx} a^x = a^x \log_e a$

5. $\frac{d}{dx} x = 1$

6. $\frac{d}{dx} \sin x = \cos x$

7. $\frac{d}{dx} \cos x = -\sin x$

8. $\frac{d}{dx} \tan x = \sec^2 x$

9. $\frac{d}{dx} \cot x = -\operatorname{cosec}^2 x$

10. $\frac{d}{dx} \sec x = \sec x \tan x$

11. $\frac{d}{dx} \operatorname{cosec} x = -\operatorname{cosec} x \cot x$

12. $\frac{d}{dx} C = 0$, where C is constant

13. $\frac{d}{dx} \log_a x = \frac{1}{x} \log_a e$

***Example* 1:** Differentiate x^{10} w.r.t. x

***Solution*:**

$$y = x^{10}$$

$$\frac{dy}{dx} = \frac{d}{dx} x^{10}$$

$$= 10\, x^{10-1}$$

$$\frac{dy}{dx} = 10x^9$$

Example 2: Differentiate $\dfrac{lx^2 + mx + n}{\sqrt{x}}$ w.r.t. x.

***Solution*:**

$$y = \frac{lx^2 + mx + n}{\sqrt{x}}$$

$$= \frac{lx^2 + mx + n}{x^{1/2}}$$

$$y = lx^{\frac{3}{2}} + mx^{\frac{1}{2}} + nx^{\frac{-1}{2}}$$

$$\frac{dy}{dx} = l\frac{d}{dx}x^{\frac{3}{2}} + m\frac{d}{dx}x^{\frac{1}{2}} + n\frac{d}{dx}x^{\frac{-1}{2}}$$

$$= l.\frac{3}{2}x^{\frac{3}{2}-1} + m.\frac{1}{2}x^{\frac{1}{2}-1} + n\left(\frac{-1}{2}\right)x^{\frac{-1}{2}-1}$$

$$\frac{dy}{dx} = \frac{3l}{2}x^{\frac{1}{2}} + \frac{m}{2}x^{\frac{-1}{2}} - \frac{n}{2}x^{\frac{-3}{2}}$$

Differentiation from First Principle $\dfrac{dy}{dx} = \lim\limits_{h \to 0}\dfrac{f(x+h) - f(x)}{h}$

***Example*:** Find the derivative of the function $y = x^2$ from first principle

***Solution*:** by the first principle

$$\frac{dy}{dx} = \lim_{h \to 0}\frac{f(x+h) - f(x)}{h}$$

Given $f(x) = x^2$

$$\frac{dy}{dx} = \lim_{h \to 0}\frac{(x+h)^2 - x^2}{h}$$

$$= \lim_{h \to 0}\frac{x^2 + h^2 + 2hx - x^2}{h}$$

$$= \lim_{h \to 0}\frac{h^2 + 2hx}{h}$$

$$= \lim_{h \to 0}(2x + h)$$

$$= 2x$$

Example: If $y = \sqrt{x}$, find $\frac{dy}{dx}$ by first principle

Solution: Given $f(x) = \sqrt{x}$

by definition $\frac{dy}{dx} = \lim_{h\to 0} \frac{f(x+h) - f(x)}{h}$

$$= \lim_{h\to 0} \frac{\sqrt{x+h} - \sqrt{x}}{h}$$

$$= \lim_{h\to 0} \frac{\sqrt{x+h} - \sqrt{x}}{h} \times \frac{(\sqrt{x+h} + \sqrt{x})}{(\sqrt{x+h} + \sqrt{x})}$$

$$= \lim_{h\to 0} \frac{x+h-x}{h(\sqrt{x+h} + \sqrt{x})}$$

$$= \lim_{h\to 0} \frac{1}{\sqrt{x+h} + \sqrt{x}} = \frac{1}{2\sqrt{x}}$$

Example: Differentiate $\sin^2 x$ from first principle

Solution: Given $f(x) = \sin^2 x$, $f(x+h) = \sin^2 (x+h)$

by definition

$$\frac{dy}{dx} = \lim_{h\to 0} \frac{\sin^2(x+h) - \sin^2 x}{h}$$

$$= \lim_{h\to 0} \frac{\sin(x+h+x)\sin(x+h-x)}{h}$$

$$(\because \sin^2 A - \sin^2 B = \sin(A+B)\sin(A-B)$$

$$= \lim_{h\to 0} \frac{\sin(2x+h)\sin h}{h}$$

$$= \lim_{h\to 0} (\sin 2x + h). \lim_{h\to 0} \frac{\sin h}{h}$$

$$= \lim_{h\to 0} \sin(2x+h) = \sin 2x$$

Example: Differentiate e^x by first principle

Solution: Given $f(x) = e^x$, $f(x+h) = e^{x+h}$

by definition

$$\frac{dy}{dx} = \underset{h\to 0}{\text{Lim}} \frac{e^{x+h} - e^x}{h}$$

$$= \underset{h\to 0}{\text{Lim}} \frac{e^x . e^h - e^x}{h}$$

$$= \lim_{h\to 0} \frac{e^x (e^h - 1)}{h}$$

$$= \underset{h\to 0}{\text{Lim}} \frac{e^x \left(1 + h + \frac{h^2}{2!} + ... - 1\right)}{h}$$

$$= \underset{h\to 0}{\text{Lim}} \frac{e^x . h\left(1 + \frac{h}{2!} + ...\right)}{h}$$

$$= \underset{h\to 0}{\text{Lim}}\ e^x \left(1 + \frac{h}{2!} + ...\right) = e^x$$

Example: Differentiate logx by first principle

Solution: Given f(x) = logx, f(x + h) = log (x + h)

By definition

$$\frac{dy}{dx} = \underset{h\to 0}{\text{Lim}} \frac{\log(x+h) - \log x}{h}$$

$$= \underset{h\to 0}{\text{Lim}} \frac{1}{h} \log\left(\frac{x+h}{x}\right) \quad \left(\because \log m - \log n = \log\frac{m}{n}\right)$$

$$= \underset{h\to 0}{\text{Lim}} \frac{1}{h} \log\left(1 + \frac{h}{x}\right)$$

$$= \underset{h\to 0}{\text{Lim}} \frac{1}{h} \left(\frac{h}{x} - \frac{1}{2}\left(\frac{h}{x}\right)^2 + ...\right)$$

$$\left(\because \log(1+x) = x - \frac{x^2}{2} + \frac{x^3}{3} + ...\right)$$

$$= \underset{h\to 0}{\text{Lim}} \left(\frac{1}{x} - \frac{1}{2}\frac{h}{x^2} + ...\right)$$

$$= \frac{1}{x}$$

Example: Differentiate sin x by first principle

Solution: Given f(x) = sin x, f (x + h) = sin (x + h)

$$= \lim_{h\to 0}\left(\frac{\sin(x+h)-\sin x}{h}\right)$$

$$= \lim_{h\to 0}\left(\frac{2\cos\frac{x+h+x}{2}.\sin\frac{x+h-x}{2}}{h}\right)$$

$$= \lim_{h\to 0}\cos\left(x+\frac{h}{2}\right).\frac{\sin\frac{h}{2}}{\frac{h}{2}}$$

$$= \lim_{h\to 0}\cos\left(x+\frac{h}{2}\right).\lim_{h\to 0}\frac{\sin\frac{h}{2}}{\frac{h}{2}}$$

$$= \cos x$$

Example: Differentiate (sec x + cosec x) w.r.t. x

Solution: Let $y = \sec x + \operatorname{cosec} x$

$$\frac{dy}{dx} = \frac{d}{dx}\sec x + \frac{d}{dx}\operatorname{cosec} x$$

$$= \sec x \tan x - \operatorname{cosec} x \cot x$$

Formula

(i) Differential coefficient of the product of two functions (product rule)

$$\frac{d}{dx}[f_1(x).f_2(x)] = f_1(x)\frac{d}{dx}f_2(x) + f_2(x)\frac{d}{dx}f_1(x)$$

(ii) Differential coefficient of the quotient of two functions (quotient rule)

$$\frac{d}{dx}\frac{f_1(x)}{f_2(x)} = \frac{f_2(x)\times\frac{d}{dx}f_1(x) - f_1(x)\frac{d}{dx}f_2(x)}{[f_2(x)]^2}$$

***Example* 4:** Differentiate $x^2 \log x$ w.r.t. x.

***Solution*:**

$$\frac{d}{dx}(x^2 . \log x) = x^2 \frac{d}{dx} \log x + \log x \frac{d}{dx} x^2$$

$$= x^2 . \frac{1}{x} + \log x . 2x$$

$$= x + 2x \log x$$

$$= x (1 + 2\log x)$$

***Example* 5:** Differentiate $\sin x . \log x$ w.r.t. x.

***Solution*:**

$$\frac{d}{dx}(\sin x . \log x) = \sin x \frac{d}{dx} \log x + \log x \frac{d}{dx} \sin x$$

$$= \sin x . \frac{1}{x} + \log x . \cos x$$

$$= \frac{\sin x}{x} + \log x . \cos x$$

***Example* 6:** Differentiate $x^3 e^x \sin x$ w.r.t. x.

***Solution*:**

$$\frac{d}{dx}(x^3 . e^x . \sin x) = x^3 \frac{d}{dx}(e^x . \sin x) + e^x \sin x \frac{d}{dx} x^3$$

$$= x^3 \left[e^x \frac{d}{dx} \sin x + \sin x \frac{d}{dx} e^x \right] + e^x \sin x . 3x^2$$

$$= x^3 \left[e^x \cos x + \sin x . e^x \right] + 3x^2 e^x \sin x$$

$$= x^3 e^x \cos x + x^3 e^x \sin x + 3x^2 e^x \sin x$$

***Example* 7:** Differentiate $\frac{e^x}{x}$ w.r.t. x.

***Solution*:**

$$\frac{d}{dx}\left(\frac{e^x}{x}\right) = \frac{x \frac{d}{dx} e^x - e^x \frac{d}{dx} x}{x^2}$$

$$= \frac{x e^x - e^x . 1}{x^2}$$

$$= \frac{e^x (x - 1)}{x^2}$$

***Example* 8:** Differentiate $\dfrac{5x^2+6x+7}{2x^2+3x+4}$ w.r.t. x.

***Solution*:**

$$\frac{d}{dx}\left(\frac{5x^2+6x+7}{2x^2+3x+4}\right)$$

$$=\frac{\left(2x^2+3x+4\right)\frac{d}{dx}\left(5x^2+6x+7\right)-\left(5x^2+6x+7\right)\frac{d}{dx}\left(2x^2+3x+4\right)}{\left(2x^2+3x+4\right)^2}$$

$$=\frac{\left(2x^2+3x+4\right)(5.2x+6.1+0)-\left(5x^2+6x+7\right)(2.2x+3.1+0)}{\left(2x^2+3x+4\right)^2}$$

$$=\frac{\left(2x^2+3x+4\right)(10x+6)-\left(5x^2+6x+7\right)(4x+3)}{\left(2x^2+3x+4\right)^2}$$

$$\frac{20x^3+12x^2+30x^2+18x+40x+24-20x^3-15x^2-24x^2-18x-28x-21}{\left(2x^2+3x+4\right)^2}$$

$$=\frac{3x^2+12x+3}{\left(2x^2+3x+4\right)^2}$$

***Example* 9:** If $y=\dfrac{x}{x+5}$, then prove that $x\dfrac{dy}{dx}=y(1-y)$.

***Solution*:**

$$y=\frac{x}{x+5}$$

$$\frac{dy}{dx}=\frac{(x+5)\frac{dy}{dx}x-x\frac{d}{dx}(x+5)}{(x+5)^2}$$

$$=\frac{(x+5).1-x(1+0)}{(x+5)^2}$$

$$=\frac{(x+5)-x}{(x+5)^2}$$

$$\frac{dy}{dx}=\frac{5}{(x+5)^2}$$

$$\therefore \quad x\frac{dy}{dx} = \frac{5x}{(x+5)^2}$$

$$= \frac{x}{x+5} \cdot \frac{5}{x+5}$$

$$= y\left(1 - \frac{x}{x+5}\right)$$

$$x\frac{dy}{dx} = y\,(1-y)$$

Differential Coefficient of a Function of Function

If y = f(z), where z = F(x), then by eliminating z, we get y = f {F(x)}

e.g. sin x^3, $e^{\sin x}$, cos 2x are the function of function.

Again, let $\quad y = f(t)$, where $t = g(x)$

Then $\quad \frac{dy}{dx} = \frac{dy}{dt} \cdot \frac{dt}{dx}$, this is known as '*chain rule*'.

***Example* 10:** Find the differential coefficient of sin x^2 w.r.t. x

***Solution*:** Let $\quad y = \sin x^2$

$$\frac{dy}{dx} = \frac{d}{dx}\sin x^2$$

Put $\quad x^2 = t$

$$\frac{dy}{dx} = \frac{d}{dx}\sin t$$

$$= \frac{d}{dt}\sin t.\ \frac{dt}{dx}$$

$$= \cos t.\ \frac{d}{dx}x^2$$

$$= \cos t.\ 2x$$

$$= 2x \cos x^2$$

***Example* 11:** Differentiate $e^{\sin x}$ w.r.t. x.

***Solution*:** Let $y = e^{\sin x}$

Put $\sin x = t$

$y = e^t$

Now,
$$\frac{dy}{dx} = \frac{dy}{dt} \cdot \frac{dt}{dx}$$
$$= \frac{d}{dt} e^t . \frac{d}{dx} \sin x$$
$$= e^t . \cos x$$
$$= e^{\sin x} . \cos x$$

***Example* 12:** Differentiate log (log sin x) w.r.t. x.

***Solution*:** Let $y = \log(\log \sin x)$

Put $\log \sin x = t$

$y = \log t$

$$\frac{dy}{dx} = \frac{dy}{dt} \cdot \frac{dt}{dx}$$
$$= \frac{d}{dt} \log t . \frac{d}{dx} \log \sin x$$
$$= \frac{1}{t} \frac{d}{dx} \log \sin x$$

Again, put $\sin x = u$

$$\frac{dy}{dx} = \frac{1}{t} . \frac{d}{du} \log u \frac{du}{dx}$$
$$= \frac{1}{t} . \frac{1}{u} . \frac{d}{dx} \sin x$$
$$= \frac{1}{\log \sin x} . \frac{1}{\sin x} . \cos x$$
$$= \frac{\cot x}{\log \sin x}$$

***Example* 13:** If $y = \log\sqrt{\dfrac{1-\cos mx}{1+\cos mx}}$, then find $\dfrac{dy}{dx}$.

***Solution*:**

$$y = \log\sqrt{\frac{1-\cos mx}{1+\cos mx}}$$

$$\frac{dy}{dx} = \frac{d}{dx}\log\sqrt{\frac{2\sin^2\frac{mx}{2}}{2\cos^2\frac{mx}{2}}}$$

$$= \frac{d}{dx}\log\tan\frac{mx}{2}$$

Put $\tan\frac{mx}{2} = t$

$$\frac{dy}{dx} = \frac{d}{dt}\log t.\ \frac{d}{dx}\tan\frac{mx}{2}$$

$$= \frac{1}{t}.\frac{d}{dx}\tan\frac{mx}{2}$$

Again put $\frac{mx}{2} = u$

$$\frac{dy}{dx} = \frac{1}{\tan\frac{mx}{2}}.\ \frac{d}{du}\tan u.\ \frac{d}{dx}\frac{mx}{2}$$

$$= \frac{1}{\tan\frac{mx}{2}}\sec^2 u.\ \frac{m}{2}$$

$$= \frac{m}{2}\ \frac{1}{\tan\frac{mx}{2}}.\sec^2\frac{mx}{2}$$

$$= \frac{m}{2}.\frac{\cos\frac{mx}{2}}{\sin\frac{mx}{2}}.\frac{1}{\cos^2\frac{mx}{2}}$$

$$= \frac{m}{\sin mx}$$

$$= m\ \text{cosec}\ mx$$

Differential coefficient of inverse Trigonometric Functions or Trigonometrical Transformation

Formula
$$\frac{d}{dx}\sin^{-1}x = \frac{1}{\sqrt{1-x^2}}$$
$$\frac{d}{dx}\cos^{-1}x = \frac{-1}{\sqrt{1-x^2}}$$
$$\frac{d}{dx}\tan^{-1}x = \frac{1}{1+x^2}$$
$$\frac{d}{dx}\cot^{-1}x = \frac{-1}{1+x^2}$$
$$\frac{d}{dx}\sec^{-1}x = \frac{1}{x\sqrt{x^2-1}}$$
$$\frac{d}{dx}\text{cosec}^{-1}x = \frac{-1}{x\sqrt{x^2-1}}$$

Differentiation by Trigonometric Transformation

***Example* 14:** Differentiate $\sin^{-1}\left(\frac{2x}{1+x^2}\right)$ w.r.t. x.

***Solution*:** Let $y = \sin^{-1}\left(\frac{2x}{1+x^2}\right)$

Put $x = \tan\theta \quad \Rightarrow \theta = \tan^{-1}x$

$$y = \sin^{-1}\left(\frac{2\tan\theta}{1+\tan^2\theta}\right)$$
$$= \sin^{-1}(\sin 2\theta) \qquad \left(\because \sin 2\theta = \frac{2\tan\theta}{1+\tan^2\theta}\right)$$
$$y = 2\theta$$
$$y = 2\tan^{-1}x$$
$$\frac{dy}{dx} = 2\frac{d}{dx}\tan^{-1}x$$
$$= \frac{2}{1+x^2}$$

***Example* 15:** If $y = \cot^{-1}\left(\frac{\sqrt{1+x^2}+1}{x}\right)$, then find $\frac{dy}{dx}$.

***Solution*:**

$$y = \cot^{-1}\left(\frac{\sqrt{1+x^2}+1}{x}\right)$$

Put $x = \tan\theta \quad \Rightarrow \theta = \tan^{-1} x$

$$y = \cot^{-1}\left(\frac{\sqrt{1+\tan^2\theta}+1}{\tan\theta}\right)$$

$$= \cot^{-1}\left(\frac{\sec\theta + 1}{\tan\theta}\right)$$

$$= \cot^{-1}\left(\frac{\frac{1}{\cos\theta}+1}{\frac{\sin\theta}{\cos\theta}}\right)$$

$$= \cot^{-1}\left(\frac{1+\cos\theta}{\sin\theta}\right)$$

$$= \cot^{-1}\left(\frac{2\cos^2\frac{\theta}{2}}{2\sin\frac{\theta}{2}\cos\frac{\theta}{2}}\right)$$

$$= \cot^{-1}\cot\frac{\theta}{2}$$

$$y = \frac{\theta}{2}$$

$$y = \frac{1}{2}\tan^{-1} x$$

$$\frac{dy}{dx} = \frac{1}{2}\frac{d}{dx}\tan^{-1} x$$

$$= \frac{1}{2}\cdot\frac{1}{1+x^2}$$

***Example* 16:** If $y = \sin^{-1}\left(2x\sqrt{1-x^2}\right)$, then find $\frac{dy}{dx}$.

***Solution*:**

$$y = \sin^{-1}\left(2x\sqrt{1-x^2}\right)$$

Put $x = \sin\theta \quad \Rightarrow \theta = \sin^{-1} x$

$$y = \sin^{-1}\left(2\sin\theta\sqrt{1-\sin^2\theta}\right)$$

$$= \sin^{-1}(2\sin\theta\cos\theta)$$

$$= \sin^{-1}\sin 2\theta$$

$$y = 2\theta$$

$$y = 2\sin^{-1} x$$

$$\frac{dy}{dx} = 2\frac{d}{dx}\sin^{-1} x$$

$$= \frac{2}{\sqrt{1-x^2}}$$

***Example* 17:** Differentiate $\sin^{-1}\left(x\sqrt{1-x} - \sqrt{x}\sqrt{1-x^2}\right)$ w.r.t. x.

***Solution*:** Let

$$y = \sin^{-1}\left(x\sqrt{1-x} - \sqrt{x}\sqrt{1-x^2}\right)$$

Put $x = \sin A,\ \sqrt{x} = \sin B \quad \Rightarrow \quad A = \sin^{-1} x, \quad B = \sin^{-1}\sqrt{x}$

$$y = \sin^{-1}\left(\sin A\sqrt{1-\sin^2 B} - \sin B\sqrt{1-\sin^2 A}\right)$$

$$= \sin^{-1}(\sin A\cos B - \cos A\sin B)$$

$$= \sin^{-1}\sin(A - B)$$

$$y = A - B$$

$$y = \sin^{-1} x - \sin^{-1}\sqrt{x}$$

$$\frac{dy}{dx} = \frac{d}{dx}\sin^{-1} x - \frac{d}{dx}\sin^{-1}\sqrt{x}$$

$$= \frac{1}{\sqrt{1-x^2}} - \frac{d}{dt}\sin^{-1} t.\ \frac{dt}{dx} \qquad \text{put } \sqrt{x} = t$$

$$= \frac{1}{\sqrt{1-x^2}} - \frac{1}{\sqrt{1-t^2}} \cdot \frac{d}{dx}\sqrt{x}$$

$$= \frac{1}{\sqrt{1-x^2}} - \frac{1}{\sqrt{1-x}} \cdot \frac{1}{2\sqrt{x}}$$

$$= \frac{1}{\sqrt{1-x^2}} - \frac{1}{2\sqrt{x-x^2}}$$

Differentiation of Implicit Functions

There may be a relation between x and y when it is not possible to express it in the form of $y = f(x)$ 'or' to solve an equation for y in terms of x. We call such a relation as implicit function.

For $\frac{dy}{dx}$ by differentiating the given relation w.r.t. x

***Example* 18:** Find $\frac{dy}{dx}$, when $x^2 + y^2 = a^2$

***Solution*:** Given $x^2 + y^2 = a^2$

differentiating both the sides w.r.t. x, we get

$$\frac{d}{dx}x^2 + \frac{d}{dx}y^2 = \frac{d}{dx}a^2$$

$$2x + 2y\frac{dy}{dx} = 0$$

$$2y\frac{dy}{dx} = -2x$$

$$\frac{dy}{dx} = \frac{-2x}{2y}$$

$$= \frac{-x}{y} \quad \text{Ans}$$

***Example* 19:** Find $\frac{dy}{dx}$, when $ax^2 + 2hxy + by^2 + 2gx + 2fy + c = 0$

***Solution*:** Given $ax^2 + 2hxy + by^2 + 2gx + 2fy + c = 0$

differentiating both the sides w.r.t. x, we get

$$a\frac{d}{dx}x^2 + 2h\frac{d}{dx}xy + b\frac{d}{dx}y^2 + 2g\frac{d}{dx}x + 2f\frac{d}{dx}y + \frac{d}{dx}c = 0$$

$$2ax + 2h\left(x\frac{dy}{dx} + y\frac{d}{dx}x\right) + 2by\frac{dy}{dx} + 2g + 2f\frac{dy}{dx} + 0 = 0$$

$$2ax + 2hx\frac{dy}{dx} + 2yh + 2by\frac{dy}{dx} + 2g + 2f\frac{dy}{dx} = 0$$

$$(2ax + 2hy + 2g) + (2hx + 2by + 2f)\frac{dy}{dx} = 0$$

$$2\,(hx + by + f)\frac{dy}{dx} = -2\,(ax + hy + g)$$

$$\frac{dy}{dx} = \frac{-(ax + hy + g)}{(hx + by + f)}$$

***Example* 20:** Find $\frac{dy}{dx}$, when $4x^2 + 9y^2 = 36$.

***Solution*:** Given $4x^2 + 9y^2 = 36$

differentiating both the sides w.r.t. x, we get

$$4.\frac{d}{dx}x^2 + 9\frac{d}{dx}y^2 = \frac{d}{dx}36$$

$$4.2x + 9.2y\frac{dy}{dx} = 0$$

$$8x + 18y\frac{dy}{dx} = 0$$

$$18y\frac{dy}{dx} = -8x$$

$$\frac{dy}{dx} = \frac{-8x}{18y}$$

$$= \frac{-4x}{9y}$$

***Example* 21:** Find $\frac{dy}{dx}$, when $\frac{1}{x}+\frac{1}{y}=1$.

***Solution*:** Given $\frac{1}{x}+\frac{1}{y}=1$

$$\frac{y+x}{xy}=1$$

$$y+x=xy$$

differentiating both the sides w.r.t. x, we get

$$\frac{dy}{dx}+\frac{d}{dx}x=x\frac{dy}{dx}+y\frac{d}{dx}x$$

$$\frac{dy}{dx}+1=x\frac{dy}{dx}+y$$

$$\frac{dy}{dx}-x\frac{dy}{dx}=y-1$$

$$\frac{dy}{dx}(1-x)=y-1$$

$$\frac{dy}{dx}=\frac{y-1}{1-x}$$

***Example* 22:** Find $\frac{dy}{dx}$, when $x^3+8xy+y^3=64$.

***Solution*:** Given $x^3+8xy+y^3=64$

Differentiating both the sides w.r.t x, we get

$$\frac{d}{dx}x^3+8\frac{d}{dx}xy+\frac{d}{dx}y^3=\frac{d}{dx}64$$

$$3x^2+8\left(x\frac{dy}{dx}+y\right)+3y^2\frac{dy}{dx}=0$$

$$3x^2+8x\frac{dy}{dx}+8y+3y^2\frac{dy}{dx}=0$$

$$(3x^2+8y)+(8x+3y^2)\frac{dy}{dx}=0$$

$$(8x+3y^2)\frac{dy}{dx}=-(3x^2+8y)$$

$$\frac{dy}{dx}=\frac{-(3x^2+8y)}{8x+3y^2}$$

Logarithmic Differentiation: If the power of a function of x is also a function of x, then it would be convenient for us if we first take logarithms w.r.t the base e and then differentiate both sides w.r.t x.

Let us recall the formulae

(i) $\log m^n = n \log m$

(ii) $\log(mn) = \log m + \log n$

(iii) $\log \frac{m}{n} = \log m - \log n$

***Example* 23:** If $y = x^x$, then prove that $\frac{dy}{dx} = x^x (1+ \log x)$

Solution: Given $y = x^x$

Taking log on both the sides, we get

$$\log y = \log x^x$$

$$\log y = x \cdot \log x$$

Differentiating both the sides w.r.t x, we get

$$\frac{d}{dx}\log y = x\frac{d}{dx}\log x + \log x\frac{d}{dx}x$$

$$\frac{1}{y}\frac{dy}{dx} = x.\frac{1}{x} + \log x.1$$

$$\frac{1}{y}\frac{dy}{dx} = 1 + \log x$$

$$\frac{dy}{dx} = y(1 + \log x)$$

$$= x^x(1 + \log x)$$

***Differentiation of Parametric equation*:** when variables x and y are expressed in terms of a third variable like t or θ, then this third variable is called parameter.

e.g. $x = g(t)$ and $y = h(t)$, these are called parametric equations.

Formula $\frac{dy}{dx} = \frac{dy/dt}{dx/dt}$

***Example* 24:** If $x = a \sec \theta$, $y = b \tan \theta$, find $\frac{dy}{dx}$.

Solution: Given $x = a \sec \theta$, $y = b \tan \theta$

$$\frac{dx}{d\theta} = a \sec\theta \tan\theta, \qquad \frac{dy}{d\theta} = b\sec^2\theta$$

$$\therefore \qquad \frac{dy}{dx} = \frac{dy/d\theta}{dx/d\theta}$$

$$= \frac{b\sec^2\theta}{a\sec\theta\tan\theta}$$

$$= \frac{b}{a}\operatorname{cosec}\theta$$

***Example* 25:** If $x = at^2$, $y = 2at$, then find $\frac{dy}{dx}$.

Solution: Given $x = at^2$, $y = 2at$

$$\frac{dx}{dt} = 2at \quad , \quad \frac{dy}{dt} = 2a$$

$$\frac{dy}{dx} = \frac{dy/dt}{dx/dt}$$

$$= \frac{2a}{2at}$$

$$= \frac{1}{t}$$

***Differentiation of Infinite Series*:** When the function is given in infinite series. In this case we use the fact that if a term is deleted from an infinite series it remains uneffected.

***Example*:** If $y = x^{x^{x^{x^{x.......\infty}}}}$, find $\frac{dy}{dx}$.

***Solution*:** here $y = x^y$

$$\Rightarrow \qquad \log y = y \log x$$

Differentiating w.r.t x, we have

$$\frac{1}{y}\frac{dy}{dx} = y\frac{d}{dx}\log x + \log x\frac{dy}{dx}$$

$$= \frac{y}{x} + \log x\frac{dy}{dx}$$

$$\frac{dy}{dx}\left(\frac{1}{y} - \log x\right) = \frac{y}{x}$$

$$\frac{dy}{dx}\left(\frac{1-y\log x}{y}\right)=\frac{y}{x}$$

$$\frac{dy}{dx}=\frac{y^2}{x(1-y\log x)}$$

Example: If $y=\sqrt{\sin x+\sqrt{\sin x+\sqrt{\sin x+.......\infty}}}$, Prove that $\frac{dy}{dx}=\frac{\cos x}{2y-1}$

Solution: Here $y=\sqrt{\sin x+y}$

$$y^2 = \sin x + y$$

$$2y\frac{dy}{dx}=\cos x+\frac{dy}{dx}$$

$$\frac{dy}{dx}(2y-1)=\cos x$$

$$\frac{dy}{dx}=\frac{\cos x}{2y-1}$$

Example: If $y=\left(\sqrt{x}\right)^{\left(\sqrt{x}\right)^{\left(\sqrt{x}\right)^{.....\infty}}}$ show that $\frac{dy}{dx}=\frac{y^2}{x(2-y\log x)}$

Solution: Here $y=\left(\sqrt{x}\right)^y$

$$\log y = y\log\sqrt{x}$$

$$\log y = \frac{y}{2}\log x$$

Differentiating w.r.t x, we get

$$\frac{1}{y}\frac{dy}{dx}=\frac{y}{2}.\frac{1}{x}+\frac{\log x}{2}\frac{dy}{dx}$$

$$\frac{dy}{dx}\left(\frac{1}{y}-\frac{\log x}{2}\right)=\frac{y}{2x}$$

$$\frac{dy}{dx}\left(\frac{2-y\log x}{2y}\right)=\frac{y}{2x}$$

$$\frac{dy}{dx}=\frac{y^2}{x(2-y\log x)}$$

Example: If $y = e^{x+e^{x+e^{x}+\ldots\ldots\infty}}$, show that $\frac{dy}{dx} = \frac{y}{1-y}$

Solution: Here $y = e^{x+y}$

Taking log on both sides

$$\log y = (x + y) \log e \qquad (\because \log e = 1)$$

$$\log y = x + y$$

Differentiating w.r.t x, we get

$$\frac{1}{y}\frac{dy}{dx} = 1 + \frac{dy}{dx}$$

$$\frac{dy}{dx}\left(\frac{1}{y} - 1\right) = 1$$

$$\frac{dy}{dx}\left(\frac{1-y}{y}\right) = 1$$

$$\frac{dy}{dx} = \frac{y}{1-y}$$

Example: If $y = \sqrt{x + \sqrt{x + \sqrt{x + \ldots\ldots\infty}}}$ then prove that $\frac{dy}{dx} = \frac{1}{2y-1}$

Solution: Here $y = \sqrt{x+y}$ .....(1)

Taking log on both sides

$$\log y = \frac{1}{2} \log(x + y)$$

Differentiating w.r.t x, we get

$$\frac{1}{y}\frac{dy}{dx} = \frac{1}{2}.\frac{1}{x+y}\left(1 + \frac{dy}{dx}\right)$$

$$\left[\frac{1}{y} - \frac{1}{2(x+y)}\right]\frac{dy}{dx} = \frac{1}{2(x+y)} \qquad .....(2)$$

From eq. (1) $x + y = y^2$

Putting in eq. (2) we get

$$\left(\frac{1}{y} - \frac{1}{2y^2}\right)\frac{dy}{dx} = \frac{1}{2y^2}$$

$$\left(\frac{2y-1}{2y^2}\right)\frac{dy}{dx}=\frac{1}{2y^2}$$

$$\frac{dy}{dx}=\frac{1}{2y-1}$$

Successive Differentiation: Differentiating the first derivative of a function we obtain second derivative, differentiating the second derivative, we get the third derivative and so on. This process of finding derivatives of higher order is known as successive differentiation.

***Example* 27:** Find second derivative of $ax^3 + bx^2 + cx + d$

Solution: Let $y = ax^3 + bx^2 + cx + d$

Differentiating w.r.t x, we get

$$\frac{dy}{dx} = 3ax^2 + 2bx + c$$

Again differentiating, we get

$$\frac{d}{dx}\left(\frac{dy}{dx}\right) = 3a\,(2x) + 2b + 0$$

$$\frac{d^2y}{dx^2} = 6ax + 2b$$

***Example* 28:** If $y = x + \tan x$, Prove that $\cos^2 x\,\frac{d^2y}{dx^2} + 2x = 2y$

Solution: Given $y = x + \tan x$

Differentiating w.r.t x, we get

$$\frac{dy}{dx} = 1 + \sec^2 x$$

Differentiating again, we get

$$\frac{d}{dx}\left(\frac{dy}{dx}\right) = 0 + 2\sec x\,.\,\sec x\tan x$$

$$\frac{d^2y}{dx^2} = 2\sec^2 x\tan x$$

$$= \frac{2\tan x}{\cos^2 x}$$

$$\cos^2 x \frac{d^2y}{dx^2} = 2 \tan x$$

$$\cos^2 x \frac{d^2y}{dx^2} = 2(y - x)$$

$$\cos^2 x \frac{d^2y}{dx^2} = 2y - 2x$$

$$\cos^2 x \frac{d^2y}{dx^2} + 2x = 2y$$

Some Practice Questions

First Principle

***Example* 1:** Differentiate cos x by first principle.

***Solution*:** Let f(x) = cos x, f(x + h) = cos(x + h)

Then by first principle

$$\frac{dy}{dx} = \lim_{h \to 0} \frac{f(x+h) - f(x)}{h}$$

$$= \lim_{h \to 0} \frac{\cos(x+h) - \cos x}{h}$$

$$= \lim_{h \to 0} \frac{2\sin\frac{x+h+x}{2}\sin\frac{x-x-h}{2}}{h}$$

$$= \lim_{h \to 0} \frac{2\sin\left(x+\frac{h}{2}\right)\sin\left(\frac{-h}{2}\right)}{h}$$

$$= -\lim_{h \to 0} \sin\left(x+\frac{h}{2}\right) . \lim_{h \to 0} \frac{\sin\frac{h}{2}}{\frac{h}{2}}$$

$$= -\sin x$$

***Example* 2:** Differentiate tan x by first principle.

***Solution*:** Let f(x) = tan x, f(x + h) = tan (x + h)

then by first principle

$$\frac{dy}{dx} = \lim_{h \to 0} \frac{f(x+h) - f(x)}{h}$$

$$= \lim_{h \to 0} \frac{\tan(x+h) - \tan x}{h}$$

$$= \lim_{h \to 0} \frac{\frac{\sin(x+h)}{\cos(x+h)} - \frac{\sin x}{\cos x}}{h}$$

$$= \lim_{h \to 0} \frac{\sin(x+h)\cos x - \cos(x+h)\sin x}{h \cos x \cos(x+h)}$$

$$= \lim_{h \to 0} \frac{\sin(x+h-x)}{h \cos x \cos(x+h)}$$

$$= \lim_{h \to 0} \frac{\sin h}{h} \cdot \lim_{h \to 0} \frac{1}{\cos x . \cos(x+h)}$$

$$= 1 \times \frac{1}{\cos x . \cos x}$$

$$= \frac{1}{\cos^2 x}$$

$$= \sec^2 x$$

***Example* 3:** Differentiate sec x by first principle.

Solution: Let f(x) = sec x, f(x + h) = sec(x + h)

Then by first principle

$$\frac{dy}{dx} = \lim_{h \to 0} \frac{f(x+h) - f(x)}{h}$$

$$= \lim_{h \to 0} \frac{\sec(x+h) - \sec x}{h}$$

$$= \lim_{h\to 0} \frac{\dfrac{1}{\cos(x+h)} - \dfrac{1}{\cos x}}{h}$$

$$= \lim_{h\to 0} \frac{\cos x - \cos(x+h)}{h\cos x\cos(x+h)}$$

$$= \lim_{h\to 0} \frac{2\sin(x+\frac{h}{2}).\ \sin\frac{h}{2}}{h\cos x\cos(x+h)}$$

$$= \lim_{h\to 0} \sin(x+\frac{h}{2}).\ \lim_{h\to 0} \frac{1}{\cos x\cos(x+h)}.\lim_{h\to 0} \frac{\sin\frac{h}{2}}{\frac{h}{2}}$$

$$= \frac{\sin x}{\cos x.\cos x} \times 1$$

$$= \sec x.\ \tan x$$

***Example* 4 :** Differentiate $\sqrt{\sin x}$ by first principle.

***Solution*:** Let $f(x) = \sqrt{\sin x}$, $f(x+h) = \sqrt{\sin(x+h)}$

Then by first principle

$$\frac{dy}{dx} = \lim_{h\to 0} \frac{f(x+h) - f(x)}{h}$$

$$= \lim_{h\to 0} \frac{\sqrt{\sin(x+h)} - \sqrt{\sin x}}{h}$$

$$= \lim_{h\to 0} \frac{\sqrt{\sin(x+h)} - \sqrt{\sin x}}{h} \times \frac{\sqrt{\sin(x+h)} + \sqrt{\sin x}}{\sqrt{\sin(x+h)} + \sqrt{\sin x}}$$

$$= \lim_{h\to 0} \frac{\sin(x+h) - \sin x}{h(\sqrt{\sin(x+h)} + \sqrt{\sin x})}$$

$$= \lim_{h\to 0} \frac{2\cos\left(x+\frac{h}{2}\right).\sin\frac{h}{2}}{h(\sqrt{\sin(x+h)} + \sqrt{\sin x})}$$

$$= \lim_{h \to 0} \frac{\cos\left(x + \frac{h}{2}\right)}{\sqrt{\sin(x+h)} + \sqrt{\sin x}} \cdot \lim_{h \to 0} \frac{\sin \frac{h}{2}}{\frac{h}{2}}$$

$$= \frac{\cos x}{\sqrt{\sin x} + \sqrt{\sin x}}$$

$$= \frac{\cos x}{2\sqrt{\sin x}}$$

***Example* 5:** Differentiate $\sin^{-1} x$ by first principle.

***Solution*:** Let $f(x) = \text{Sin}^{-1} x = y, \quad f(x+h) = \sin^{-1}(x+h) = y + k$

$$\Rightarrow x = \sin y \qquad x + h = \sin(y+k)$$

Then by first principle

$$\frac{dy}{dx} = \lim_{h \to 0} \frac{f(x+h) - f(x)}{h}$$

$$= \lim_{h \to 0} \frac{\sin^{-1}(x+h) - \sin^{-1} x}{h}$$

$$= \lim_{h \to 0} \frac{y + k - y}{(x+h) - x}$$

$$= \lim_{k \to 0} \frac{k}{\sin(y+k) - \sin y} \qquad (\text{as } h \to 0 \Rightarrow k \to 0)$$

$$= \lim_{k \to 0} \frac{k}{2\cos\left(y + \frac{k}{2}\right) . \sin \frac{k}{2}}$$

$$= \lim_{k \to 0} \frac{1}{\cos\left(y + \frac{k}{2}\right)} \cdot \lim_{k \to 0} \frac{\frac{k}{2}}{\sin \frac{k}{2}}$$

$$= \frac{1}{\cos y} \times 1 = \frac{1}{\cos y}$$

$$= \frac{1}{\sqrt{1 - \sin^2 y}}$$

$$= \frac{1}{\sqrt{1 - x^2}}$$

***Example* 6:** Differentiate $\sin\sqrt{x}$ by first principle.

***Solution*:** Let $f(x) = \sin\sqrt{x}$, $f(x+h) = \sin\sqrt{x+h}$

Then by first principle

$$\frac{dy}{dx} = \lim_{h\to 0}\frac{f(x+h)-f(x)}{h}$$

$$= \lim_{h\to 0}\frac{\sin\sqrt{(x+h)} - \sin\sqrt{x}}{h}$$

$$= \lim_{h\to 0}\frac{2\cos\left(\dfrac{\sqrt{x+h}+\sqrt{x}}{2}\right)\sin\left(\dfrac{\sqrt{x+h}-\sqrt{x}}{2}\right)}{h}$$

$$= \lim_{h\to 0}\frac{2\cos\left(\dfrac{\sqrt{x+h}+\sqrt{x}}{2}\right)}{h} . \sin\left(\frac{\sqrt{x+h}-\sqrt{x}}{2}\right)\times\frac{\sqrt{x+h}+\sqrt{x}}{\sqrt{x+h}+\sqrt{x}}$$

$$= \lim_{h\to 0}\frac{2\cos\left(\dfrac{\sqrt{x+h}+\sqrt{x}}{2}\right)}{\sqrt{x+h}+\sqrt{x}} \cdot \left(\frac{\sin\dfrac{x+h-x}{2}}{h}\right)$$

$$= \lim_{h\to 0}\frac{\cos\left(\dfrac{\sqrt{x+h}+\sqrt{x}}{2}\right)}{\sqrt{x+h}+\sqrt{x}} . \lim_{h\to 0}\frac{\sin\dfrac{h}{2}}{\dfrac{h}{2}}$$

$$= \frac{\cos\left(\dfrac{\sqrt{x}+\sqrt{x}}{2}\right)}{\sqrt{x}+\sqrt{x}}\times 1$$

$$= \frac{\cos\sqrt{x}}{2\sqrt{x}}$$

***Example* 7:** Differentiate sin x^2 by first principle.

***Solution*:** Let $f(x) = \sin x^2$, $f(x+h) = \sin (x+h)^2$

Then by first principle

$$\frac{dy}{dx} = \lim_{h\to0}\frac{f(x+h)-f(x)}{h}$$

$$= \lim_{h\to0}\frac{\sin (x+h)^2 - \sin x^2}{h}$$

$$= \lim_{h\to0}\frac{2\cos\frac{(x+h)^2+x^2}{2}.\sin\frac{(x+h)^2-x^2}{2}}{h}$$

$$= \lim_{h\to0}\frac{2\cos\frac{2hx+h^2+2x^2}{2}.\sin\frac{h^2+2hx}{2}}{h}$$

$$= \lim_{h\to0}\cos\left(x^2+\frac{h^2+2hx}{2}\right)\lim_{h\to0}\frac{\sin\left(\frac{h^2+2hx}{2}\right)}{\frac{h^2+2hx}{2}}\times(h+2x)$$

$$= \cos x^2 \times 1 \times 2x$$

$$= 2x\cos x^2$$

***Example* 8:** Differentiate sin 2x by first principle.

***Solution*:** Let $f(x) = \sin 2x$

by first principle,

$$\frac{dy}{dx} = \lim_{h\to0}\frac{f(x+h)-f(x)}{h}$$

$$= \lim_{h\to0}\frac{\sin 2(x+h) - \sin 2x}{h}$$

$$= \lim_{h\to0}\frac{2\cos\frac{2x+2h+2x}{2}\ \sin\frac{2x+2h-2x}{2}}{h}$$

$$= \lim_{h\to0}\frac{2\cos(2x+h)\ .\sin h}{h}$$

$$= 2 \lim_{h\to 0} \cos(2x+h) . \lim_{h\to 0} \frac{\sin h}{h}$$

$$= 2 \cos 2x \times 1$$

$$= 2 \cos 2x$$

Simple Differentiation

Example 9: Differentiate (i) $\log x^2 - 3e^x$ (ii) $\left(x+\frac{1}{x}\right)^2$ (iii) $\left(x-\frac{1}{x}\right)\left(x^2-\frac{1}{x^2}\right)$

Solution: (i) $$\frac{d}{dx}(\log x^2 - 3e^x) = \frac{d}{dx}(\log x^2) - \frac{d}{dx}(3e^x)$$

$$= \frac{d}{dx}(2 \log x) - 3\frac{d}{dx}e^x$$

$$= \frac{2}{x} - 3e^x$$

(ii) $$\frac{d}{dx}\left(x+\frac{1}{x}\right)^2 = \frac{d}{dx}\left(x^2+2+\frac{1}{x^2}\right)$$

$$= \frac{d}{dx}(x^2+2+x^{-2})$$

$$= 2x + 0 - 2x^{-2-1}$$

$$= 2x - 2x^{-3}$$

$$= 2x - \frac{2}{x^3}$$

(iii) $$\left(x-\frac{1}{x}\right)\left(x^2-\frac{1}{x^2}\right) = x^3 - \frac{1}{x} - x + \frac{1}{x^3}$$

$$= x^3 - x^{-1} - x + x^{-3}$$

$$= \frac{d}{dx}x^3 - \frac{d}{dx}x^{-1} - \frac{d}{dx}x + \frac{d}{dx}x^{-3}$$

$$= 3x^2 - (-1)x^{-1-1} - 1 - 3x^{-3-1}$$

$$= 3x^2 + x^{-2} - 1 - 3x^{-4}$$

$$= 3x^2 + \frac{1}{x^2} - 1 - \frac{3}{x^4}$$

***Example* 10:** If $y = \dfrac{2 - 3\cos x}{\sin x}$, find $\dfrac{dy}{dx}$ at $x = \dfrac{\pi}{4}$

***Solution*:** Let $y = \dfrac{2 - 3\cos x}{\sin x}$

$$= \frac{2}{\sin x} - 3\frac{\cos x}{\sin x}$$

$$y = 2\,\text{cosec}\,x - 3\cot x$$

differentiate w.r.t. x, we get

$$\frac{dy}{dx} = 2\frac{d}{dx}\,\text{cosec}\,x - 3\frac{d}{dx}\cot x$$

$$\frac{dy}{dx} = -2\,\text{cosec}\,x\cot x + 3\,\text{cosec}^2 x$$

Put $x = \dfrac{\pi}{4}$, we get $\dfrac{dy}{dx} = -2\,\text{cosec}\dfrac{\pi}{4}\cot\dfrac{\pi}{4} + 3\left(\text{cosec}\dfrac{\pi}{4}\right)^2$

$$= -2 \times \sqrt{2} \times 1 + 3\left(\sqrt{2}\right)^2$$

$$= -2\sqrt{2} + 3 \times 2$$

$$= 6 - 2\sqrt{2}$$

***Example* 11:** If $y = \sqrt{x} + \dfrac{1}{\sqrt{x}}$, show that $2x\dfrac{dy}{dx} + y = 2\sqrt{x}$

***Solution*:** Let $y = \sqrt{x} + \dfrac{1}{\sqrt{x}}$

$$y = x^{\frac{1}{2}} + x^{\frac{-1}{2}}$$

differentiate w.r.t. x, we get

$$\frac{dy}{dx} = \frac{1}{2}x^{\frac{1}{2}-1} - \frac{1}{2}x^{\frac{-1}{2}-1}$$

$$2\frac{dy}{dx} = x^{\frac{-1}{2}} - x^{\frac{-3}{2}}$$

$$= \frac{1}{\sqrt{x}} - \frac{1}{x\sqrt{x}}$$

multiply by x, we get

$$2x\frac{dy}{dx} = \frac{x}{\sqrt{x}} - \frac{1}{\sqrt{x}}$$

$$= \sqrt{x} - \frac{1}{\sqrt{x}}$$

$$= 2\sqrt{x} - \left(\sqrt{x} + \frac{1}{\sqrt{x}}\right)$$

$$= 2\sqrt{x} - y$$

$$2x\frac{dy}{dx} + y = 2\sqrt{x}$$

***Example* 12:** Find $\frac{dy}{dx}$ when $y = e^x \cos x$.

***Solution*:** Let $y = e^x \cos x$

differentiate w.r.t. x, we have

$$\frac{dy}{dx} = e^x \frac{d}{dx}\cos x + \cos x \frac{d}{dx} e^x$$

$$= e^x (-\sin x) + \cos x.\, e^x$$

$$= e^x (\cos x - \sin x)$$

***Example* 13:** Find $\frac{dy}{dx}$ f(x) for $f(x) = x^{-2} \sin 3x \,.\, \log x$

***Solution*:** Let $f(x) = x^{-2} \,.\, \sin 3x \,.\, \log x$

differentiating w.r.t. x, we get

$$\frac{dy}{dx} f(x) = x^{-2} \sin 3x \frac{d}{dx} \log x + x^{-2} \log x \frac{d}{dx} \sin 3x + \sin 3x.\, \log x \frac{d}{dx} x^{-2}$$

$$= x^{-2} \sin 3x \frac{1}{x} + x^{-2} \log x.\, 3 \cos 3x + \sin 3x \log x\, (-2)x^{-3}$$

$$= x^{-3} \sin 3x + 3x^{-2} \log x \cos 3x - 2x^{-3} \sin 3x \log x$$

$$= x^{-2} (x^{-1} \sin 3x + 3 \cos x.\, \log x - 2x^{-1} \sin 3x \log x)$$

***Example* 14:** Find $\frac{dy}{dx}$ if $y = (ax^2 + bx + c)(px + q)$

***Solution*:** Let $y = (ax^2 + bx + c)(px + q)$

$$\Rightarrow \quad \frac{dy}{dx} = (ax^2 + bx + c)\frac{dy}{dx}(px + q) + (px + q)\frac{dy}{dx}(ax^2 + bx + c)$$

$$= (ax^2 + bx + c)\,p + (px + q)(2ax + b)$$

$$= apx^2 + bpx + cp + 2apx^2 + pbx + 2aqx + bq$$

$$= 3apx^2 + 2pbx + 2aqx + cp + bq$$

***Example* 15:** Find $\frac{dy}{dx}$ if $y = x \log x \tan x$

***Solution*:** Let $y = x \log x. \tan x$

$$\Rightarrow \quad \frac{dy}{dx} = x \log x \frac{d}{dx} \tan x + x \tan x \frac{d}{dx} \log x + \log x \tan x \frac{d}{dx} x$$

$$= x \log x \,.\, \sec^2 x + x \tan x \,.\, \frac{1}{x} + \log x \tan x$$

$$= x \log x \,.\, \sec^2 x + \tan x + \log x. \tan x$$

***Example* 16:** Find the derivative of $x^3 e^x \log x$

***Solution*:** Let $y = x^3.e^x. \log x$

differentiating w.r.t. x, we get

$$\Rightarrow \quad \frac{dy}{dx} = x^3 e^x \frac{d}{dx} \log x + x^3 \log x \frac{d}{dx} e^x + e^x \log x \frac{d}{dx} x^3$$

$$= x^3 e^x \,.\, \frac{1}{x} + x^3 \log x.\, e^x + e^x \,.\, \log x.\, 3x^2$$

$$= x^2 e^x (1 + x \log x + 3 \log x)$$

***Example* 17:** Find $\frac{dy}{dx}$ when $y = \frac{x + \tan x}{x^2 - 1}$

***Solution*:** Let $y = \frac{x + \tan x}{x^2 - 1}$

$$\Rightarrow \quad \frac{dy}{dx} = \frac{(x^2-1)\frac{d}{dx}(x+\tan x) - (x+\tan x)\frac{d}{dx}(x^2-1)}{(x^2-1)^2}$$

$$= \frac{(x^2-1)(1+\sec^2 x) - (x+\tan x)\,2x}{(x^2-1)^2}$$

$$= \frac{(x^2-1)(\sec^2 x+1) - 2x^2 - 2x\tan x}{(x^2-1)^2}$$

***Example* 18:** Find $\frac{dy}{dx}$ when $y = \frac{ax^2+b}{\sin x + \cos x}$

***Solution*:** Let $y = \frac{ax^2+b}{\sin x + \cos x}$

$$\Rightarrow \quad \frac{dy}{dx} = \frac{(\sin x + \cos x)\frac{d}{dx}(ax^2+b) - (ax^2+b)\frac{d}{dx}(\sin x + \cos x)}{(\sin x + \cos x)^2}$$

$$= \frac{(\sin x + \cos x)2ax - (ax^2+b)(\cos x + \sin x)}{(\sin x + \cos x)^2}$$

***Example* 19:** Differentiate w.r.t x, $\frac{e^x + e^{-x}}{e^x - e^{-x}}$

***Solution*:** Let $y = \frac{e^x + e^{-x}}{e^x - e^{-x}}$

$$\Rightarrow \quad \frac{dy}{dx} = \frac{(e^x - e^{-x})\frac{d}{dx}(e^x + e^{-x}) - (e^x + e^{-x})\frac{d}{dx}(e^x - e^{-x})}{(e^x - e^{-x})^2}$$

$$= \frac{(e^x - e^{-x})(e^x - e^{-x}) - (e^x + e^{-x})(e^x + e^{-x})}{(e^x - e^{-x})^2}$$

$$= \frac{e^{2x} - 2e^x e^{-x} + e^{-2x} - e^{2x} - e^{-2x} - 2e^x e^{-x}}{(e^x - e^{-x})^2}$$

$$= \frac{-4}{(e^x - e^{-x})^2} \qquad (\because e^x . e^{-x} = 1)$$

***Example* 20:** Find $\frac{dy}{dx}$ if $y = \frac{e^x + \tan x}{\cot x - x^n}$

***Solution*:** Let $y = \frac{e^x + \tan x}{\cot x - x^n}$

$$\Rightarrow \frac{dy}{dx} = \frac{(\cot x - x^n)\frac{d}{dx}(e^x + \tan x) - (e^x + \tan x)\frac{d}{dx}(\cot x - x^n)}{(\cot x - x^n)^2}$$

$$= \frac{(\cot x - x^n)(e^x + \text{see}^2 x) - (e^x + \tan x)(-\text{cosec}^2 x - nx^{n-1})}{(\cot x - x^n)^2}$$

***Example* 21:** Find $\frac{dy}{dx}$ when $y = \frac{5x^2 + 6x + 7}{2x^2 + 3x + 4}$

***Solution*:** Let $y = \frac{5x^2 + 6x + 7}{2x^2 + 3x + 4}$

$\Rightarrow$

$$\frac{dy}{dx} = \frac{(2x^2 + 3x + 4)\frac{d}{dx}(5x^2 + 6x + 7) - (5x^2 + 6x + 7)\frac{d}{dx}(2x^2 + 3x + 4)}{(2x^2 + 3x + 4)^2}$$

$$= \frac{(2x^2 + 3x + 4)\ (10x + 6) - (5x^2 + 6x + 7)\ (4x + 3)}{(2x^2 + 3x + 4)^2}$$

$$= \frac{20x^3 + 12x^2 + 30x^2 + 18x + 40x + 24 - 20x^3 - 15x^2 - 24x^2 - 18x - 28x - 21}{(2x^2 + 3x + 4)^2}$$

$$= \frac{3x^2 + 12x + 3}{(2x^2 + 3x + 4)^2}$$

Function of a function

***Example* 22:** Find $\frac{dy}{dx}$, when $y = \sin e^x$

***Solution*:** Let $y = \sin e^x$

Put $e^x = t \Rightarrow y = \sin t$

$$\Rightarrow \quad \frac{dy}{dx} = \frac{dy}{dt} . \frac{dt}{dx}$$

$$= \frac{d}{dt} \sin t. \frac{d}{dx} e^x$$

$$= \cos t\, e^x$$

$$= e^x \cos e^x \qquad \text{Ans}$$

***Example* 23:** Find $\frac{dy}{dx}$, when $y = \cos\sqrt{x}.\ \log \sin x$

***Solution*:** Let $y = \cos\sqrt{x}.\ \log \sin x$

$$\Rightarrow \quad \frac{dy}{dx} = \cos\sqrt{x}. \frac{dy}{dx} \log \sin x + \log \sin x \frac{dy}{dx} \cos\sqrt{x}$$

Put $\sin x = t$, $\sqrt{x} = u$

$$= \cos\sqrt{x}. \frac{d}{dt} \log t . \frac{d}{dx} \sin x + \log \sin x. \frac{d}{du} \cos u. \frac{d}{dx}\sqrt{x},$$

$$= \cos\sqrt{x}. \frac{1}{t} \cos x + \log \sin x . (-\sin\sqrt{x}). \frac{1}{2\sqrt{x}}$$

$$= \frac{\cos x \cos \sqrt{x}}{\sin x} - \frac{\sin \sqrt{x} \log \sin x}{2 \sqrt{x}}$$

$$= \cos \sqrt{x} \cot x - \frac{\sin \sqrt{x} \log \sin x}{2\sqrt{x}}$$

***Example* 24:** Find $\frac{dy}{dx}$, when $y = \tan\left(\frac{1}{x}\right)$

***Solution*:** Let $y = \tan\left(\frac{1}{x}\right)$

Put $\frac{1}{x} = t \Rightarrow y = \tan t$

$$\Rightarrow \quad \frac{dy}{dx} = \frac{dy}{dt} . \frac{dt}{dx}$$

$$= \frac{d}{dt} \tan t \frac{d}{dx} . x^{-1}$$

$$= \sec^2 t. (-1)x^{-2} \qquad = -\frac{\sec^2\left(\frac{1}{x}\right)}{x^2}$$

***Example* 25:** Find $\frac{dy}{dx}$ when $y = e^{3x^2}$.

***Solution*:** Let $y = e^{3x^2}$

Put $3x^2 = t \Rightarrow y = e^t$

$$\Rightarrow \frac{dy}{dx} = \frac{dy}{dt} \cdot \frac{dt}{dx}$$

$$= \frac{d}{dt} e^t . \frac{d}{dx} 3x^2$$

$$= e^t . 6x$$

$$= 6x\ e^{3x^2}$$

***Example* 26:** If $y = \sin^2 (\log \sec x)$, find $\frac{dy}{dx}$.

***Solution*:** Let $y = \sin^2 (\log \sec x)$

Put $\sin (\log \sec x) = t$ and $\log \sec x = u$, $\sec x = v$

$$y = t^2$$

$$\Rightarrow \frac{dy}{dx} = \frac{dy}{dt} \cdot \frac{dt}{dx}$$

$$= \frac{d}{dt} t^2 . \frac{d}{dx} \sin (\log \sec x)$$

$$= 2t . \frac{d}{du} \sin u \frac{d}{dx} \log \sec x$$

$$= 2 \sin (\log \sec x) . \cos u . \frac{d}{dv} \log v \frac{d}{dx} \sec x$$

$$= 2 \sin (\log \sec x) . \cos (\log \sec x) . \frac{1}{v} \sec x \tan x$$

$$= \sin 2 (\log \sec x) . \frac{1}{\sec x} . \sec x \tan x$$

$$= \tan x \sin 2 (\log \sec x)$$

***Example* 27:** If $y = \sin^{-1} (x^2 + 2x)$, find $\frac{dy}{dx}$.

***Solution*:** Let $y = \sin^{-1} (x^2 + 2x)$

Put $x^2 + 2x = t \Rightarrow y = \sin^{-1} t$

$$\Rightarrow \frac{dy}{dx} = \frac{dy}{dt} \cdot \frac{dt}{dx}$$

$$= \frac{d}{dt}\sin^{-1} t \cdot \frac{d}{dx}(x^2 + 2x)$$

$$= \frac{1}{\sqrt{1+t^2}} \cdot (2x + 2)$$

$$= \frac{2x+2}{\sqrt{1-(x^2+2x)^2}} \quad \text{Ans}$$

***Example* 28:** If $y = \log \cos x$, find $\frac{dy}{dx}$.

***Solution*:** Let $y = \log \cos x$

Put $\cos x = t \Rightarrow y = \log t$

$$\Rightarrow \frac{dy}{dx} = \frac{dy}{dt} \cdot \frac{dt}{dx}$$

$$= \frac{d}{dt}\log t \cdot \frac{d}{dx}\cos x$$

$$= \frac{1}{t} \cdot (-\sin x)$$

$$= \frac{-\sin x}{\cos x} = -\tan x$$

***Example* 29:** Find $\frac{dy}{dx}$ when $y = \sec(3x + 2) + (3x+1)^{\frac{1}{3}}$

***Solution*:** Let $y = \sec(3x + 2) + (3x+1)^{\frac{1}{3}}$

Put $3x + 2 = t, \quad (3x + 1) = u$

$$\Rightarrow \frac{dy}{dx} = \frac{dy}{dt} \cdot \frac{dt}{dx} + \frac{dy}{du}\frac{du}{dx}$$

$$= \frac{d}{dt} \sec t \ \frac{d}{dx}(3x+2) + \frac{d}{du} u^{\frac{1}{3}} . \frac{d}{dx}(3x+1)$$

$$= \sec t. \tan t. 3 + \frac{1}{3} u^{\frac{1}{3}-1} . 3$$

$$= 3 \sec (3x+2) \tan (3x+2) + (3x+1)^{\frac{-2}{3}}$$

***Example* 30:** Find $\frac{dy}{dx}$ when $y = \sqrt{\sin x}$

***Solution*:** Let $y = \sqrt{\sin x}$

Put $\sin x = t, \quad \Rightarrow \quad y = \sqrt{t}$

$$\Rightarrow \quad \frac{dy}{dx} = \frac{dy}{dt} . \frac{dt}{dx}$$

$$= \frac{d}{dt} t^{\frac{1}{2}} . \frac{d}{dx} \sin x$$

$$= \frac{1}{2} t^{\frac{-1}{2}} . \cos x$$

$$= \frac{\cos x}{2\sqrt{\sin x}}$$

Implicit form

***Example* 31:** Find $\frac{dy}{dx}$ when $x^3 + y^3 + 3xy + 1 = 0$

***Solution*:** differentiating w.r.t. x, we get

$$\frac{dy}{dx} x^3 + \frac{d}{dx} y^3 + 3 \frac{d}{dx} xy + \frac{d}{dx} 1 = 0$$

$$3x^2 + 3y^2 \frac{dy}{dx} + 3\left(x\frac{dy}{dx} + y\right) = 0$$

$$x^2 + y^2 \frac{dy}{dx} + x \frac{dy}{dx} + y = 0$$

$$\frac{dy}{dx}(y^2 + x) = -(x^2 + y)$$

$$\frac{dy}{dx} = \frac{-(x^2 + y)}{(x + y^2)}$$

***Example* 32:** Find $\frac{dy}{dx}$ when (i) $\tan(x + y) + \tan(x - y) = 1$

(ii) $x = y \log(xy)$

(iii) $x^3 + y^3 = \sin(x + y)$

***Solution*:** (i) $\tan(x + y) + \tan(x - y) = 1$

differentiating w.r.t. x, we have

$$\sec^2(x + y)\left(1 + \frac{dy}{dx}\right) + \sec^2(x - y)\left(1 - \frac{dy}{dx}\right) = 0$$

$$\frac{dy}{dx}[\sec^2(x + y) - \sec^2(x - y)] + \sec^2(x + y) + \sec^2(x - y) = 0$$

$$\frac{dy}{dx} = -\left(\frac{\sec^2(x + y) + \sec^2(x - y)}{\sec^2(x + y) - \sec^2(x - y)}\right)$$

(ii) $x = y \log(xy)$

differentiating w.r.t. x, we get

$$1 = y.\frac{1}{xy}\left(y + x\frac{dy}{dx}\right) + \log(xy)\frac{dy}{dx}$$

$$1 = \frac{y}{x} + \frac{dy}{dx} + \log(xy)\frac{dy}{dx}$$

$$\frac{dy}{dx}(1 + \log xy) = 1 - \frac{y}{x}$$

$$\frac{dy}{dx} = \frac{x - y}{x(1 + \log xy)}$$

(iii) $x^3 + y^3 = \sin(x + y)$

differentiating w.r.t. x, we get

$$3x^2 + 3y^2 \frac{dy}{dx} = \cos(x+y)\left(1 + \frac{dy}{dx}\right)$$

$$3x^2 - \cos(x+y) = \frac{dy}{dx}(\cos(x+y) - 3y^2)$$

$$\frac{dy}{dx} = \frac{3x^2 - \cos(x+y)}{\cos(x+y) - 3y^2}$$

***Example* 33:** If $e^x + e^y = e^{x+y}$ then show that $\frac{dy}{dx} = -e^{y-x}$

***Solution*:** Let $e^x + e^y = e^{x+y}$

differentiating w.r.t. x, we have

$$e^x + e^y \frac{dy}{dx} = e^{x+y}\left(1 + \frac{dy}{dx}\right)$$

$$e^x - e^{x+y} = (e^{x+y} - e^y)\frac{dy}{dx}$$

$$e^x - e^x - e^y = (e^x + e^y - e^y)\frac{dy}{dx}$$

$$-e^y = e^x \frac{dy}{dx}$$

$$\frac{dy}{dx} = -\frac{e^y}{e^x}$$

$$= -e^y \cdot e^{-x}$$

$$\frac{dy}{dx} = -e^{y-x}$$

***Example* 34:** If $x\sqrt{1+y} + y\sqrt{1+x} = 0$, prove that $(1+x)^2 \frac{dy}{dx} + 1 = 0$

***Solution*:**

$$x\sqrt{1+y} + y\sqrt{1+x} = 0$$

$$\Rightarrow \quad x\sqrt{1+y} = -y\sqrt{1+x}$$

Squaring on both the sides, we get

$$x^2 (1 + y) = y^2 (1 + x)$$

$$x^2 - y^2 + x^2y - y^2x = 0$$

$$(x - y)(x + y) + xy(x - y) = 0$$

$$(x - y)(x + y + xy) = 0$$

$$\Rightarrow \quad x = y \quad \text{or} \quad x + y + xy = 0$$

But x = y does not satisfy the equation

Hence $\quad x + y + xy = 0$

$$y(1 + x) = -x$$

$$y = \frac{-x}{1+x}$$

differentiating w.r.t. x, we get

$$\frac{dy}{dx} = \frac{-[(1+x)\frac{d}{dx}x - x\frac{d}{dx}(1+x)]}{(1+x)^2}$$

$$= \frac{-(1+x)+x}{(1+x)^2}$$

$$\frac{dy}{dx} = \frac{-1}{(1+x)^2}$$

$$(1 + x)^2 \frac{dy}{dx} + 1 = 0 \qquad \text{proved}$$

Parametric Equations

***Example* 35:** Find $\frac{dy}{dx}$ when x = 2 cos 3t, y = 2 sin 3t

***Solution*:** Since $\quad x = 2 \cos 3t, \qquad y = 2 \sin 3t$

$$\frac{dx}{dt} = -2.3 \sin 3t \qquad \frac{dy}{dt} = 2.3 \cos 3t$$

$$= -6 \sin 3t \qquad = 6 \cos 3t$$

$$\therefore \quad \frac{dy}{dx} = \frac{\frac{dy}{dt}}{\frac{dx}{dt}} = \frac{6 \cos 3t}{-6 \sin 3t} = -\cot 3t$$

***Example* 36:** If $x = a(\theta - \sin\theta)$, $y = a(1 - \cos\theta)$, find $\frac{dy}{dx}$

***Solution*:** We have $x = a(\theta - \sin\theta)$, $y = a(1 - \cos\theta)$

$$\frac{dx}{d\theta} = a(1 - \cos\theta) \qquad \frac{dy}{d\theta} = a\sin\theta$$

$$\therefore \quad \frac{dy}{dx} = \frac{\frac{dy}{d\theta}}{\frac{dx}{d\theta}} = \frac{a\sin\theta}{a(1-\cos\theta)} = \frac{\sin\theta}{1-\cos\theta}$$

***Example* 37:** Find $\frac{dy}{dx}$ at $t = \frac{\pi}{2}$ when $x = 2\cos t - 2\cos 2t$ and $y = 2\sin t - \sin 2t$

***Solution*:** We have $x = 2\cos t - \cos 2t$, $y = 2\sin t - \sin 2t$

$$\frac{dx}{dt} = -2\sin t + 2\sin 2t \qquad \frac{dy}{dt} = 2\cos t - 2\cos 2t$$

$$\therefore \quad \frac{dy}{dx} = \frac{\frac{dy}{dt}}{\frac{dx}{dt}} = \frac{2(\cos t - \cos 2t)}{2(\sin 2t - \sin t)}$$

at $t = \frac{\pi}{2}$, $\frac{dy}{dx} = \frac{\cos\frac{\pi}{2} - \cos 2\times\frac{\pi}{2}}{\sin 2\times\frac{\pi}{2} - \sin\frac{\pi}{2}}$

$$= \frac{0-(-1)}{0-1} = -1$$

***Example* 38:** Find $\frac{dy}{dx}$, when $x = a\left(\cos t + \log\tan\frac{t}{2}\right)$, $y = a\sin t$

***Solution*:** When $x = a\left(\cos t + \log\tan\frac{t}{2}\right)$, $y = a\sin t$

$$\therefore \quad \frac{dx}{dt} = a\left(-\sin t + \frac{1}{\tan\frac{t}{2}}.\sec^2\frac{t}{2}.\frac{1}{2}\right)$$

$$= a\left(-\sin t + \frac{1}{2\sin\frac{t}{2}\cos\frac{t}{2}}\right)$$

$$= a\left(-\sin t + \frac{1}{\sin t}\right)$$

$$\frac{dy}{dt} = a\cos t$$

$$\therefore \quad \frac{dy}{dx} = \frac{\frac{dy}{dt}}{\frac{dx}{dt}} = \left(\frac{a\cos t}{a\left(\frac{-\sin^2 t + 1}{\sin t}\right)}\right)$$

$$= \frac{\cos t\ \sin t}{\cos^2 t} = \tan t$$

***Example* 39:** Find $\frac{dy}{dx}$, when $x = a\cos\theta$ and $y = b\sin\theta$

***Solution*:** When $x = a\cos\theta$, $\quad y = b\sin\theta$

$$\frac{dx}{d\theta} = -a\sin\theta, \qquad \frac{dy}{d\theta} = b\cos\theta$$

$$\frac{dy}{dx} = \frac{\frac{dy}{d\theta}}{\frac{dx}{d\theta}} = \left(\frac{b\cos\theta}{-a\sin\theta}\right)$$

$$= \frac{-b}{a}\cot\theta$$

***Example* 40:** Find $\frac{dy}{dx}$ when $x = \log t + \sin t$, $y = e^t + \cos t$

***Solution*:** When $x = \log t + \sin t$, $\quad y = e^t + \cos t$

$$\frac{dx}{dt} = \frac{1}{t} + \cos t, \qquad \frac{dy}{dt} = e^t - \sin t$$

$$= \frac{1 + t\cos t}{t}$$

$$\therefore \quad \frac{dy}{dx} = \frac{\frac{dy}{dt}}{\frac{dx}{dt}} = \left(\frac{e^t - \sin t}{\frac{1 + t\cos t}{t}}\right)$$

$$= \frac{t(e^t - \sin t)}{1 + t\cos t}$$

Logarithm Differentiation

***Example* 41:** Find the differential coefficient of $(\sin x)^{\log x}$ w.r.t. x.

***Solution*:** Let $y = (\sin x)^{\log x}$

Taking log on both the sides

$$\log y = \log x . \log \sin x$$

differentiating w.r.t. x, we get

$$\frac{d}{dx} \log y = \log x \frac{d}{dx} \log \sin x + \log \sin x \frac{d}{dx} \log x$$

$$\frac{1}{y}\frac{dy}{dx} = \log x . \frac{1}{\sin x} \cos x + \log \sin x . \frac{1}{x}$$

$$= \log x \ \cot x + \frac{\log \sin x}{x}$$

$$\therefore \quad \frac{dy}{dx} = y\left[\log x \ \cot x + \frac{\log \sin x}{x}\right]$$

$$= (\sin x)^{\log x}\left[\log x \ \cot x + \frac{\log \sin x}{x}\right]$$

***Example* 41:** Differentiate w.r.t. x the function $x^{\log x} + (\sin x)^x$.

***Solution*:** Let $y = x^{\log x} + (\sin x)^x$

$y = u + v$ where $u = x^{\log x}$, $v = (\sin x)^x$

$$\frac{dy}{dx} = \frac{du}{dx} + \frac{dv}{dx} \quad \text{.....(i)}$$

Now $u = x^{\log x}$

Taking log on both sides, $\log u = \log x . \log x$

$$\Rightarrow \quad \frac{1}{u}\frac{du}{dx} = \log x \times \frac{1}{x} + \log x \times \frac{1}{x}$$

$$= \frac{2 \log x}{x}$$

$$\frac{du}{dx} = u\left(\frac{2 \log x}{x}\right)$$

$$= 2\, x^{\log x} . \frac{\log x}{x}$$

Now, $v = (\sin x)^x$

Taking log on both sides, $\log v = x \log \sin x$

$$\Rightarrow \quad \frac{1}{v}\frac{dv}{dx} = x \frac{1}{\sin x} \cos x + \log \sin x$$

$$= x \cot x + \log \sin x$$

$$\frac{dv}{dx} = v\,[x \cot x + \log \sin x]$$

$$= (\sin x)^x\ [x \cot x + \log \sin x]$$

Substituting the values of $\frac{du}{dx}$ and $\frac{dv}{dx}$ in (i), we get

$$\frac{dy}{dx} = 2\, x^{\log x} . \frac{\log x}{x} + (\sin x)^x\,[x \cot x + \log \sin x]$$

***Example* 43:** If $x^y = e^{x-y}$, show that $\frac{dy}{dx} = \frac{\log x}{(1+\log x)^2}$

***Solution*:** $x^y = e^{x-y}$

Taking log on both sides, $\log y = \log x = (x - y) \log_e$

$$y \log x = (x - y) \log e$$

$$y \log x = x - y \qquad \text{.....(i)}$$

($\because \log e = 1$)

$$\Rightarrow \quad y . \frac{1}{x} + \log x \frac{dy}{dx} = 1 - \frac{dy}{dx}$$

$$\Rightarrow \quad \frac{dy}{dx}(1 + \log x) = 1 - \frac{y}{x}$$

$$\frac{dy}{dx} = \frac{x - y}{x(1 + \log x)} \qquad \text{.....(ii)}$$

Now from eq. (i)

$$y \log x + y = x$$

$$y(\log x + 1) = x$$

$$y = \frac{x}{1 + \log x}$$

Putting the value of y in (ii), we get

$$\frac{dy}{dx} = \frac{x - \dfrac{x}{1 + \log x}}{x(1 + \log x)}$$

$$= \frac{x + x \log x - x}{x(1 + \log x)^2}$$

$$\frac{dy}{dx} = \frac{\log x}{(1 + \log x)^2}$$

***Example* 44:** Find $\frac{dy}{dx}$ when $y = (\sin x)^{\cos x} + (\cos x)^{\sin x}$

***Solution*:**

$$y = (\sin x)^{\cos x} + (\cos x)^{\sin x}$$

$$y = u + v \qquad \text{where } u = (\sin x)^{\cos x},\ v = (\cos x)^{\sin x}$$

$$\Rightarrow \quad \frac{dy}{dx} = \frac{du}{dx} + \frac{dv}{dx} \qquad \text{.....(i)}$$

Now $u = (\sin x)^{\cos x}$

Taking log on both sides, $\log u = \cos x . \log (\sin x)$

$$\Rightarrow \quad \frac{1}{u}\frac{du}{dx} = \cos x . \frac{1}{\sin x} \cos x + \log(\sin x)(-\sin x)$$

$$= \cot x . \cos x - \sin x \log \sin x$$

$$\frac{du}{dx} = u\,[\cot x.\ \cos x - \sin x \log \sin x]$$

$$= (\sin x)^{\cos x}\,[\cot x \cos x - \sin x \log \sin x]$$

again $\quad v = (\cos x)^{\sin x}$

Taking log on both sides,

$$\log v = \sin x \log (\cos x)$$

$$\Rightarrow \quad \frac{1}{v}\frac{dv}{dx} = \sin x \frac{1}{\cos x}(-\sin x) + \log(\cos x) \,.\, \cos x$$

$$= -\tan x \,.\, \sin x + \cos x \log(\cos x)$$

$$\frac{dv}{dx} = v\,[\cos x.\ \log(\cos x) - \tan x \sin x]$$

$$= (\cos x)^{\sin x}\,[\cos x \log(\cos x) - \tan x \sin x]$$

Substituting values of $\frac{du}{dx}$ and $\frac{dv}{dx}$ in eq. (i) we get

$$\frac{dy}{dx} = (\sin x)^{\cos x}\,[\cot x \cos x - \sin x \log(\sin x)] + (\cos x)^{\sin x}$$

$$[\cos x \log(\cos x) - \tan x \sin x]$$

***Example* 45:** If $y = (\sin x)^{(\sin x)^{\sin x \ldots \infty}}$ then prove that $\frac{dy}{dx} = \frac{y^2 \cot x}{1 - y \log \sin x}$

***Solution*:** we can write $y = (\sin x)^y$

Taking log on both sides, $\log y = y \log \sin x$

$$\Rightarrow \quad \frac{1}{y}\frac{dy}{dx} = y.\frac{1}{\sin x}\cos x + \log \sin x \frac{dy}{dx}$$

$$\Rightarrow \quad \left(\frac{1}{y} - \log \sin x\right)\frac{dy}{dx} = y \cot x$$

$$\left(\frac{1 - y \log \sin x}{y}\right)\frac{dy}{dx} = y \cot x$$

$$\frac{dy}{dx} = \frac{y^2 \cot x}{1 - y \log \sin x}$$

Example **46:** If $y = \sqrt{\tan x + \sqrt{\tan x + \sqrt{\tan x + \ldots \infty}}}$ then prove that $\frac{dy}{dx} = \frac{\sec^2 x}{2y - 1}$

***Solution*:** we can write $y = \sqrt{\tan x + y}$ .....(i)

Taking log on both sides, $\log y = \frac{1}{2} \log (\tan x + y)$

$$\Rightarrow \quad \frac{1}{y}\frac{dy}{dx} = \frac{1}{2} \cdot \frac{1}{\tan x + y}\left(\sec^2 x + \frac{dy}{dx}\right)$$

From eq. (i) $\tan x + y = y^2$

$$\Rightarrow \quad \frac{1}{y}\frac{dy}{dx} = \frac{1}{2y^2}\left(\sec^2 x + \frac{dy}{dx}\right)$$

$$\Rightarrow \quad \left(\frac{1}{y} - \frac{1}{2y^2}\right)\frac{dy}{dx} = \frac{\sec^2 x}{2y^2}$$

$$\Rightarrow \quad \left(\frac{2y-1}{2y^2}\right)\frac{dy}{dx} = \frac{\sec^2 x}{2y^2}$$

$$\Rightarrow \quad \frac{dy}{dx} = \frac{\sec^2 x}{2y - 1}$$

Function w.r.t a function

Example **47:** Find the derivative of the following

(i) $\sin x^2$ w.r.t. x^2 (ii) $\sin (\cos x)$ w.r.t. $\cos x$

***Solution*:** (i) Let $y = \sin x^2$, $u = x^2$

$$\frac{dy}{dx} = \cos x^2 \,.\, 2x \qquad \frac{du}{dx} = 2x$$

$$\therefore \quad \frac{dy}{du} = \frac{\frac{dy}{dx}}{\frac{du}{dx}} = \frac{2x \cos x^2}{2x}$$

$$= \cos x^2$$

(ii) Let $y = \sin(\cos x)$ $\qquad$ $u = \cos x$

$$\frac{dy}{dx} = \cos(\cos x)(-\sin x) \qquad \frac{du}{dx} = -\sin x$$

$$\Rightarrow \quad \frac{dy}{du} = \frac{\frac{dy}{dx}}{\frac{du}{dx}} = \frac{-\sin x.\cos(\cos x)}{-\sin x}$$

$$= \cos(\cos x)$$

***Example* 48:** Differentiate $\tan^{-1}\frac{2x}{1-x^2}$ w.r.t. $\sin^{-1}\frac{2x}{1+x^2}$

***Solution*:** Let $y = \tan^{-1}\frac{2x}{1-x^2}$ $\qquad$ $u = \sin^{-1}\frac{2x}{1+x^2}$

Put $x = \tan\theta$ $\qquad$ $u = \sin^{-1}\left(\frac{2\tan\theta}{1+\tan^2\theta}\right)$

$$\Rightarrow \quad \theta = \text{Tan}^{-1} x$$

$$y = \tan^{-1}\left(\frac{2\tan\theta}{1+\tan^2\theta}\right)$$

$$= \tan^{-1}(\tan 2\theta) \qquad = \sin^{-1}(\sin 2\theta)$$

$$= 2\theta \qquad = 2\theta$$

$$y = 2\tan^{-1} x \qquad u = 2\tan^{-1} x$$

$$\frac{dy}{dx} = \frac{2}{1+x^2} \qquad \frac{du}{dx} = \frac{2}{1+x^2}$$

$$\therefore \quad \frac{dy}{du} = \frac{\frac{dy}{dx}}{\frac{du}{dx}} = \frac{\frac{2}{(1+x^2)}}{\frac{2}{(1+x^2)}} = 1$$

Successive Differentiation

***Example* 49:** If $y = ae^{mx} + be^{-mx}$, show that $\frac{d^2y}{dx^2} = m^2y$

***Solution*:** Let $y = ae^{mx} + be^{-mx}$

differentiating w.r.t. x, we get

$$\frac{dy}{dx} = am\, e^{mx} - bm\, e^{-mx}$$

differentiating again w.r.t. x, we have

$$\frac{d^2y}{dx^2} = am^2\, e^{mx} + bm^2\, e^{-mx}$$

$$= m^2\,(ae^{mx} + be^{-mx})$$

$$\frac{d^2y}{dx^2} = m^2y$$

***Example* 50:** If y = A cos nx + B sin nx, show that $\frac{d^2y}{dx^2} + n^2y = 0$

***Solution*:** Let y = A cos nx + B sin nx

differentiating w.r.t. x

$$\frac{dy}{dx} = -\,An \sin nx + Bn \cos nx$$

differentiating again w.r.t. x

$$\frac{d^2y}{dx^2} = -An^2 \cos nx - Bn^2 \sin nx$$

$$= -n^2\,(A \cos nx + B \sin nx)$$

$$= -n^2\,y$$

$$\frac{d^2y}{dx^2} + n^2y = 0$$

***Example* 51:** Find $\frac{d^2y}{dx^2}$ when y = log x

***Solution*:** Let y = log x

differentiating w.r.t. x $\frac{dy}{dx} = \frac{1}{x}$

differentiating again w.r.t. x $\frac{d^2y}{dx^2} = -x^{-2}$

$$= \frac{-1}{x^2}$$

***Example* 52:** Find $\frac{d^2y}{dx^2}$ when y = sin x + tan 2x

***Solution*:** Let y = sin x + tan 2x

differentiating w.r.t. x, we get, $\frac{dy}{dx} = \cos x + 2\sec^2 2x$

differentiating again w.r.t. x, $\frac{d^2y}{dx^2} = -\sin x + 2.2 \sec 2x . \sec 2x \tan 2x . 2$

$= -\sin x + 8\sec^2 2x . \tan 2x$

***Example* 53:** When y = a sin x + b cos x, find $\frac{d^2y}{dx^2}$

***Solution*:** Let y = a sin x + b cos x

differentiating w.r.t. x, $\frac{dy}{dx} = a\cos x - b\sin x$

differentiating again w.r.t. x, $\frac{d^2y}{dx^2} = -a\sin x - b\cos x$

$= -(a\sin x + b\cos x)$

$= -y$

$$\frac{d^2y}{dx^2} + y = 0$$

***Example* 54:** If y = A cos (log x) + B sin (log x) prove that $x^2\frac{d^2y}{dx^2} + x\frac{dy}{dx} + y = 0$

***Solution*:** Let y = A cos (log x) + B sin (log x)

differentiating w.r.t. x, $\frac{dy}{dx} = -A\sin(\log x).\frac{1}{x} + B\cos(\log x).\frac{1}{x}$

$x\frac{dy}{dx} = -A\sin(\log x) + B\cos(\log x)$

differentiating again w.r.t. x, we have

$$x\frac{d^2y}{dx^2} + \frac{dy}{dx} = -A\cos(\log x).\frac{1}{x} - B\sin(\log x).\frac{1}{x}$$

$$x^2 \frac{d^2y}{dx^2} + x \frac{dy}{dx} = -[A \cos(\log x) + B \sin(\log x)]$$

$$= -y$$

$$\therefore \quad x^2 \frac{d^2y}{dx^2} + x \frac{dy}{dx} + y = 0$$

***Example* 55:** If $y = Ae^{px} + Be^{qx}$ show that $\frac{d^2y}{dx^2} - (p+q)\frac{dy}{dx} + pqy = 0$

***Solution*:** Let $y = Ae^{px} + Be^{qx}$

differentiating w.r.t. x,

$$\frac{dy}{dx} = Ape^{px} + Bqe^{qx} \quad .....(i)$$

differentiating again w.r.t. x, we have

$$\frac{d^2y}{dx^2} = Ap^2 e^{px} + Bq^2 e^{qx} \quad .....(i)$$

Now $\frac{d^2y}{dx^2} - (p+q)\frac{dy}{dx} + pqy$

$$= Ap^2 e^{px} + Bq^2 e^{qx} - (p+q)(Ape^{px} + Bqe^{qx}) + pq(Ae^{px} + Be^{qx})$$

$$= Ap^2 e^{px} + Bq^2 e^{qx} - Ap^2 e^{px} - Bq^2 e^{qx} - Apqe^{pq} - Bpqe^{qx} + pqAe^{px} + pqBe^{qx}$$

$$= 0$$

***Example* 56:** If $y = \tan^{-1} x$, show that $(1+x^2)\frac{d^2y}{dx^2} + 2x\frac{dy}{dx} = 0$

***Solution*:** Let $y = \tan^{-1} x$

differentiating w.r.t. x, we get

$$\frac{dy}{dx} = \frac{1}{1+x^2}$$

$$\Rightarrow \quad (1+x^2)\frac{dy}{dx} = 1$$

differentiating again w.r.t. x,

$$(1+x^2)\frac{d^2y}{dx^2} + 2x\frac{dy}{dx} = 0$$

Exercises

Find the differential coefficient of the following

1. $\sqrt{x^{-3}}$, $9x^{-8}$
2. $5x^3 - 4x^2 + 3x + 1$
3. $\dfrac{(x-4)(x+2)}{x}$
4. $\dfrac{(3x-1)(4x^{\frac{1}{2}}+5)}{x^{\frac{1}{2}}}$
5. $\dfrac{2x^3+6x^2+7}{\sqrt{x}}$
6. $x^5 - 2\log x$
7. $e^x + x^n$
8. $(x\sin x + 7x^2 - 3) \cdot \dfrac{1}{x}$
9. $\dfrac{a+b\sqrt{x}}{\sqrt[3]{x}}$
10. $\dfrac{x e^x - 1}{x}$
11. $\dfrac{x^2\log x - x + x^2 . e^x}{x^2}$
12. $3e^x + 7\log x + x^{-3}$
13. $e^{ax}\sin bx$
14. $3\log x \sin x$
15. $2\sqrt{x}\log x$
16. $ae^x \sin x + bx^n \cos x$
17. $(x^2+1)(e^x + 2\sin x)$
18. $\sqrt{x}(1-x)(x^2+2)$
19. $(1 - 2\tan x)(5 + 4\sin x)$
20. $\dfrac{2x^3+6x+9}{x^2+5}$
21. $\dfrac{1-\cos x}{1+\cos x}$
22. $\dfrac{x^n}{\log_a x}$
23. $\dfrac{\sin x - x\cos x}{x\sin x + \cos x}$
24. $\dfrac{x . \cos x}{\log x}$
25. $\dfrac{1-\tan x}{1+\tan x}$
26. $\dfrac{\sqrt{x}+\sqrt{a}}{\sqrt{a}-\sqrt{x}}$
27. $\dfrac{e^x + \tan x}{\cot x - x^n}$
28. $\dfrac{e^x + 5x^3\log x}{\log x}$

29. $\dfrac{1}{(x+a)(x+b)(x+c)}$

30. $\dfrac{1+\log x}{1-\log x}$

31. $\left(\sqrt{x}+\dfrac{1}{\sqrt{x}}\right)^2$

32. $(2x-3)\sqrt{7+x^2}$

33. $\sqrt{\dfrac{1-x}{1+x}}$

34. $\dfrac{1}{x+\sqrt{1+x^2}}$

35. $\dfrac{a^2+x^2}{\sqrt{a^2-x^2}}$

36. $\tan x^3$

37. $\log \tan x$

38. $\sin(\tan x)$

39. $\sqrt{\log x}$

40. e^{ax+b}

41. $\log\sqrt{\sin e^x}$

42. $\log\left(\dfrac{ax+b}{px+q}\right)$

43. $\dfrac{\sqrt{\sin x}}{\sin\sqrt{x}}$

44. $\cot(x^2 \sin 2x)$

45. $\sqrt{\dfrac{1+\sin x}{1-\sin x}}$

46. $\log\sqrt{\dfrac{1+\sin x}{1-\sin x}}$

47. $\log\dfrac{x+\sqrt{x^2-1}}{x-\sqrt{x^2-1}}$

48. $\log\left(\sqrt{x+1}+\sqrt{x-1}\right)$

49. $\sqrt{\operatorname{cosec} x \ \cot x}$

50. $\log\log\log x$

51. $\dfrac{e^{\sin x}}{\sin x^n}$

52. $\sqrt{\sec\sqrt{x}}$

53. If $y=\tan^{-1}\dfrac{\sqrt{1-x^2}}{x}$, find $\dfrac{dy}{dx}$

54. Differentiate $\sin^{-1}(3x-4x^3)$

55. If $y=\tan^{-1}\sqrt{\dfrac{1-\cos x}{1+\cos x}}$, find $\dfrac{dy}{dx}$ differentiate the following:

56. $\sin^{-1}\dfrac{2x}{1+x^2}$

57. $\tan^{-1}\dfrac{x}{\sqrt{1+x^2}}$

58. $\sin^{-1}\left(\dfrac{1-x^2}{1+x^2}\right)$

59. $\cos^{-1}\left(\dfrac{1-x^2}{1+x^2}\right)$

60. $2\tan^{-1}\sqrt{\dfrac{1-x}{1+x}}$

61. $\sin^{-1}\left(2x\sqrt{1-x^2}\right)$

62. $\cos^{-1}(4x^3-3x^2)$

63. $\tan^{-1}\left(\dfrac{a+x}{1-ax}\right)$

64. $\cos^{-1}\left(\dfrac{x-x^{-1}}{x+x^{-1}}\right)$

65. $\sin^{-1}\left(x\sqrt{1-x}-\sqrt{x}\sqrt{1-x^2}\right)$

66. If $y = 1+\dfrac{x}{1!}+\dfrac{x^2}{2!}+\dfrac{x^3}{3!}+\ldots$ Then show that $\dfrac{dy}{dx}=y$

67. $\dfrac{3^x}{x+\tan x}$

68. $\sin x^o$ $\left(\text{Hint}: x^o = \dfrac{\pi}{180}\right)$

Differentiate by first principle:

69. x^7

70. $\cos(ax+b)$

71. $\tan ax$

72. $\cos^2 x$

73. $\cos^{-1} x$

74. e^{3x}

75. $\sin x^2$

76. $\cos\sqrt{x}$

77. $\tan 3x$

78. $x^{\frac{-3}{2}}$

79. If $\sin y = x\sin(a+y)$, prove that $\dfrac{dy}{dx}=\dfrac{\sin^2(a+y)}{\sin a}$

80. If $\dfrac{1}{2}(e^y - e^{-y}) = x$, prove that $\dfrac{dy}{dx}=\dfrac{1}{\sqrt{1+x^2}}$

81. If $y=\sqrt{\sin x+\sqrt{\sin x+\sqrt{\sin x+\ldots\infty}}}$ prove that $(2y-1)\dfrac{dy}{dx}=\cos x$.

82. If $y\sqrt{1-x^2}+x\sqrt{1-y^2}=1$, prove that $\dfrac{dy}{dx}+\sqrt{\dfrac{1-y^2}{1+x^2}}=0$

Find $\dfrac{dy}{dx}$ by implicit differentiation when.

83. $xy = c^2$

84. $\frac{x^2}{a^2} + \frac{y^2}{b^2} = 1$

85. $x^{\frac{2}{3}} + y^{\frac{2}{3}} = a^{\frac{2}{3}}$

86. $xy^3 - yx^3 = x$

87. $y \sec x + \tan x + x^2 y = 0$

88. $\tan(x + y) + \tan(x - y) = 1$

89. If $(x + y)^{m+n} = x^m y^n$, prove that $\frac{dy}{dx} = \frac{y}{x}$

90. Find the value of $\frac{dy}{dx}$ at (2, –2) for the curve $x^2 + xy + y^2 = 4$

91. If $y = \sqrt{\log x \sqrt{\log x + \sqrt{\log x + \ldots\infty}}}$ show that $(2y - 1)\frac{dy}{dx} = \frac{1}{x}$

92. If $y = a^{x+y}$, show that $\frac{dy}{dx} = \frac{y \log a}{1 - y \log a}$

93. If $\cos(xy) = x$, show that $\frac{dy}{dx} + \frac{1 + y\sin(xy)}{x\sin(xy)} = 0$

94. If $x = y \log(xy)$, find $\frac{dy}{dx}$

95. If $x^3 + y^3 = \sin(x + y)$, find $\frac{dy}{dx}$

96. If $x^3 y^2 = \log(x + y) + \sin e^x$, find $\frac{dy}{dx}$

97. If $x^a . y^b = (x + y)^{a+b}$, prove that $\frac{dy}{dx} = \frac{y}{x}$ provided that $ay \neq bx$

Find $\frac{dy}{dx}$ of the following.

98. $(\log x)^x$

99. $(2x + 3)^{x-5}$

100. $(\cos x)^{\log x}$

101. e^{x^x}

102. $(2 - x)\left(\frac{3 - x}{1 + x}\right)^{\frac{1}{2}}$

103. $x^2 e^x \sin x$

104. $x \sin^{-1} x + (\log x)^x$

105. $(\tan x)^{\cot x} + (\cot x)^{\sin x}$

106. $\frac{1}{(x + a)(x + b)(x + c)}$

107. $\tan x \tan 2x \tan 3x \tan 4x$

108. x^{x^x}

109. $y = x^{1+x+x^2}$

110. $\dfrac{(x+1)(x+2)(x+3)}{(x-1)(x-2)(x-3)}$

111. $\dfrac{(x+2)^{\frac{5}{2}}}{(x+4)^{\frac{1}{3}}(x+1)^{\frac{3}{2}}}$ Find $\dfrac{dy}{dx}$ for the following

112. $x = a\sec^2\theta,\ y = b\tan^2\theta$

113. $x = a\sin^3\theta,\ y = a\cos^3\theta$

114. $x = \sqrt{\sin 2\theta},\ y = \sqrt{\cos 2\theta}$

115. $x = at^4,\ y = bt^3$

116. $x = \dfrac{3\,at}{1+t^3},\ y = \dfrac{3\,at^2}{1+t^3}$

117. $x = \sin^{-1}\dfrac{2u}{1+u^2},\ y = \tan^{-1}\dfrac{2u}{1-u^2}$

118. $x = \log t,\ y = e^t + \cos t$ Find the derivative of the following

119. $\sin x^2$ w.r.t. x^2

120. $x^{\sin x}$ w.r.t. $(\sin x)^x$

121. $e^{\tan x}$ w.r.t. $\tan x$

122. $\tan^{-1}\dfrac{\sqrt{1+x^2}-1}{x}$ w.r.t. $\tan^{-1}x$

123. If $x = e^{\cos 2t}$ and $y = e^{\sin 2t}$, prove that $\dfrac{dy}{dx} = \dfrac{-y\log x}{x\log y}$

124. If $x = ae^t(\sin t + \cos t),\ y = ae^t(\sin t - \cos t)$ prove that $\dfrac{dy}{dx} = \tan t$

125. If $x = a\left(t+\dfrac{1}{t}\right)$ and $y = a\left(t-\dfrac{1}{t}\right)$ prove that $\dfrac{dy}{dx} = \dfrac{x}{y}$.

126. Find second derivative of $ax^3 + bx^2 + cx + d$

127. If $y = a\cos px + b\sin px$, prove that $\dfrac{d^2y}{dx^2} + p^2y = 0$

128. If $y = \log(1 + \cos x)$ prove that $y_1 y_2 + y_3 = 0$

$$\left[\text{Hint: } y_1 = \frac{dy}{dx},\ y_2 = \frac{d^2y}{dx^2},\ y_3 = \frac{d^3y}{dx^3}\right]$$

129. If $y = \tan x + \sec x$, prove that $\dfrac{d^2y}{dx^2} = \dfrac{\cos x}{(1-\sin x)^2}$

130. If $y = x^2 \log x$, prove that $\frac{d^3y}{dx^3} = \frac{2}{x}$

131. If $y = e^{ax} \sin bx$, prove that $y_2 - 2ay_1 + (a^2 + b^2)y = 0$

132. If $y = e^{m\cos^{-1}x}$, prove that $(1-x^2)\, y_2 - xy_1 - m^2y = 0$

133. If $y = \sin(\sin x)$, prove that $y_2 + y_1 \tan x + y \cos^2 x = 0$

134. If $y = t^2 - 1$, $x = 2t$ prove that $\frac{d^2y}{dx^2} = \frac{1}{2}$

135. If $y = a \cos(\log x) + b \sin(\log x)$ prove that $x^2y_2 + xy_1 + y = 0$

136. If $y = a \sin(\log x)$ prove that $x^2y_2 + xy_1 + y = 0$

137. If $y = (\sin^{-1} x)^2$, prove that $(1-x^2)\frac{d^2y}{dx^2} - x\frac{dy}{dx} = 2$

138. If $y = x + \tan x$, prove that $\cos^2 x \frac{d^2y}{dx^2} - 2y + 2x = 0$

139. If $y = \frac{\sin^{-1}x}{\sqrt{1-x^2}}$, prove that $(1-x^2)\frac{d^2y}{dx^2} - 3x\frac{dy}{dx} - y = 0$

140. If $y = Ae^{-kt} \cos(pt + c)$ show that $\frac{d^2y}{dt^2} - 2k\frac{dy}{dt} - n^2y = 0$, where $n^2 = p^2 + k^2$.

141. If $y = e^{\tan^{-1}x}$, prove that $(1 - x^2)\, y_2 - xy_1 - 2 = 0$

142. If $y = \frac{\log x}{x}$, prove that $\frac{d^2y}{dx^2} + \frac{2\log x - 3}{x^3}$

143. If $y = \tan^{-1} x$, prove that $\frac{d^2y}{dx^2} + \frac{2x}{(1+x^2)^2} = 0$

144. If $y = \sin(x + y)$; prove that $\frac{d^2y}{dx^2} + y\left(1 + \frac{dy}{dx}\right)^3 = 0$

145. If $\frac{x^2}{a^2} + \frac{y^2}{b^2} = 1$, show that $\frac{d^2y}{dx^2} = \frac{-b^4}{a^2y^3}$.

Answers

1. $\frac{-3}{2}x^{\frac{-5}{2}}$, $-72x^{-9}$

2. $15x^2 - 8x + 3$

3. $1 + \frac{8}{x^2}$

4. $12 + \frac{15}{2}x^{\frac{-1}{2}} + \frac{5}{2}x^{\frac{-3}{2}}$

5. $5x^{\frac{3}{2}} + 9x^{\frac{1}{2}} - \frac{7}{2x^{\frac{3}{2}}}$

6. $5x^4 - \frac{2}{x}$

7. $e^x + nx^{n-1}$

8. $\cos x + 7 + \frac{3}{x^2}$

9. $\frac{-3}{2}ax^{\frac{-5}{2}} - \frac{b}{x^2}$

10. $e^x + \frac{1}{x^2}$

11. $\frac{x + x^2e^x + 1}{x^2}$

12. $3e^x + \frac{7}{x} - 3x^{-4}$

13. $e^{ax}(a \sin bx + b \cos bx)$

14. $3\left(\frac{1}{x}\sin x + \cos x \log x\right)$

15. $\frac{1}{\sqrt{x}}(2 + \log x)$

16. $ae^x(\cos x + \sin x) + bx^{n-1}(-x \sin x + n \cos x)$

17. $e^x(x^2 + 2x + 1) + 2 \cos x (x^2 + 1) + 4x \sin x$

18. $2x^{\frac{3}{2}}(1 - x) + \frac{1}{2\sqrt{x}}(1 - 3x)(x^2 + 2)$

19. $4(\cos x - 2 \sin x) - 2 \sec^2 x (5 + 4 \sin x)$

20. $\frac{2(x^4 + 12x^2 - 9x + 15)}{(x^2 + 5)^2}$

21. $\tan\frac{x}{2} \sec^2\frac{x}{2}$

22. $\frac{nx^{n-1}\log_a x - x^{n-1}\log_a e}{(\log_a x)^2}$

23. $\frac{x^2}{(x \sin x + \cos x)^2}$

24. $\frac{\log x(\cos x - x \sin x) - \cos x}{(\log x)^2}$

25. $\frac{-2}{1 + \sin 2x}$

26. $\dfrac{\sqrt{a}}{\sqrt{x}\left(\sqrt{a}-\sqrt{x}\right)^2}$

27. $\dfrac{(e^x+\sec x^2)(\cot x-x^n)+(\operatorname{cosec}^2 x+nx^{n-1})(e^x+\tan x)}{(\cot x - x^n)^2}$

28. $\dfrac{(e^x+15x^2\log x+5x^2)\log x-x^{-1}(e^x+5x^3\log x)}{(\log x)^2}$

29. $\dfrac{3x^2+2(a+b+c)x+(ab+bc+ca)}{(x+a)^2(x+b)^2(x+c)^2}$

30. $\dfrac{2}{x(1-\log x)^2}$

31. $\left(\sqrt{x}+\dfrac{1}{\sqrt{x}}\right)\left(\dfrac{1}{\sqrt{x}}-\dfrac{1}{x\sqrt{x}}\right)$

32. $x(2x-3)(7+x^2)^{\frac{-1}{2}}+2\sqrt{7+x^2}$

33. $-\dfrac{1}{(1+x)(1-x^2)^{\frac{1}{2}}}$

34. $-\dfrac{1}{(x+\sqrt{1+x^2})\sqrt{1+x^2}}$

35. $\dfrac{x(3a^2-x^2)}{(a^2-x^2)^{\frac{3}{2}}}$

36. $3x^2\sec^2 x^3$

37. $\cot x.\ \sec^2 x$

38. $\cos(\tan x)\sec^2 x$

39. $\dfrac{1}{2x\sqrt{\log x}}$

40. $a\,e^{ax+b}$

41. $\dfrac{e^x\cos e^x}{2\sin e^x}$

42. $\dfrac{a}{ax+b}-\dfrac{p}{px+q}$

43. $\dfrac{\sqrt{x}\cos x\sin\sqrt{x}-\sin x\cos\sqrt{x}}{2\sqrt{x}\sqrt{\sin x}\sin^2\sqrt{x}}$

44. $-\operatorname{cosec}^2(x^2\sin 2x)(2x\sin 2x+2x^2\cos 2x)$

45. $\dfrac{1}{1-\sin x}$

46. $\sec x$

47. $\dfrac{2}{\sqrt{x^2-1}}$

48. $\dfrac{1}{2\sqrt{x^2-1}}$

49. $\dfrac{-\operatorname{cosec} x\,(\operatorname{cosec}^2 x + \cot^2 x)}{2\sqrt{\operatorname{cosec} x.\cot x}}$

50. $\dfrac{1}{x \log x.\log \log x}$

51. $\dfrac{\sin x^n.e^{\sin x}.\cos x - e^{\sin x}.\cos x^n.nx^{n-1}}{\sin^2 x^n}$

52. $\dfrac{\sec\sqrt{x}\tan\sqrt{x}}{4\sqrt{x}\sqrt{\sec\sqrt{x}}}$

53. $\dfrac{-1}{\sqrt{1-x^2}}$

54. $\dfrac{3}{\sqrt{1-x^2}}$

55. $\dfrac{1}{2}$

56. $\dfrac{2}{1+x^2}$

57. $\dfrac{1}{(1+2x^2)\sqrt{1+x^2}}$

58. $\dfrac{-2}{1+x^2}$

59. $\dfrac{2}{1+x^2}$

60. $\dfrac{-1}{\sqrt{1-x^2}}$

61. $\dfrac{2}{\sqrt{1+x^2}}$

62. $\dfrac{-3}{\sqrt{1-x^2}}$

63. $\dfrac{1}{1+x^2}$

64. $\dfrac{-2}{1+x^2}$

65. $\dfrac{1}{\sqrt{1-x^2}} - \dfrac{1}{2\sqrt{x-x^2}}$

67. $\dfrac{3^x\,[(x+\tan x)\log 3 - (1+\sec^2 x)]}{(x+\tan x)^2}$

68. $\dfrac{\pi}{180}\cos x^{o}$

69. $7x^6$

70. $-a\sin(ax+b)$

71. $a\sec^2 ax$

72. $-\sin 2x$

73. $\dfrac{-1}{\sqrt{1-x^2}}$

74. $3e^{3x}$

75. $2x\cos x^2$

76. $\dfrac{-\sin\sqrt{x}}{2\sqrt{x}}$

77. $3\sec^2 3x$

78. $\dfrac{-3}{2}x^{\frac{-5}{2}}$

83. $\dfrac{-y}{x}$

84. $\frac{-b^2}{a^2} \cdot \frac{x}{y}$

85. $-\left(\frac{y}{x}\right)^{\frac{1}{3}}$

86. $\frac{1+3x^2y-y^3}{3xy^2-x^3}$

87. $\frac{-y\sec x\tan x+\sec^2 x+2xy}{(\sec x+x^2)}$

88. $\frac{\sec^2(x+y)+\sec^2(x-y)}{\sec^2(x-y)-\sec^2(x+y)}$

90. $\frac{-2x+y}{x+2y}$, 1

94. $\frac{x-y}{x(1+\log xy)}$

95. $\frac{\cos(x+y)-3x^2}{3y^2-\cos(x+y)}$

96. $\frac{\frac{1}{x+y}+e^x\cos e^x-3x^2y^2}{2x^3y-\frac{1}{x+y}}$

98. $(\log x)^x\left[\log\log x+\frac{1}{\log x}\right]$

99. $(2x+3)^{x-5}\left[\frac{2(x-5)}{2x+3}+\log(2x+3)\right]$

100. $(\cos x)^{\log x}\left[\frac{1}{x}.\log\cos x-\tan x\log x\right]$

101. $e^{x^x}.x^x\ (1+\log x)$

102. $\frac{x^2-7}{(1+x)^{\frac{3}{2}}\sqrt{3-x}}$

103. $x^2 e^x \sin x\left[\frac{2}{x}+1+\cot x\right]$

104. $x^{\sin^{-1}x}\left[\frac{\sin^{-1}x}{x}+\frac{\log x}{\sqrt{1-x^2}}\right]+(\log x)^x\left[\frac{1}{\log x}+\log\log x\right]$

105. $(\tan x)^{\cot x}\,\text{cosec}^2 x\,(1-\log\tan x)+(\cot x)^{\tan x}\,.\,\sec^2 x\,(\log\cot x-1)$

106. $\frac{-1}{(x+a)(x+b)(x+c)}\left[\frac{1}{x+a}+\frac{1}{x+b}+\frac{1}{x+c}\right]$

107. $2\tan x\tan 2x\tan 3x\tan 4x\,(\text{cosec}\,2x+2\,\text{cosec}\,4x+3\,\text{cosec}\,6x+4\,\text{cosec}\,8x)$

108. $x^{x^x}\left[x^x(1+\log x \log x + x^{x-1}\right]$

109. $x^{1+x+x^2}\left[(2x+1)\log x + \frac{1}{x} + 1 + x\right]$

110. $\frac{(x+1)(x+2)(x+3)}{(x-1)(x-2)(x-3)}\left[\frac{2}{1-x^2}+\frac{4}{4-x^2}+\frac{6}{9-x^2}\right]$

111. $\frac{(x+2)^{\frac{5}{2}}}{(x+4)^{\frac{1}{3}}(x+1)^{\frac{3}{2}}}\left[\frac{5}{2(x+2)}-\frac{1}{3(x+4)}-\frac{3}{2(x+1)}\right]$

112. $\frac{b}{a}$

113. $-\cot\theta$

114. $-(\tan 2\theta)^{\frac{3}{2}}$

115. $\frac{3b}{4at}$

116. $\frac{t(2-t^3)}{1-2t^3}$

117. 1

118. $t(e^t - \sin t)$

119. $\cos x^2$

120. $\frac{x^{\sin x}\left(\frac{\sin x}{x}+\cos x \log x\right)}{(\sin x)^x (x\cot x + \log \sin x)}$

121. $\frac{e^{\tan x}}{\cos^3 x}$

122. $\frac{1}{2}$

126. $6ax + 2b$

CHAPTER 8

INTEGRAL CALCULUS

Introduction

Integration of a function is the inverse process of differentiation. If F(x) is the integral of f(x) then we write $\int f(x)dx = F(x)$ 'or' $\int f(x)dx = F(x) + c$.

Here,

herehere, the symbol $\int$ stands for integral and dx after f(x) indicates that x is the variable of integration. The function f(x) which is to be integrated is called the integrand and C is an arbitrary constant.

Standard Integral Formulae

1. $\int x^n dx = \dfrac{x^{n+1}}{n+1}$
2. $\int \dfrac{1}{x} dx = \log_e x$
3. $\int e^x dx = e^x$
4. $\int a^x dx = \dfrac{a^x}{\log_e a}$

5. $\int \sin x dx = -\cos x$

6. $\int \cos x dx = \sin x$

7. $\int \sec^2 x dx = \tan x$

8. $\int \operatorname{cosec}^2 x dx = -\cot x$

9. $\int \sec x \tan x dx = \sec x$

10. $\int \operatorname{cosec} x \cot x \, dx = -\operatorname{cosec} x$

11. $\int 1 \, dx = x$

12. $\int 0 \, dx = c$

13. $\int \frac{1}{\sqrt{1-x^2}} dx = \sin^{-1} x$

14. $\int \frac{1}{1+x^2} dx = \tan^{-1} x$

15. $\int \frac{1}{x\sqrt{x^2-1}} dx = \sec^{-1} x$

Integral of Algebraic Functions

***Example* 1:** Integrate x^{99} w.r.t x.

***Solution*:**

$$\int x^{99} dx = \frac{x^{99+1}}{99+1}$$

$$= \frac{1}{100} x^{100}$$

***Example* 2:** Integrate $x^{\frac{3}{2}}$ w.r.t x.

***Solution*:**

$$\int x^{\frac{3}{2}} dx = \frac{x^{\frac{3}{2}+1}}{\frac{3}{2}+1}$$

$$= \frac{x^{\frac{5}{2}}}{\frac{5}{2}} = \frac{2}{5}x^{\frac{5}{2}}$$

***Example* 3:** Integrate $\frac{x^3 + 3x^2 + 4}{\sqrt{x}}$ w.r.t x.

***Solution*:**

$$\int \frac{x^3 + 3x^2 + 4}{\sqrt{x}}\,dx = \int \frac{x^3}{x^{\frac{1}{2}}}\,dx + 3\int \frac{x^2}{x^{\frac{1}{2}}}\,dx + 4\int \frac{1}{\sqrt{x}}\,dx$$

$$= \int x^{\frac{5}{2}}dx + 3\int x^{\frac{3}{2}}dx + 4\int x^{\frac{-1}{2}}\,dx$$

$$= \frac{x^{\frac{5}{2}+1}}{\frac{5}{2}+1} + 3.\frac{x^{\frac{3}{2}+1}}{\frac{3}{2}+1} + 4.\frac{x^{\frac{-1}{2}+1}}{\frac{-1}{2}+1}$$

$$= \frac{2}{7}x^{\frac{7}{2}} + \frac{6}{5}x^{\frac{5}{2}} + 8x^{\frac{1}{2}}$$

***Example* 4:** Integrate the following w.r.t x.

(i) $\frac{2}{x} + 3e^x$ (ii) $\frac{1}{x} + \cos x$

(iii) $\sin x - \frac{2}{x} + e^x$ (iv) $\frac{x^4 + x^2 - 1}{2(x^2 + 1)}$

***Solution*:**

(i) $$\int \left(\frac{2}{x} + 3e^x\right) dx = 2\int \frac{1}{x}\,dx + 3\int e^x\,dx$$

$$= 2\log x + 3e^x$$

(ii) $$\int \left(\frac{1}{x} + \cos x\right) dx = \int \frac{1}{x}\,dx + \int \cos x\,dx$$

$$= \log x + \sin x$$

(iii) $$\int\left(\sin x-\frac{2}{x}+e^x\right)dx=\int \sin x dx-2\int\frac{1}{x}dx+\int e^x dx$$

$$=-\cos x-2\log x+e^x$$

(iv) $$\int\frac{x^4+x^2-1}{2(x^2+1)}dx=\int\frac{x^2(x^2+1)-1}{2(x^2+1)}dx$$

$$=\frac{1}{2}\int\frac{x^2(x^2+1)}{x^2+1}dx-\frac{1}{2}\int\frac{1}{x^2+1}dx$$

$$=\frac{1}{2}\int x^2\,dx-\frac{1}{2}\int\frac{1}{1+x^2}dx$$

$$=\frac{1}{2}.\frac{x^3}{3}-\frac{1}{2}\tan^{-1}x$$

$$=\frac{x^3}{6}-\frac{1}{2}\tan^{-1}x$$

Integral of Trigonometrical Functions

***Example* 5:** Evaluate $\int\left(\frac{\sec x}{\sec x-\tan x}\right)dx$

***Solution*:**

$$\int\frac{\sec x}{\sec x-\tan x}dx=\int\frac{\sec x}{\sec x-\tan x}\times\frac{\sec x+\tan x}{\sec x+\tan x}dx$$

$$=\int\frac{\sec^2 x+\sec x\tan x}{\sec^2 x-\tan^2 x}dx$$

$$=\int\frac{\sec^2 x+\sec x\tan x}{1}dx$$

$$=\int\sec^2 x dx+\int\sec x\tan x dx$$

$$=\tan x+\sec x$$

***Example* 6:** Evaluate $\int\frac{\sin^2 x-\cos^2 x}{\sin^2 x\cos^2 x}dx$

***Solution*:**

$$\int\frac{\sin^2 x-\cos^2 x}{\sin^2 x\cos^2 x}dx=\int\left(\frac{\sin^2 x}{\sin^2 x\cos^2 x}-\frac{\cos^2 x}{\sin^2 x\cos^2 x}\right)dx$$

$$= \int \left(\frac{1}{\cos^2 x} - \frac{1}{\sin^2 x} \right) dx$$

$$= \int \sec^2 x\, dx - \int \operatorname{cosec}^2 x\, dx$$

$$= \tan x + \cot x$$

Integration by Substitution

When we have to integrate such functions which are not in any standard form, then we substitute a proper function of variable t for the variable x.

Let $\quad I = \int f(x)dx$

Put $\quad x = \phi(t)$

$\therefore \quad \frac{dx}{dt} = \phi'(t)$

$\therefore \quad I = \int f(x)\, \phi'(t)\, dt$

$\therefore \quad \int f(x)\, dx = \int f[\phi(t)]\phi'(t)\, dt$

Some Important Results

(i) $\int \sin(ax+b)\, dx = \frac{-1}{a} \cos(ax+b)$

(ii) $\int \cos(ax+b)\, dx = \frac{1}{a} \sin(ax+b)$

(iii) $\int \sec^2(ax+b)\, dx = \frac{1}{a} \tan(ax+b)$

(iv) $\int e^{ax} dx = \frac{1}{a} e^{ax}$

(v) $\int \frac{1}{ax+b} = \frac{1}{a} \log(ax+b)$

(vi) $\int (ax+b)^n dx + \frac{1}{a(n+1)} (ax+b)^{n+1}$

(vii) $\int \sec(ax+b) \tan(ax+b)\, dx = \frac{1}{a} \sec(ax+b)$

(viii) $\int \sin ax dx = \frac{-1}{a} \cos ax$

(ix) $\int \cos ax\, dx = \frac{1}{a} \sin ax$

(x) $\int \sec^2 ax\, dx = \frac{1}{a} \tan ax$

(xi) $\int \sec ax \tan ax\, dx = \frac{1}{a} \sec ax$

(xii) $\int \operatorname{cosec} ax \cot ax\, dx = \frac{-1}{a} \operatorname{cosec} ax$

***Example* 6:** Evaluate $\int (3x+2)^5 dx$

***Solution*:** Let $I = \int (3x+2)^5 dx$

Put $3x + 2 = t$

$\Rightarrow \quad 3\frac{d}{dx}x + \frac{d}{dx}2 = \frac{dt}{dx}$

$3 = \frac{dt}{dx}$

$\Rightarrow \quad dx = \frac{1}{3}dt$

$\therefore \quad I = \int t^5 . \frac{1}{3} dt$

$= \frac{1}{3}\int t^5 dt$

$= \frac{1}{3}\frac{t^{5+1}}{5+1} = \frac{1}{18}t^6 = \frac{1}{18}(3x+2)^6$

***Example* 7:** Evaluate $\int \frac{1}{1-4x} dx$

***Solution*:** Let $I = \int \frac{1}{1-4x} dx$

Put $1 - 4x = t$

$-4\, dx = dt \quad \Rightarrow \quad dx = \frac{-1}{4} dt$

$\therefore \quad I = \int \frac{1}{t} . \left(-\frac{1}{4}\right) dt$

$$= \frac{-1}{4} \int \frac{1}{t} dt$$

$$= \frac{-1}{4} \log t$$

$$= \frac{-1}{4} \log (1 - 4x)$$

***Example* 8:** Evaluate $\int \cos x \cos 2x \cos 3x \, dx$

***Solution*:** Let $I = \int \cos x \cos 2x \cos 3x \, dx$

Now, $\cos x \cos 2x \cos 3x = \frac{1}{2} .2 \cos 3x \cos x \cos 2x$

$$= \frac{1}{2} [\cos 4x + \cos 2x] \cos 2x$$

$$= \frac{1}{2} [\cos 4x \cos 2x + \cos^2 2x]$$

$$= \frac{1}{4} [\cos 6x + \cos 2x] + \frac{1}{2} \left[\frac{1 + \cos 4x}{2}\right]$$

$$= \frac{1}{4} [\cos 6x + \cos 2x + \cos 4x + 1]$$

$\therefore$ $$I = \frac{1}{4} \int [\cos 6x + \cos 4x + \cos 2x + 1] dx$$

$$= \frac{1}{4} \left[\frac{\sin 6x}{6} + \frac{\sin 4x}{4} + \frac{\sin 2x}{2} + x\right]$$

***Example* 9:** Evaluate $\int x^2 \sin x^3 dx$

***Solution*:** Let $I = \int x^2 \sin x^3 dx$

Put $x^3 = t \Rightarrow \quad 3x^2 \, dx = dt \quad \Rightarrow \quad dx = \frac{1}{3x^2} dt$

$\therefore$ $$I = \int x^2 \sin t . \frac{1}{3x^2} dt$$

$$= \frac{1}{3}\int \sin t \, dt$$

$$= \frac{-1}{3}\cos t$$

$$= \frac{-1}{3}\cos x^3$$

Integration of $\frac{f'(x)}{f(x)}$

$$\int \frac{f'(x)}{f(x)} dx = \log f(x)$$

Integration of tan x Let

$$I = \int \tan x \, dx$$

$$= \int \frac{\sin x}{\cos x} dx$$

Put $\cos x = t \Rightarrow -\sin x \, dx = dt$

$$dx = \frac{-dx}{\sin x}$$

$$\therefore \quad I = \int \frac{\sin x}{\cos x}\left(\frac{-dt}{\sin x}\right)$$

$$= -\int \frac{1}{t} dt$$

$$= -\log t$$

$$= -\log \cos x = \log\left(\frac{1}{\cos x}\right)$$

$$= \log(\sec x)$$

$$\therefore \quad \int \tan x \, dx = \log(\sec x) = -\log(\cos x)$$

Integration of cot x

Let $I = \int \cot x \, dx$

$$= \int \frac{\cos x}{\sin x} dx$$

Put $\sin x = t \Rightarrow \cos x\, dx = dt$

$\therefore \quad I = \int \frac{1}{t}\, dt$

$= \log t$

$= \log(\sin x)$

$\therefore \quad \int \cot x\, dx = \log(\sin x)$

Integration of cosec x

Let $I = \text{cosec } x$

$$= \int \frac{1}{\sin x}\, dx$$

$$= \int \frac{1}{2\sin\frac{x}{2}\cos\frac{x}{2}}\, dx$$

Divide by $\cos^2 \frac{x}{2}$, we get

$$I = \int \frac{\frac{1}{\cos^2\frac{x}{2}}}{\frac{2\sin\frac{x}{2}}{\cos\frac{x}{2}}}\, dx$$

$$= \int \frac{\sec^2\frac{x}{2}}{2\tan\frac{x}{2}}\, dx$$

Put $\tan\frac{x}{2} = t \Rightarrow 2\sec^2\frac{x}{2}\, dx = dt$

$\therefore \quad I = \int \frac{1}{t}\, dt$

$= \log t = \log(\tan\frac{x}{2})$

$\therefore \quad \int \text{cosec } x\, dx = \log\left(\tan\frac{x}{2}\right)$ or $\int \text{cosec } x\, dx = \log(\text{cosec } x - \cot x)$

Integration of sec x

Let $I = \int \sec x \, dx$

$$= \int \operatorname{cosec}\left(\frac{\pi}{2} + x\right) dx$$

Put $\frac{\pi}{2} + x = t \quad \Rightarrow \quad dx = dt$

$\therefore \quad I = \int \operatorname{cosec} t \, dt$

$$= \log \tan \frac{t}{2}$$

$$= \log \tan \frac{1}{2}\left(\frac{\pi}{2} + x\right)$$

$$= \log \tan \left(\frac{\pi}{4} + \frac{x}{2}\right)$$

$\therefore \quad \int \sec x dx = \log \tan\left(\frac{\pi}{4} + \frac{x}{2}\right)$

or $\quad \int \sec x dx = \log(\sec x + \tan x)$

Indefinite Integrals of Additional Standard Form

Formula

(i) $\int \frac{dx}{\sqrt{a^2 - x^2}} = \sin^{-1}\frac{x}{a}$

(ii) $\int \frac{dx}{a^2 + x^2} = \frac{1}{a}\tan^{-1}\frac{x}{a}$

(iii) $\int \frac{dx}{x^2 - a^2} = \frac{1}{2a}\log\frac{x-a}{x+a}, x > a$

(iv) $\int \frac{dx}{a^2 - x^2} = \frac{1}{2a}\log\frac{a+x}{a-x}, x < a$

(v) $\int \frac{dx}{x\sqrt{x^2 - a^2}} = \frac{1}{a}\sec^{-1}\frac{x}{a}$

(vi) $\int \frac{dx}{\sqrt{x^2 + a^2}} = \log(x + \sqrt{x^2 + a^2})$

(viii) $\int \sqrt{a^2 - x^2}\,dx = \frac{1}{2}\left[x\sqrt{a^2 - x^2} + a^2 \sin^{-1}\frac{x}{a}\right]$

(ix) $\int \sqrt{a^2 + x^2}\,dx = \frac{1}{2}\left[x\sqrt{a^2 + x^2} + a^2 \log(x + \sqrt{a^2 + x^2})\right]$

(x) $\int \sqrt{x^2 - a^2}\,dx = \frac{1}{2}\left[x\sqrt{x^2 - a^2} - a^2 \log(x + \sqrt{x^2 - a^2})\right]$

***Example* 10:** Evaluate $\int \frac{1}{\sqrt{4 - x^2}}dx$

***Solution*:** $\int \frac{1}{\sqrt{4 - x^2}}dx = \int \frac{1}{\sqrt{(2)^2 - x^2}}dx$

$$= \sin^{-1}\frac{x}{2} \qquad \left(\because \int \frac{dx}{\sqrt{a^2 - x^2}} = \sin^{-1}\frac{x}{a}\right)$$

***Example* 11:** Evaluate $\int \frac{dx}{\sqrt{9 - 25x^2}}$

***Solution*:** $\int \frac{dx}{\sqrt{9 - 25x^2}} = \frac{1}{5}\int \frac{dx}{\sqrt{\frac{9}{25} - x^2}}$

$$= \frac{1}{5}\int \frac{dx}{\sqrt{\left(\frac{3}{5}\right)^2 - x^2}}$$

$$= \frac{1}{5}\sin^{-1}\frac{5}{3}x$$

***Example* 12:** Evaluate $\int \sqrt{2 - 3x^2}\,dx$

***Solution*:** $\int \sqrt{2 - 3x^2}\,dx = \sqrt{3}\int \sqrt{\frac{2}{3} - x^2}\,dx$

$$= \sqrt{3}\int \sqrt{\left(\frac{\sqrt{2}}{\sqrt{3}}\right)^2 - x^2}\,dx$$

$$= \sqrt{3}.\frac{1}{2}\left[x\sqrt{\frac{2}{3}-x^2} + \left(\frac{\sqrt{2}}{\sqrt{3}}\right)^2 \sin^{-1}\frac{x}{\frac{\sqrt{2}}{\sqrt{3}}}\right]$$

(by $\int\sqrt{a^2-x^2}\,dx$)

$$= \frac{\sqrt{3}}{2}\left[x\sqrt{\frac{2}{3}-x^2} + \frac{2}{3}\sin^{-1}\frac{\sqrt{3}x}{\sqrt{2}}\right]$$

Integration by Parts: The integration of a product of two functions can be easily evaluated by the rule of integration by parts.

Method $$\int u.v\,dx = u\int v\,dx - \int\left[\frac{du}{dx}\int v\,dx\right]dx$$

***Example* 13:** Solve $\int x\,e^x dx$

***Solution*:** Let $I = \int x\,e^x dx$

Integrating by parts, we get

$$I = x\int e^x\,dx - \int\left[\frac{d}{dx}x\int e^x\,dx\right]dx$$

$$= xe^x - \int 1.e^x dx$$

$$= x\,e^x - e^x \quad = e^x(x-1)$$

***Example* 14:** Solve $\int x^2.\sin x\,dx$

***Solution*:** Let $I = \int x^2.\sin x\,dx$

Integrating by parts, we get

$$I = x^2\int \sin xdx - \int\left[\frac{d}{dx}x^2\int \sin xdx\right]dx$$

$$= x^2(-\cos x) + \int 2x \cos x \, dx$$

$$= -x^2 \cos x + 2\left\{x\int \cos x dx - \int\left[\frac{d}{dx}x\int \cos x dx\right]dx\right\}$$

$$= -x^2 \cos x + 2\left\{x \sin x - \int \sin x \, dx\right\}$$

$$= -x^2 \cos x + 2x \sin x + 2 \cos x$$

***Example* 15:** Solve $\int x \cos^2 x \, dx$

***Solution*:** Let $I = \int x \cos^2 x \, dx$

$$= \int x\left(\frac{1+\cos 2x}{2}\right)dx$$

$$= \frac{1}{2}\int x(1+\cos x)dx$$

$$= \frac{1}{2}\int x dx + \frac{1}{2}\int x \cos x \, dx$$

Integrating by parts, we get

$$I = \frac{1}{2}.\frac{x^2}{2} + \frac{1}{2}\left\{x\int \cos x dx - \int\left[\frac{d}{dx}x\int \cos x \, dx\right]dx\right\}$$

$$= \frac{x^2}{4} + \frac{1}{2}\left\{x \sin x - \int \sin x \, dx\right\}$$

$$= \frac{x^2}{4} + \frac{1}{2}\{x \sin x + \cos x\}$$

***Example* 16:** Solve $\int x^2 e^{-x} dx$

***Solution*:** Let $I = \int x^2 e^{-x} dx$

Integrating by parts we get

$$I = x^2\int e^{-x}dx - \int\left[\frac{d}{dx}x^2\int e^{-x}dx\right]dx$$

$$= -x^2e^{-x} + \int 2x\,e^{-x}dx$$

$$= -x^2e^{-x} + 2\left\{x\int e^{-x}dx - \int\left[\frac{d}{dx}x\int e^{-x}dx\right]dx\right\}$$

$$= -x^2\,e^{-x} + 2x(-e^{-x}) + 2\int e^{-x}dx$$

$$= -x^2\,e^{-x} - 2x\,e^{-x} - 2e^{-x} \qquad = -e^{-x}(x^2 + 2x + 2)$$

***Example* 17:** Solve $\int x\sin^3 x dx$

***Solution*:** Let $I = \int x\sin^3 x\,dx$

$$= \int \frac{x}{4}(3\sin x - \sin 3x)dx \qquad (\because \sin 3A = 3\sin A - 4\sin^3 A)$$

$$= \frac{3}{4}\int x\sin x\,dx - \frac{1}{4}\int x\sin 3x\,dx$$

Integrating by parts, we get

$$I = \frac{3}{4}\left\{x\int \sin x\,dx - \int\left[\frac{d}{dx}\int \sin x\,dx\right]dx\right\} - \frac{1}{4}\left\{x\int \sin 3x\,dx - \int\left[\frac{d}{dx}x\int \sin 3x\,dx\right]dx\right\}$$

$$= \frac{3}{4}\left[-x\cos x + \int \cos x dx\right] - \frac{1}{4}\left[-x\frac{\cos 3x}{3} + \int\frac{\cos 3x}{3}dx\right]$$

$$= \frac{3}{4}[-x\cos x + \sin x] - \frac{1}{4}\left[-\frac{x\cos 3x}{3} + \frac{\sin 3x}{9}\right]$$

$$= \frac{1}{4}\left[-3x\cos x + 3\sin x\frac{x\cos 3x}{3} - \frac{\sin 3x}{9}\right]$$

***Example* 18:** Solve $\int \log_e x\,dx$

***Solution*:** Let $I = \int \log_e x\,dx$

$$= \int \log_e x\ .\ 1dx$$

Integrating by parts, we get

$$I = \log_e x \int 1\,dx - \int \left[\frac{d}{dx} \log_e x \int 1\,dx \right] dx$$

$$= \log_e x.x - \int \frac{1}{x}.x\,dx$$

$$= x \log_e x - \int 1\,dx$$

$$= x \log_e x - x$$

$$= x(\log_e x - 1)$$

Formulae

(i) $\int e^{ax} \sin bx dx = \frac{e^{ax}}{a^2 + b^2}(a \sin bx - b \cos bx)$

(ii) $\int e^{ax} \sin bx\,dx = \frac{e^{ax}}{\sqrt{a^2 + b^2}} \sin(bx - \tan^{-1} \frac{b}{a})$

(iii) $\int e^{ax} \cos bx\,dx = \frac{e^{ax}}{a^2 + b^2}(a \cos bx + b \sin bx)$

(iv) $\int e^{ax} \cos bx\,dx = \frac{e^{ax}}{\sqrt{a^2 + b^2}} \cos(bx - \tan^{-1} \frac{b}{a})$

***Example* 19:** Solve $\int e^{3x} \sin 4x dx$

***Solution*:** Let $I = \int e^{3x} \sin 4x dx$

by compare $\int e^{ax} \sin bx\,dx$, $a = 3, b = 4$

by using formula $\int e^{ax} \sin bx\,dx = \frac{e^{ax}}{a^2 + b^2}(a \sin bx - b \cos bx)$

$\therefore$ $I = \int e^{3x} \sin 4x\,dx = \frac{e^{3x}}{9 + 16}(3 \sin 4x - 4 \cos 4x)$

$$= \frac{e^{3x}}{25}(3 \sin 4x - 4 \cos 4x)$$

Integration by Partial Fraction: Let $y=\frac{f(x)}{g(x)}$, where f(x) and g(x) are polynomial functions. If the degree of g(x) is more than that of f(x), we call such a function a partial fraction.

Rules

1. When the denominator of given function contains non repeated linear factor. Then,
$$\frac{f(x)}{(x+a)(x+b)(x+c)}=\frac{A}{x+a}+\frac{B}{x+b}+\frac{C}{x+c}$$
2. When the denominator contains repeated linear factors. Then,
$$\frac{f(x)}{(x+a)(x+b)^n}=\frac{A}{x+a}+\frac{B}{x+b}+\frac{C}{(x+b)^2}+....+\frac{N}{(x+b)^n}$$
3. When the denominator contains a quadratic factor. Then,
$$\frac{f(x)}{(x+a)(x^2+b^2)}=\frac{A}{x+a}+\frac{Bx+C}{x^2+b^2}$$
4. When the denominator contains repeated quadratic factors. Then
$$\frac{f(x)}{(x^2+a^2)^2(x+b)}=\frac{Ax+B}{x^2+a^2}+\frac{cx+D}{(x^2+a^2)^2}+\frac{E}{x+b}$$

***Example* 20:** Find the value of $\int\frac{x^2}{(x+1)(x+2)(x+3)}dx$

***Solution*:** Let $I=\frac{x^2}{(x+1)(x-2)(x+3)}=\frac{A}{x+1}+\frac{B}{x-2}+\frac{C}{x+3}$

$x^2 = A(x-2)(x+3)+B(x+1)(x+3)+(x+1)(x-2)$ .....(1)

Put $x+1=0 \Rightarrow x=-1$, in eqn (1), we get

$1 = A(-3)2 \Rightarrow 1=-6A \quad \therefore A=\frac{-1}{6}$

Put $x-2=0 \Rightarrow x=2$ in eqn (1)

$4 = B(3)(5) \Rightarrow B=\frac{4}{15}$

Put $x+3=0 \Rightarrow x=-3$ in equn (1)

$$9 = c(-2)(-5) \Rightarrow \quad c = \frac{9}{10}$$

$$\therefore \int \frac{x^2}{(x+1)(x-2)(x+3)} dx = \frac{-1}{6}\int \frac{1}{x+1} dx + \frac{4}{15}\int \frac{1}{x-2} dx + \frac{9}{10}\int \frac{1}{x+3} dx$$

$$= \frac{-1}{6}\log(x+1) + \frac{4}{15}\log(x-2) + \frac{9}{10}\log(x+3)$$

Definite Integral

If $\int f(x)dx = F(x)$

Then $\int_a^b f(x)dx = F(b) - F(a)$

Integral $\int_a^b f(x)dx$ is called the definite integral of f(x) between the limits a and b. a is called lower limit and b is upper limit. The interval (a, b) is called the range of integration.

***Example* 21:** Evaluate the following

(i) $\int_1^2 x^3 dx$ (ii) $\int_1^4 \sqrt{x}\, dx$

(iii) $\int_0^{\frac{\pi}{2}} \sin x\, dx$ (iv) $\int_{-1}^1 2^x\, dx$

***Solution*:**

(i) $$\int_{-1}^2 x^3\, dx = \left[\frac{x^4}{4}\right]_1^2 = \frac{2^4}{4} - \frac{1^4}{4}$$

$$= \frac{16}{4} - \frac{1}{4} = \frac{15}{4}$$

(ii) $$\int_1^4 \sqrt{x} dx = \int_1^4 x^{\frac{1}{2}} dx$$

$$= \left[\frac{x^{\frac{1}{2}+1}}{\frac{1}{2}+1}\right]_1^4 = \frac{2}{3}\left[x^{\frac{3}{2}}\right]_1^4$$

$$= \frac{2}{3}\left[4^{\frac{2}{3}} - 1^{\frac{3}{2}}\right]$$

$$= \frac{2}{3}[8-1] = \frac{14}{3}$$

(iii) $$\int_0^{\frac{\pi}{2}} \sin x dx = [-\cos x]_0^{\frac{\pi}{2}}$$

$$= -\left[\cos\frac{\pi}{2} - \cos 0\right]$$

$$= -(0-1) = 1$$

(iv) $$\int_{-1}^{1} 2^x dx = \left[\frac{2^x}{\log_e 2}\right]_{-1}^{1}$$

$$= \frac{1}{\log_e 2}\left[2^1 - 2^{-1}\right]$$

$$= \frac{1}{\log_e 2}\left[2 - \frac{1}{2}\right] = \frac{3}{2\log_e 2}$$

***Example* 22:** Evaluate the following

(i) $$\int_1^{\sqrt{3}} \frac{dx}{1+x^2}$$ (ii) $$\int_0^a \frac{dx}{\sqrt{a^2-x^2}}$$

***Solution*:** (i) $$\int_1^{\sqrt{3}} \frac{dx}{1+x^2} = \left[\tan^{-1} x\right]_1^{\sqrt{3}}$$

$$= \tan^{-1}\sqrt{3} - \tan_1^{-1}$$

$$= \frac{\pi}{3} - \frac{\pi}{4} = \frac{\pi}{12}$$

(ii) $$\int_0^a \frac{dx}{\sqrt{a^2-x^2}} = \left[\sin^{-1}\frac{x}{a}\right]_0^a$$

$$= \sin^{-1}\frac{a}{a} - \sin^{-1}\frac{0}{a}$$

$$= \sin^{-1} 1 - \sin^{-1} 0$$

$$= \frac{\pi}{2} - 0$$

$$= \frac{\pi}{2}$$

Properties of Definite Integrals

1. $\int_a^b f(x)\,dx = \int_a^b f(t)\,dt$

2. $\int_a^b f(x)\,dx = -\int_b^a f(x)\,dx$

3. $\int_a^b f(x)\,dx = -\int_a^c f(x)\,dx + \int_c^b f(x)\,dx$

4. $\int_0^a f(x)\,dx = \int_0^a f(a-x)\,dx$

5. $\int_{-a}^a f(x)\,dx = 2\int_0^a f(x)\,dx$, if f(x) is even function of x.

6. $\int_{-a}^a f(x)\,dx = 0$, if f(x) is odd function of x.

7. $\int_a^{2a} f(x)\,dx = 2\int_0^a f(x)\,dx$, if f(2a – x) f(x)

8. $\int_a^{2a} f(x)\,dx = 0$, if f(2a – x) = –f(x)

***Example* 23:** Evaluate $\int_{\frac{-\pi}{2}}^{\frac{\pi}{2}} \sin^5 x\,dx$

***Solution*:** Here $f(x) = \sin^5 x$

$\Rightarrow \quad f(-x) = \sin^5(-x) = -\sin^5 x$

Since function f(x) is odd function, then by property 6

$$\int_{-\frac{\pi}{2}}^{\frac{\pi}{2}} \sin^5 x\, dx = 0 \qquad \left(\because \int_{-a}^{a} f(x)dx = 0, f(x) \text{is odd}\right)$$

***Example* 24:** Prove that $\int_0^{\frac{\pi}{2}} \sin x\, dx = \int_0^{\frac{\pi}{2}} \cos x\, dx$

***Solution*:**

$$\int_0^{\frac{\pi}{2}} \sin x\, dx = \int_0^{\frac{\pi}{2}} \sin\left(\frac{\pi}{2} - x\right) dx \text{, by property} \qquad (4)$$

$$= \int_0^{\frac{\pi}{2}} \cos x\, dx$$

***Example* 25:** Prove that $\int_{-\frac{\pi}{2}}^{\frac{\pi}{2}} x \sin x\, dx = 2\int_0^{\frac{\pi}{2}} x \sin x\, dx$

***Solution*:** Let $f(x) = x \sin x$

$$f(-x) = (-x) \sin(-x)$$

$$= x \sin x$$

$$= f(x)$$

Hence function is even, then by property $\int_{-a}^{a} f(x)\, dx = 2\int_0^{a} f(x)\, dx$

$$\therefore \quad \int_{-\frac{\pi}{2}}^{\frac{\pi}{2}} x \sin x\, dx = 2\int_0^{\frac{\pi}{2}} x \sin x\, dx$$

***Example* 26:** Prove that $\int_0^{\frac{\pi}{2}} \frac{\sqrt{\sin x}}{\sqrt{\cos x} + \sqrt{\sin x}} dx = \frac{\pi}{4}$

Solution: Let $I = \int_0^{\frac{\pi}{2}} \frac{\sqrt{\sin x}}{\sqrt{\cos x} + \sqrt{\sin x}} dx$ ....(1)

$$= \int_0^{\frac{\pi}{2}} \frac{\sqrt{\sin\left(\frac{\pi}{2} - x\right)}}{\sqrt{\cos(\frac{\pi}{2} - x)} + \sqrt{\sin(\frac{\pi}{2} - x)}} dx$$

$$\left(\because \int_0^a f(x)dx = \int_0^a f(a-x)dx\right)$$

$$I = \int_0^{\frac{\pi}{2}} \frac{\sqrt{\cos x}}{\sqrt{\sin x} + \sqrt{\cos x}} dx \qquad ..... (2)$$

Adding (1) and (2), we get

$$2I = \int_0^{\frac{\pi}{2}} \frac{\sqrt{\sin x} + \sqrt{\cos x}}{\sqrt{\sin x} + \sqrt{\cos x}} dx$$

$$= \int_0^{\frac{\pi}{2}} 1dx$$

$$= [x]_0^{\frac{\pi}{2}}$$

$$= \frac{\pi}{2} - 0$$

$$2I = \frac{\pi}{2}$$

$$I = \frac{\pi}{4}$$ proved

Some Practice Questions

***Example* 1:** Evaluate

(i) $\int\left(x+\frac{1}{x}\right)^3 dx$ (ii) $\int(x^{\frac{1}{2}}+x^{\frac{1}{3}}+2x^{\frac{3}{5}})dx$

(iii) $\int\frac{x^4+1}{x^2+1}dx$

***Solution*:** (i) $\left(x+\frac{1}{x}\right)^3 dx=\int(x^3+3x+\frac{3}{x}+\frac{1}{x^3})dx$

$$=\int x^3dx+3\int xdx+3\int\frac{1}{x}dx+\int\frac{1}{x^3}dx$$

$$=\frac{x^4}{4}+\frac{3x^2}{2}+3\log x-\frac{x^2}{2}$$

$$=\frac{x^4}{4}+\frac{3}{2}x^2+3\log x-\frac{1}{2}x^{-2}$$

(ii) $\int(x^{\frac{1}{2}}+x^{\frac{1}{3}}+2x^{\frac{3}{5}})\,dx=\int x^{\frac{1}{2}}\,dx+\int x^{\frac{1}{3}}\,dx+2\int x^{\frac{3}{5}}dx$

$$=\frac{x^{\frac{1}{2}+1}}{\frac{1}{2}+1}+\frac{x^{\frac{1}{3}+1}}{\frac{1}{3}+1}+2\times\frac{x^{\frac{3}{5}+1}}{\frac{3}{5}+1}$$

$$=\frac{x^{\frac{3}{2}}}{\frac{3}{2}}+\frac{x^{\frac{4}{3}}}{\frac{4}{3}}+2\times\frac{x^{\frac{8}{5}}}{\frac{8}{5}}$$

$$=\frac{2}{3}x^{\frac{3}{2}}+\frac{3}{4}x^{\frac{4}{3}}+\frac{10}{8}x^{\frac{8}{5}}$$

(iii) $\int\frac{x^4+1}{x^2+1}dx=\int(x^2-1+\frac{2}{x^2+1})\,dx$

$$=\int x^2\,dx-\int 1\,dx+2\int\frac{1}{x^2+1}dx$$

$$=\frac{x^3}{3}-x+2\tan^{-1}x$$

***Example* 2:** Evaluate

(i) $\int \frac{\cos x}{\sin^2 x} dx$ (ii) $\int (4e^{3x} + 1) dx$

(iii) $\int \frac{1 + 2\sin x}{\cos^2 x} dx$

***Solution*:**

(i) $\int \frac{\cos x}{\sin^2 x} dx = \int \frac{\cos x}{\sin x} . \frac{1}{\sin x} dx$

$= \int \cot x . \operatorname{cosec} x \, dx$

$= -\operatorname{cosec} x$

(ii) $\int (4e^{3x} + 1) dx = 4\int e^{3x} dx + \int 1 dx$

$= 4.\frac{e^{3x}}{3} + x$

(iii) $\int \frac{1 + 2\sin x}{\cos^2 x} dx = \int \frac{1}{\cos^2 x} dx + 2\int \frac{\sin x}{\cos^2 x} dx$

$= \int \sec^2 x \, dx + 2\int \tan x \sec x \, dx$

$= \tan x + 2 \sec x$

***Example* 3:** Evaluate

(i) $\int \frac{dx}{\sqrt{x + a} + \sqrt{x + b}}$ (ii) $\int (3 \operatorname{cosec}^2 x + 2 \sin 3x) dx$

(iii) $\int \sin 2x \cos 3x \, dx$

***Solution*:** (i) $\int \frac{dx}{\sqrt{x + a} + \sqrt{x + b}} = \int \frac{\sqrt{x + a} - \sqrt{x + b}}{\left(\sqrt{x + a} + \sqrt{x + b}\right)\left(\sqrt{x + a} - \sqrt{x + b}\right)} dx$

$= \int \frac{\sqrt{x + a} - \sqrt{x + b}}{x + a - x - b} dx$

$= \frac{1}{a - b}\int \sqrt{x + a} \, dx - \frac{1}{a - b}\int \sqrt{x + b} \, dx$

$$= \frac{1}{a-b}\left[\frac{(x+a)^{\frac{1}{2}+1}}{\frac{1}{2}+1} - \frac{(x+b)^{\frac{1}{2}+1}}{\frac{1}{2}+1}\right]$$

$$= \frac{1}{a-b}\left[\frac{(x+a)^{\frac{3}{2}}}{\frac{3}{2}} - \frac{(x+b)^{\frac{3}{2}}}{\frac{3}{2}}\right]$$

$$= \frac{2}{3(a-b)}[(x+a)^{\frac{3}{2}} - (x+b)^{\frac{3}{2}}]$$

(ii) $\int (3\operatorname{cosec}^2 x + 2\sin 3x)dx = 3\int \operatorname{cosec}^2 x dx + 2\int \sin 3x dx$

$$= -3 \cot x - 2\ \frac{\cos 3x}{3}$$

$$= -3 \cot x - \frac{2}{3} \cos 3x$$

(iii) $\int \sin 2x \cos 3x\ dx = \frac{1}{2}\int 2 \sin 2x \cos 3x\ dx$

$$= \frac{1}{2}\int [\sin (2x+3x) + \sin(2x - 3x)]\,dx$$

$$= \frac{1}{2}\int [\sin 5x - \sin x]\,dx$$

$$= \frac{1}{2}\left[\int \sin 5x\ dx - \int \sin x\ dx\right]$$

$$= \frac{1}{2}\left[\frac{-\cos 5x}{5} + \cos x\right]$$

***Example* 4:** Evaluate

(i) $\int \sin^3 x\ dx$ (ii) $\int \frac{\cos 2\theta - \cos 2\alpha}{\cos\theta - \cos\alpha}\ d\theta$ (iii) $\int \sin^2 x \cos x\ dx$

***Solution*:**

(i) $\int \sin^3 x\ dx = \int \frac{1}{4}(3\sin x - \sin 3x)dx$

($\because \sin 3A = 3 \sin A - 4\sin^3 A$)

$$= \frac{1}{4}\left[3\int \sin x dx - \int \sin 3x\ dx\right]$$

$$= \frac{-3}{4}\cos x + \frac{1}{4} \times \frac{\cos 3x}{3}$$

$$= \frac{1}{12}\cos 3x - \frac{3}{4}\cos x$$

(ii) $\int \frac{\cos 2\theta - \cos 2\alpha}{\cos\theta - \cos\alpha} d\theta$

$$= \int \frac{2\cos^2\theta - 1 - 2\cos^2\alpha + 1}{\cos\theta - \cos\alpha} d\theta$$

$$= 2\int \frac{\cos^2\theta - \cos^2\alpha}{\cos\theta - \cos\alpha} d\theta$$

$$= 2\int \frac{(\cos\theta + \cos\alpha)(\cos\theta - \cos\alpha)}{\cos\theta - \cos\alpha} d\theta$$

$$= 2\int (\cos\theta + \cos\alpha)\, d\theta$$

$$= 2\,[\sin\theta + \theta\cos\alpha]$$

(iii) $I = \int \sin^2 x \cos x \, dx$

Put $\sin x = t$

$\cos x \, dx = dt$

$\therefore \quad I = \int t^2 dt$

$$= \frac{t^3}{3}$$

$$= \frac{1}{3}\sin^3 x$$

***Example* 5:** Evaluate

(i) $\int \frac{x^2 \tan^{-1} x^3}{1 + x^6} dx$ (ii) $\int x^{x-1} \sin x^x dx$ (iii) $\int \frac{1}{e^x + e^{-x}} dx$

***Solution*:**

(i) $\int \frac{x^2 \tan^{-1} x^3}{1 + x^6} dx$

Put $x^3 = t$ $\Rightarrow$ $3x^2\,dx = dt$

$$\therefore \quad \int \frac{x^2 \tan^{-1} x^3}{1+(x^3)^2}\,dx = \frac{1}{3}\int \frac{\tan^{-1} t}{1+t^2}\,dt$$

Put $\tan^{-1}t = 4$ $\Rightarrow$ $\frac{1dt}{1+t^2} = du$

$$\therefore \quad \frac{1}{3}\int \frac{\tan^{-1} t}{1+t^2}\,dt = \frac{1}{3}\int u\,du$$

$$= \frac{1}{3} \times \frac{4^2}{2}$$

$$= \frac{4^2}{6}$$

$$= \frac{1}{6}\left(\tan^{-1} t\right)^2$$

(ii) $\int x^{x-1} \sin x^n\,dx$

Let $x^n = t$ $\Rightarrow$ $nx^{n-1}\,dx = dt$

$$\therefore \quad \int x^{n-1} \sin x^n\,dx = \frac{1}{n}\int \sin t\,dt$$

$$= \frac{-1}{n}\cos t$$

$$= \frac{-1}{n}\cos x^n$$

(iii) $\int \frac{1}{e^x + e^{-x}}\,dx = \int \frac{e^x}{e^{2x}+1}\,dx$

Let $e^x = t$ $\Rightarrow$ $e^x\,dx = dt$

$$\therefore \quad \int \frac{e^x}{e^{2x}+1} = \int \frac{dt}{t^2+1}$$

$$= \tan^{-1} t$$

$$= \tan^{-1}(e^x)$$

***Example* 6:** Evaluate

(i) $\int \frac{1}{x} \log x \, dx$ (ii) $\int \cos^2 x \, dx$

(iii) $\int \frac{xdx}{x^2+1}$ (iv) $\int \frac{\sin(\tan^{-1} x)}{1+x^2} dx$

***Solution*:**

(i) $\int \frac{1}{x} \log x \, dx$

Let $\log x = t \Rightarrow \frac{1}{x} dx = dt$

$$\therefore \int \frac{1}{x} \log x \, dx = \int t dt = \frac{t^2}{2} = \frac{1}{2}(\log x)^2$$

(ii) $$\int \cos^2 x \, dx = \int \frac{1+\cos 2x}{2} dx$$

$$= \frac{1}{2}\left[\int 1 dx + \int \cos 2x \, dx\right]$$

$$= \frac{1}{2}\left[x + \frac{\sin 2x}{2}\right]$$

(iii) $\int \frac{xdx}{x^2+1}$

Let $x^2 + 1 = t \Rightarrow 2x \, dx = dt$

$$\therefore \int \frac{xdx}{x^2+1} = \frac{1}{2}\int \frac{1}{t} dt = \frac{1}{2} \log t = \frac{1}{2} \log (x^2 + 1)$$

(iv) $\int \frac{\sin(\tan^{-1} x)}{1+x^2} dx$

Let $\tan^{-1} x = t \Rightarrow \Rightarrow \frac{1}{1+x^2} dx = dt$

$$\therefore \int \frac{\sin(\tan^{-1} x)}{1+x^2} dx = \int \sin t \, dt$$

$$= -\cos t$$
$$= -\cos(\tan^{-1}x)$$

***Example* 7:** Evaluate

(i) $\int \frac{e^x - e^{-x}}{e^x + e^{-x}} dx$ (ii) $\int \sin 3x \cos 5x dx$

***Solution*:**

(i) $\int \frac{e^x - e^{-x}}{e^x + e^{-x}} dx$

Let $e^x + e^{-x} = t \quad \Rightarrow \quad (e^x - e^{-x})\, dx = dt$

$$\therefore \quad \int \frac{e^x - e^{-x}}{e^x + e^{-x}} dx = \int \frac{1}{t} dt$$

$$= \log t = \log(e^x + e^{-x})$$

(ii) $$\int \sin 3x \cos 5x\, dx = \frac{1}{2}\int 2\sin 3x \cos 5x\, dx$$

$$= \frac{1}{2}\int [\sin(3x + 5x) + \sin(3x - 5x)] dx$$

$$= \frac{1}{2}\int [\sin 8x - \sin 2x] dx$$

$$= \frac{1}{2}\left[\frac{-\cos 8x}{8} + \frac{\cos 2x}{2}\right]$$

$$= \frac{1}{16}\int [4\cos 2x - \cos 8x]$$

***Example* 8:** Integrate

(i) $\int \frac{1}{x \log x} dx$ (ii) $\int \sin(e^x).e^x dx$ (iii) $\int \frac{x^2 dx}{x^3 + 1}$

***Solution*:**

(i) $\int \frac{1}{x \log x} dx$

Let $\log x = t \Rightarrow \frac{1}{x}\,dx = dt$

$$\therefore \int \frac{1}{x\log x}\,dx = \int \frac{1}{t}\,dt$$

$$= \log t = \log(\log x)$$

(ii) $\int \sin(e^x).e^x dx$

Let $e^x = t \Rightarrow e^x\,dx = dt$

$$\therefore \int \sin(e^x).\,e^x dx = \int \sin t\,dt$$

$$= -\cos t = -\cos(e^x)$$

(iii) $\int \frac{x^2}{x^3+1}\,dx$

Let $x^3 + 1 = t \Rightarrow 3x^2\,dx = dt$

$$\therefore \int \frac{x^2 dx}{x^3+1} = \frac{1}{3}\int \frac{1}{t}\,dt$$

$$= \frac{1}{3}\log t$$

$$= \frac{1}{3}\log(x^3+1)$$

***Example* 9:** Integrate

(i) $\int \frac{1+\cos x}{x+\sin x}\,dx$ (ii) $\int \frac{\sec x}{\log(\sec x+\tan x)}\,dx$

(iii) $\int \frac{e^x-\sin x}{e^x+\cos x}\,dx$ (iv) $\int \frac{dx}{\sqrt{1+\cos 2x}}$

(v) $\int \frac{e^x(1+x)}{\cos^2(xe^x)}\,dx$ (vi) $\int xe^{x^2}\,dx$

***Solution*:**

(i) $\int \frac{1+\cos x}{x+\sin x}\,dx$

Let $x + \sin x = t \Rightarrow (1 + \cos x)\,dx = dt$

$$\therefore \quad \int \frac{1+\cos x}{x+\sin x}dx = \int \frac{dt}{t}$$

$$= \log t = \log(x + \sin x)$$

(ii) $\int \frac{\sec x}{\log(\sec x + \tan x)}dx$

Let $\log(\sec x + \tan x) = t \Rightarrow \frac{1}{(\sec x + \tan x)}$

$(\sec x \tan x + \sec^2 x)\, dx = dt$

$\Rightarrow \quad \sec x\, dx = dt$

$$\therefore \quad \int \frac{\sec x\, dx}{\log(\sec x + \tan x)} = \int \frac{dt}{t}$$

$= \log t = \log \log(\sec x + \tan x)$

(iii) $\int \frac{e^x - \sin x}{e^x + \cos x}dx$

Let $e^x + \cos x = t \Rightarrow (e^x - \sin x)\, dx = dt$

$$\therefore \quad \int \frac{e^x - \sin x}{e^x + \cos x}dx = \int \frac{dt}{t}$$

$$= \log t = \log(e^x + \cos x)$$

(iv) $\int \frac{dx}{\sqrt{1+\cos 2x}} = \int \frac{dx}{\sqrt{2\cos^2 x}}$

$$= \frac{1}{\sqrt{2}}\int \frac{dx}{\cos x}$$

$$= \frac{1}{\sqrt{2}}\int \sec x\, dx$$

$$= \frac{1}{\sqrt{2}} \log \tan \left(\frac{\pi}{4} + \frac{x}{2}\right)$$

(v) $\int \frac{e^x(1+x)}{\cos^2(xe^x)}dx$

Let $xe^x = t \Rightarrow (xe^x + e^x)\, dx = dt$

$\Rightarrow \quad e^x(x + 1)\, dx = dt$

$$\therefore \quad \int \frac{e^x(1+x)}{\cos^2(xe^x)}dx = \int \frac{dt}{\cos^2 t}$$

$$= \int \sec^2 t\, dt$$

$$= \tan t$$

$$= \tan(xe^x)$$

(vi) $\int xe^{x^2}dx$

Let $\quad x^2 = t \Rightarrow \quad 2x\,dx = dt$

$$\therefore \quad \int xe^{x^2}dx = \frac{1}{2}\int e^t dt$$

$$= \frac{1}{2}e^t = \frac{1}{2}e^{x^2}$$

***Example* 10:** Evaluate

(i) $\int \frac{dx}{\sqrt{x^2-9}}$ (ii) $\int \frac{dx}{\sqrt{9-4x^2}}$

(iii) $\int \sqrt{\frac{a+x}{a-x}}\,dx$ (iv) $\int \frac{dx}{9x^2-4}$

(v) $\int \sqrt{2-3x^2}dx$

***Solution*:**

(i) $$\int \frac{dx}{\sqrt{x^2-9}} = \int \frac{dx}{\sqrt{x^2-3^2}}$$

$$= \log\left[x+\sqrt{x^2-9}\right]$$

(ii) $$\int \frac{dx}{\sqrt{9-4x^2}} = \frac{1}{4}\int \frac{dx}{\frac{9}{4}-x^2}$$

$$= \frac{1}{4}\int \frac{dx}{\sqrt{\left(\frac{3}{2}\right)^2 - x^2}}$$

$$= \frac{1}{2}\sin^{-1}\frac{x}{\frac{3}{2}}$$

$$= \frac{1}{2}\sin^{-1}\frac{2x}{3}$$

(iii) $\int \sqrt{\frac{a+x}{a-x}}\, dx$

Put $x = a\cos\theta \quad \Rightarrow \quad dx = -a\sin\theta\, d\theta$

$$\therefore \quad \int \sqrt{\frac{a+x}{a-x}}\, dx = \int \sqrt{\frac{a+a\cos\theta}{a-a\cos\theta}}(-a\sin\theta)\, d\theta$$

$$= -a\int \sqrt{\frac{1+\cos\theta}{1-\cos\theta}}.\sin\theta\, d\theta$$

$$= -a\int \sqrt{\frac{2\cos^2\frac{\theta}{2}}{2\sin^2\frac{\theta}{2}}}.\sin\theta\, d\theta$$

$$= -a\int \frac{\cos\frac{\theta}{2}}{\sin\frac{\theta}{2}}.2\sin\frac{\theta}{2}\cos\frac{\theta}{2}\, d\theta$$

$$= -2a\int \cos^2\frac{\theta}{2}\, d\theta$$

$$= -a\int (1+\cos\theta)\, d\theta$$

$$= -a\int (\theta+\sin\theta)$$

$$= -a\left[\cos^{-1}\frac{x}{a} + \frac{\sqrt{a^2-x^2}}{a}\right]$$

(iv) $\int \frac{dx}{9x^2-4} = \frac{1}{9}\int \frac{dx}{x^2-\frac{4}{9}}$

$$= \frac{1}{9}\int \frac{dx}{x^2 - \left(\frac{2}{3}\right)^2}$$

$$= \frac{1}{9}\frac{1}{2\times\frac{2}{3}}\log\frac{x-\frac{2}{3}}{x+\frac{2}{3}} \qquad \left(\because \int\frac{dx}{x^2-a^2} = \frac{1}{2a}\log\frac{x-a}{x+a}\right)$$

$$= \frac{1}{12}\log\frac{3x-2}{3x+2}$$

(v) $$\int\sqrt{2-3x^2}\,dx = \sqrt{3}\int\sqrt{\frac{2}{3}-x^2}\,dx$$

$$= \sqrt{3}\int\sqrt{\left(\frac{2}{3}\right)^2 - x^2}\,dx$$

$$\left[\because \sqrt{a^2-x^2} = \frac{x}{2}\sqrt{a^2-x^2} + \frac{1}{2}a^2\sin^{-1}\frac{x}{a}\right]$$

$$= \sqrt{3}\left[\frac{x}{2}\sqrt{\frac{2}{3}-x^2} + \frac{1}{2}\times\frac{2}{3}\sin^{-1}\frac{x}{\sqrt{\frac{2}{3}}}\right]$$

$$\left[\because \sqrt{a^2-x^2} = \frac{x}{2}\sqrt{a^2-x^2} + \frac{1}{2}a^2\sin^{-1}\frac{x}{a}\right]$$

$$= \sqrt{3}\left[\frac{x}{2}\sqrt{\frac{2-3x^2}{3}} + \frac{1}{3}\sin^{-1}x\sqrt{\frac{3}{2}}\right]$$

$$= \frac{x}{2}\sqrt{2-3x^2} + \frac{1}{\sqrt{3}}\sin^{-1}x\frac{\sqrt{3}}{2}$$

***Example* 11:** Evaluate

(i) $\int x^2\sin 3x\,dx$ (ii) $\int x\cos x\,dx$

(iii) $\int x^2\cos x\,dx$ (iv) $\int xe^x dx$

(v) $\int x \sin dx$

***Solution*:**

(i) $$\int x^2 \sin 3x\, dx = x^2 \int \sin 3x dx - \int \left[\frac{d}{dx} x^2 \int \sin 3x dx \right] dx$$

$$= -x^2 \frac{\cos 3x}{3} + \int 2x \frac{\cos 3x}{3} dx$$

$$= \frac{-x^2 \cos 3x}{3} + \frac{2}{3} \left\{ x \int \cos 3x dx - \int \left[\frac{d}{dx} x \int \cos 3x dx \right] dx \right\}$$

$$= \frac{-x^2 \cos 3x}{3} + \frac{2}{3} \left\{ x \frac{\sin 3x}{3} - \int \frac{\sin 3x}{3} dx \right\}$$

$$= \frac{-x^2 \cos 3x}{3} + \frac{2}{9} x \sin 3x + \frac{2}{9} \frac{\cos 3x}{3}$$

$$= \frac{-1}{3} x^2 \cos 3x + \frac{2}{9} x \sin 3x + \frac{2}{27} \cos 3x$$

(ii) $$\int x \cos x dx = x \int \cos x dx - \int \left[\frac{d}{dx} x \int \cos x dx \right] dx$$

$$= x \sin x - \int \sin x\, dx$$

$$= x \sin x + \cos x$$

(iii) $$\int x^2 \cos x dx = x^2 \int \cos x dx - \int \left[\frac{d}{dx} x^2 \int \cos x dx \right] dx$$

$$= x^2 \sin x - \int 2x \sin x\, dx$$

$$= x^2 \sin x - 2 \left\{ x \int \sin x\, dx - \int \left[\frac{d}{dx} x \int \sin x\, dx \right] dx \right\}$$

$$= x^2 \sin x - 2 \left\{ -x \cos x + \int \cos x\, dx \right\}$$

$$= x^2 \sin x + 2x \cos x - 2 \sin x$$

(iv) $$\int x e^x dx = x \int e^x dx - \int \left[\frac{d}{dx} x \int e^x dx \right] dx$$

$$= x e^x - \int e^x dx$$

$$= x\,e^x - e^x$$

$$= e^x\,(x - 1)$$

(v) $\int x \sin dx = x \int \sin x dx - \int \left[\frac{d}{dx} x \int \sin x dx \right] dx$

$$= -x \cos x + \int \cos x dx$$

$$= -x \cos x + \sin x$$

***Example* 12:** Evaluate

(i) $\int (2 - x)(x + 1)\,dx$ (ii) $\int (x + \frac{1}{x})^2\,dx$

(iii) $\int (5 - 4x)(1 + x)\,dx$

***Solution*:**

(i) $\int (2 - x)(x + 1)\,dx = \int (2x + 2 - x^2 - x) dx$

$$= \int (x + 2 - x^2)\,dx$$

$$= \frac{x^2}{2} + 2x - \frac{x^3}{3}$$

(ii) $\int (x + \frac{1}{x})^2\,dx = \int (x^2 + \frac{1}{x^2} + 2) dx$

$$= \frac{x^3}{3} + \frac{x^{-1}}{-1} + 2x$$

$$= \frac{x^3}{3} - \frac{1}{x} + 2x$$

(iii) $\int (5 - 4x)(1 + x)\,dx = \int (5 + 5x - 4x - 4x^2)\,dx$

$$= \int (5 + x - 4x^2)\,dx$$

$$= 5x + \frac{x^2}{2} - \frac{4x^3}{3}$$

***Example* 13:** Evaluate

(i) $\int \frac{\cot x}{\log(\sin x)} dx$ (ii) $\int \frac{\sin x \cos x}{a\cos^2 x + b\sin^2 x} dx$

(iii) $\int \frac{1}{x+\sqrt{x}} dx$ (iv) $\int \frac{x}{1-3x^2} dx$

(v) $\int \frac{\sin\sqrt{x}}{\sqrt{x}} dx$ (vi) $\int \frac{x}{x^2-1} dx$

***Solution*:**

(i) $\int \frac{\cot x}{\log(\sin x)} dx$

Let $\log(\sin x) = t \Rightarrow \cot x\, dx = dt$

$$\therefore \int \frac{\cot x dx}{\log(\sin x)} = \int \frac{dt}{t}$$

$$= \log t$$

$$= \log \log(\sin x)$$

(ii) $\int \frac{\sin x \cos x}{a\cos^2 x + b\sin^2 x} dx$

Let a $\cos2 x + b \sin2 x = t$

$\Rightarrow$ $(-2a \sin x \cos x + 2b \sin x \cos x)\, dx = dt$

$\Rightarrow$ $2(b - a) \sin x \cos x\, dx = dt$

$$\therefore \int \frac{\sin x \cos x}{a\cos^2 x + b\sin^2 x} = \frac{1}{2(b-a)}\int \frac{dt}{t}$$

$$= \frac{1}{2(b-a)} \log t$$

$$= \frac{1}{2(b-a)} \log(a\cos^2 x + b\sin^2 x)$$

(iii) $\int \frac{1}{x+\sqrt{x}} dx$

Let $\sqrt{x} = t \quad \Rightarrow \quad x = t^2 \quad \Rightarrow \quad dx = 2t\,dt$

$$\therefore \quad \int \frac{1}{x+\sqrt{x}}\,dx = \int \frac{2t\,dt}{t^2+t}$$

$$= 2\int \frac{1}{t+1}\,dt$$

$$= 2 \log (t + 1)$$

$$= 2 \log (\sqrt{x} + 1)$$

(iv) $\int \frac{x}{1-3x^2}\,dx$

Let $1 - 3x^2 = t \quad \Rightarrow \quad -6x\,dx = dt$

$$\therefore \quad \int \frac{x}{1-3x^2}\,dx = \frac{-1}{6}\int \frac{1}{t}\,dt$$

$$= \frac{-1}{6}\log t \qquad = \frac{-1}{6}\log (1-3x^2)$$

(v) $\int \frac{\sin\sqrt{x}}{\sqrt{x}}\,dx$

Let $\sqrt{x} = t \quad \Rightarrow \quad \frac{1}{2\sqrt{x}}\,dx = dt$

$$\therefore \quad \int \frac{\sin\sqrt{x}}{\sqrt{x}}\,dx = 2\int \sin t\;dt$$

$$= -2 \cos t$$

$$= -2 \cos \sqrt{x}$$

(vi) $\int \frac{x}{x^2-1}\,dx$

Let $x^2 - 1 = t \quad \Rightarrow \quad 2x\,dx = dt$

$$\therefore \quad \int \frac{x\,dx}{x^2-1} = \frac{1}{2}\int \frac{dt}{t}$$

$$= \frac{1}{2}\log t$$

$$= \frac{1}{2} \log (x^2 - 1)$$

***Example* 14:** Evaluate

(i) $\int \sin^2 2x \, dx$ (ii) $\int \sqrt{1 + \sin x} \, dx$

(iii) $\int \cos^2 2x \, dx$

***Solution*:**

(i) $$\int \sin^2 2x \, dx = \int \frac{1 - \cos 4x}{2} dx \qquad (\because \cos 4x = 1 - 2\sin^2 2x)$$

$$= \frac{1}{2}\left[\int (1 - \cos 4x) \, dx\right]$$

$$= \frac{1}{2}\left[x - \frac{\sin 4x}{4}\right]$$

(ii) $$\int \sqrt{1 + \sin x} \, dx = \int \sqrt{\sin^2 \frac{x}{2} + \cos^2 \frac{x}{2} + 2 \sin \frac{x}{2} \cos \frac{x}{2}}$$

$$= \int \sqrt{\left(\sin \frac{x}{2} + \cos \frac{x}{2}\right)^2} \, dx$$

$$= \int \sqrt{\left(\sin \frac{x}{2} + \cos \frac{x}{2}\right)} \, dx$$

$$= -\frac{\cos \frac{x}{2}}{\frac{1}{2}} + \frac{\sin \frac{x}{2}}{\frac{1}{2}}$$

$$= 2\left(\sin \frac{x}{2} - \cos \frac{x}{2}\right)$$

(iii) $$\int \cos^2 2x \, dx = \int \frac{1 + \cos 4x}{2} dx$$

$$= \frac{1}{2} \int (1 + \cos 4x) \, dx$$

$$= \frac{1}{2}\left[x + \frac{\sin 4x}{4}\right]$$

***Example* 15:** Evaluate

(i) $\int x \log x dx$ (ii) $\int \tan^{-1} x\, dx$

(iii) $\int \log x\, dx$ (iv) $\int \log(1+x^2)\, dx$

***Solution*:**

(i) $$\int x \log x\, dx = \log x \int x\, dx - \int\left[\frac{d}{dx}\log x \int x\, dx\right] dx$$

$$= \log x \,.\, \frac{x^2}{2} - \int \frac{1}{x}.\frac{x^2}{2}\, dx$$

$$= \frac{x^2}{2}\log x - \frac{1}{2}\int x\, dx$$

$$= \frac{x^2}{2}\log x - \frac{x^2}{4}$$

(ii) $$\int \tan^{-1} x\, dx = \tan^{-1} x \int 1 dx - \int\left[\frac{d}{dx}\tan^{-1} x \int 1\, dx\right] dx$$

$$= \tan^{-1} x\,.\,x - \int \frac{1}{1+x^2}.\, x\, dx$$

$$= x \tan^{-1} x - \frac{1}{2}\int \frac{2x}{1+x^2}\, dx$$

$$= x \tan^{-1} x - \frac{1}{2}\log(1+x^2)$$

(iii) $$\int \log x\, dx = \log x \int 1\, dx - \int\left[\frac{d}{dx}\log x \int 1\, dx\right] dx$$

$$= x \log x - \int \frac{1}{x}.\, x\, dx$$

$$= x \log x - x$$

$$= x (\log x - 1)$$

(iv) $$\int \log(1+x^2)\,dx = \log(1+x^2)\int 1dx - \int\left[\frac{d}{dx}\log(1+x^2)\int 1dx\right]dx$$

$$= x \log (1+x^2) - \int \frac{1}{1+x^2}.2x.x\,dx$$

$$= x \log (1+x^2) - 2\int \frac{x^2}{x^2+1}dx$$

$$= x \log (1+x^2) - 2\int\left(1-\frac{1}{x^2+1}\right)dx$$

$$= x \log (1+x^2) - 2(x - \tan^{-1} x)$$

$$= x \log(1+x^2) - 2x + 2\tan^{-1} x$$

***Example* 16:** Evaluate

(i) $\int e^{mx} \sin x \cos x\,dx$ (ii) $\int e^{3x} \cos^3 x\,dx$

(iii) $\int e^x (\sin x + \cos x)\,dx$

***Solution*:**

(i) $$\int e^{mx} \sin x \cos x\,dx$$

$$= \frac{1}{2}\int e^{mx}(2\sin x \cos x)\,dx$$

$$= \frac{1}{2}\int e^{mx} \sin 2x\,dx$$

$$= \frac{1}{2}\cdot\frac{e^{mx}}{\sqrt{m^2+4}} \sin\left(2x - \tan^{-1}\frac{2}{m}\right)$$

(ii) $$\int e^{3x} \cos^3 x\,dx$$

$$= \int e^{3x}\left(\frac{\cos 3x + 3\cos x}{4}\right)dx$$

$$(\because \cos 3x = 4\cos^3 x - 3\cos x)$$

$$= \frac{1}{4}\int e^{3x}\cos 3x\, dx + \frac{3}{4}\int e^{3x}\cos x\, dx$$

$$= \frac{1}{4}\cdot\frac{e^{3x}}{\sqrt{9+9}}\cos\left(3x - \tan^{-1}\frac{3}{3}\right) + \frac{3}{4}\cdot\frac{e^{3x}}{\sqrt{9+1}}\cos\left(x - \tan^{-1}\frac{1}{3}\right)$$

$$= \frac{e^{3x}}{4\sqrt{18}}\cos(3x - \tan^{-1}1) + \frac{3e^{3x}}{4\sqrt{10}}\cos\left(x - \tan^{-1}\frac{1}{3}\right)$$

(iii) $\int e^x(\sin x + \cos x)\, dx = \int e^x \sin x\, dx + \int e^x \cos x\, dx$

$$= \int e^x \sin x\, dx + e^x\int \cos x\, dx - \int\left[\frac{d}{dx}e^x \int \cos x\, dx\right]dx$$

$$= \int e^x \sin x\, dx + e^x \sin x - \int e^x \sin x\, dx$$

$$= e^x \sin x$$

Integration by Partial Fractions

***Example* 17:** Evaluate

(i) $\int \frac{x+1}{(x-1)(x+2)}dx$ (ii) $\int \frac{3x}{x^2 - x - 2}dx$

(iii) $\int \frac{x-1}{(x-3)(x-2)}dx$ (iv) $\int \frac{2x+1}{(2x+3)(3x-4)}dx$

(v) $\int \frac{2x}{(x-1)(x+3)}dx$ (vi) $\int \frac{x}{x^2 + x - 6}dx$

***Solution*:**

(i) $\int \frac{x+1}{(x-1)(x+2)}dx$

Let $\frac{x+1}{(x-1)(x+2)} = \frac{A}{x-1} + \frac{B}{(x+2)}$

$$= \frac{A(x+2) + B(x-1)}{(x-1)(x+2)}$$

$$x + 1 = A(x+2) + B(x-1) \quad \text{.... (1)}$$

Put $x + 2 = 0 \Rightarrow x = -2$ in equation (1) we get

$$-1 = -3B \Rightarrow B = \frac{1}{3}$$

Put $x - 1 = 0 \Rightarrow x = 1$ in (i) we get

$$2 = 3A \Rightarrow A = \frac{2}{3}$$

$$2 = 3A \Rightarrow A = \frac{2}{3}$$

$$\therefore \quad \frac{x+1}{(x-1)(x+2)} = \frac{2}{3}.\frac{1}{x-1} + \frac{1}{3}.\frac{1}{x+2}$$

$$\int \frac{x+1}{(x-1)(x+2)} dx = \frac{2}{3}\int \frac{1}{x-1} dx + \frac{1}{3}\int \frac{1}{x+2} dx$$

$$= \frac{2}{3} \log (x-1) + \frac{1}{3} \log (x+2)$$

(ii) $$\int \frac{3x}{x^2 - x - 2} dx = \int \frac{3x}{(x-2)(x+1)} dx$$

Let $$\frac{3x}{(x-2)(x+1)} = \frac{A}{x-2} + \frac{B}{x+1}$$

$$= \frac{A(x+1) + B(x-2)}{(x-2)(x+1)}$$

$3x = A(x + 1) + B(x - 2)$ ..... (1)

Put $x = -1$, and $x = 2$ in (1) we get

$-3 = -3B \Rightarrow B = 1$

$6 = 3A \Rightarrow A = 2$

$$\therefore \quad \frac{3x}{(x-2)(x+1)} = \frac{2}{x-2} + \frac{1}{x+1}$$

$$\int \frac{3x}{(x-2)(x+1)} dx = 2\int \frac{1}{x-2} dx + \int \frac{1}{x+1}$$

$$= 2 \log (x - 2) + \log (x + 1)$$

(iii) $\int \frac{x-1}{(x-3)(x-2)} dx$

Let $\frac{x-1}{(x-3)(x-2)} = \frac{A}{x-3} + \frac{B}{x-2}$

$$= \frac{A(x-2)+B(x-3)}{(x-3)(x-2)}$$

$x - 1 = A(x-2) + B(x-3)$ ..... (1)

Put x = 2, 3 in (1) we get

$1 = -B \quad \Rightarrow \quad B = -1$

$2 = A \quad \Rightarrow \quad A = 2$

$\therefore \quad \frac{x-1}{(x-3)(x-2)} = \frac{2}{x-3} - \frac{1}{x-2}$

$$\int \frac{x-1}{(x-3)(x-2)} dx = 2\int \frac{1}{x-3} dx - \int \frac{1}{x-2} dx$$

$$= 2 \log (x-3) - \log (x-2)$$

(iv) $\int \frac{2x+1}{(2x+3)(3x-4)} dx$

Let $\frac{2x+1}{(2x+3)(3x-4)} = \frac{A}{2x+3} + \frac{B}{3x-4}$

$$= \frac{A(3x-4)+B(2x+3)}{(2x+3)(3x-4)}$$

$2x + 1 = A(3x-4) + B(2x+3)$ ..... (1)

Put $x = \frac{4}{3}, \frac{-3}{2}$ in (i), we have

$B = \frac{11}{17}, A = \frac{4}{17}$

$\therefore \quad \frac{2x+1}{(2x+3)(3x-4)} = \frac{4}{17}\frac{1}{2x+3} + \frac{11}{17}.\frac{1}{3x-4}$

$$\int \frac{2x+1}{(2x+3)(3x-4)} dx = \frac{4}{17}\int \frac{1}{2x+3} dx + \frac{11}{17}\int \frac{1}{3x-4} dx$$

$$= \frac{4}{17}\log(2x+3)\times\frac{1}{2} + \frac{11}{17}\log(3x-4)\times\frac{1}{3}$$

$$= \frac{4}{34}\log(2x+3) + \frac{11}{51}\log(3x-4)$$

(v) $\int \frac{2x}{(x-1)(x+3)}dx$

Let $\frac{2x}{(x-1)(x+3)} = \frac{A}{x-1} + \frac{B}{x+3}$

$$= \frac{A(x+3)+B(x-1)}{(x-1)(x+3)}$$

$2x = A(x+3) + B(x-1)$

Put x = 1, –3 in (1), we get

$A = \frac{1}{2}, \quad B = \frac{3}{2}$

$\therefore \quad \frac{2x}{(x-1)(x+3)} = \frac{1}{2(x-1)} + \frac{3}{2(x+3)}$

$$\int \frac{2x}{(x-1)(x+3)}dx = \frac{1}{2}\int\frac{1}{x-1}dx + \frac{2}{3}\int\frac{1}{x+3}dx$$

$$= \frac{1}{2}\log(x-1) + \frac{2}{3}\log(x+3)$$

(vi) $\int \frac{x}{x^2+x-6}dx = \int \frac{x}{(x+3)(x-2)}dx$

Let $\frac{x}{(x+3)(x-2)} = \frac{A}{x+3} + \frac{B}{x-2}$

$$= \frac{A(x-2)+B(x+3)}{(x+3)(x-2)}$$

$x = A(x-2) + B(x+3)$ .... (1)

Put x = 2, –3 in (1) we have

$A = \frac{3}{5}, \; B = \frac{2}{5}$

$$\therefore \quad \frac{x}{x^2+x-6} = \frac{3}{5(x+3)} + \frac{2}{5(x-2)}$$

$$\int \frac{x}{x^2+x-6}\,dx = \frac{3}{5}\int \frac{1}{x+3}\,dx + \frac{2}{5}\int \frac{1}{x-2}\,dx$$

$$= \frac{3}{5}\log(x+3) + \frac{2}{5}\log(x-2)$$

***Example* 18:** Evaluate

(i) $\int \frac{x}{(1+x)(1-x)^2}\,dx$ (ii) $\int \frac{x}{(x-1)(x+1)^2}\,dx$

***Solution*:** (i) $\int \frac{x}{(1+x)(1-x)^2}\,dx$

Let $\frac{x}{(1+x)(1-x)^2} = \frac{A}{1+x} + \frac{B}{1-x} + \frac{C}{(1-x)^2}$

$$x = A(1-x)^2 + B(1+x)(1-x) + C(1+x) \quad \text{.... (1)}$$

Put $x = -1, 1$ in (1) we have

When $x = 1$, $c = \frac{1}{2}$

$x = -1$, $A = \frac{-1}{4}$

$x = 0$ $B = \frac{-1}{4}$

$$\therefore \quad \frac{x}{(1+x)(1-x)^2} = \frac{-1}{4(1+x)} - \frac{1}{4(1-x)} + \frac{1}{2(1-x)^2}$$

$$\int \frac{x}{(1+x)(1-x)^2}\,dx = \frac{-1}{4}\int \frac{dx}{1+x} - \frac{1}{4}\int \frac{1}{1-x}\,dx + \frac{1}{2}\int \frac{1}{(1-x)^2}\,dx$$

$$= \frac{-1}{4}\log(1+x) + \frac{1}{4}\log(1-x) + \frac{1}{2}.\frac{1}{1-x}$$

$$= \frac{1}{4}[\log(1-x) - \log(1+x)] + \frac{1}{2(1-x)}$$

$$=\frac{1}{4}\log\left(\frac{1-x}{1+x}\right)+\frac{1}{2(1-x)}$$

(ii) $\int \frac{x}{(x-1)(x+1)^2}dx$

Let $\frac{x}{(x-1)(x+1)^2}=\frac{A}{x-1}+\frac{B}{x+1}+\frac{C}{(x+1)^2}$

$$x = A(x+1)^2 + B(x-1)(x+1) + C(x-1) \quad .....(1)$$

Put $x = -1, 1$ in (1) we get

When $x = 1,\quad A=\frac{1}{4}$

$x=-1,\quad C=\frac{1}{2}$

$x=0,\quad B=\frac{-1}{4}$

$$\therefore \int \frac{x}{(x-1)(x+1)^2}dx=\frac{1}{4}\int \frac{1}{x-1}dx-\frac{1}{4}\int \frac{1}{x+1}dx+\frac{1}{2}\int \frac{1}{(x+1)^2}dx$$

$$=\frac{1}{4}\log(x-1)-\frac{1}{4}\log(x+1)-\frac{1}{2(x+1)}$$

$$=\frac{1}{4}\log\frac{x-1}{x+1}-\frac{1}{2(x+1)}$$

***Example* 19:** Evaluate

(i) $\int \frac{x}{(2x+1)(x^2+1)}dx$ (ii) $\int \frac{dx}{(x^2+1)(x-1)}$

(iii) $\int \frac{x+1}{(2x+3)(x^2+1)}dx$

***Solution*:**

(i) $\int \frac{x}{(2x+1)(x^2+1)}dx$

Let $\dfrac{x}{(2x+1)(x^2+1)} = \dfrac{A}{2x+1} + \dfrac{Bx+c}{x^2+1}$

$x = A(x^2 + 1) + (Bx + c)(2x + 1)$

$x = x^2(A + 2B) + x(B + 2C) + A + C$ ….. (1)

Putting $x = \dfrac{-1}{2}$, $A = \dfrac{-2}{5}$

On comparing coefficients of x^2, x and we get

$A + 2B = 0 \Rightarrow B = \dfrac{-A}{2} = \dfrac{-1}{2}\left(\dfrac{-2}{5}\right) = \dfrac{1}{5}$

$B + 2C = 1$

$A + C = 0 \Rightarrow C = -A \Rightarrow C = \dfrac{2}{5}$

$$\therefore \int \frac{x}{(2x+1)(x^2+1)}dx = \frac{-2}{5}\int \frac{1}{2x+1}dx + \int \frac{\frac{1}{5}x + \frac{2}{5}}{x^2+1}dx$$

$$= \frac{-2}{5}\int \frac{1}{2x+1}dx + \frac{1}{5}\int \frac{x}{x^2+1}dx + \frac{2}{5}\int \frac{1}{x^2+1}dx$$

$$= \frac{-2}{5}\log(2x+1)\times\frac{1}{2} + \frac{1}{10}\log(x^2+1) + \frac{2}{5}\tan^{-1}x$$

$$= \frac{-1}{5}\log(2x+1) + \frac{1}{10}\log(x^2+1) + \frac{2}{5}\tan^{-1}x + c$$

(ii) $\displaystyle\int \frac{dx}{(x^2+1)(x-1)}$

Let $\dfrac{1}{(x^2+1)(x-1)} = \dfrac{Ax+B}{x^2+1} + \dfrac{C}{x-1}$

$1 = (Ax + B)(x - 1) + C(x^2 + 1)$ ….. (1)

Putting $x = 1,\ c = \dfrac{1}{2}$

From (1) $1 = x^2(A + C) + x(B - A) + (C - B)$

On comparing the coefficient, we get

$$A + C = 0 \quad \Rightarrow \quad A = -C = \frac{-1}{2}$$

$$B - A = 0 \quad \Rightarrow \quad B = A = \frac{-1}{2}$$

$$\therefore \int \frac{dx}{(x^2+1)(x-1)} = \frac{-1}{2}\int \frac{x}{x^2+1}dx - \frac{1}{2}\int \frac{1}{x^2+1}dx + \frac{1}{2}\int \frac{1}{x-1}dx$$

$$= \frac{-1}{4}\log(x^2+1) - \frac{1}{2}\tan^{-1}x + \frac{1}{2}\log x + c$$

(iii) $\displaystyle\int \frac{x+1}{(2x+3)(x^2+1)}dx$

Let $\displaystyle\frac{x+1}{(2x+3)(x^2+1)} = \frac{A}{2x+3} + \frac{Bx+c}{x^2+1}$

$$x + 1 = A(x^2 + 1) + (Bx + C)(2x + 3) \qquad \text{..... (1)}$$

Putting $x = \dfrac{-3}{2}$, we get $\quad A = \dfrac{-2}{13}$

From (1) $x + 1 = x^2(A + 2B) + x(3B + 2C) + (A + 3C)$

on comparing the coefficient, we get

$$A + 2B = 0 \quad \Rightarrow \quad B = \frac{-A}{2} = \frac{-1}{2}\times\left(\frac{-2}{13}\right) = \frac{1}{13}$$

$$A + 3C = 1 \quad \Rightarrow \quad C = \frac{1-A}{3} = \frac{1+\frac{2}{13}}{3} = \frac{15}{39}$$

$$\therefore \int \frac{x+1}{(2x+3)(x^2+1)}dx = \frac{-2}{13}\int \frac{1}{2x+3}dx + \frac{1}{13}\int \frac{x}{x^2+1}dx + \frac{15}{39}\int \frac{1}{x^2+1}dx$$

$$= \frac{-1}{13}\log(2x+3) + \frac{1}{26}\log(x^2+1) + \frac{15}{39}\tan^{-1}x + c$$

Definite Integrals

***Example* 20:** Evaluate

(i) $\int_0^1 \frac{(\tan^{-1} x)^2}{1+x^2} dx$ (ii) $\int_0^1 x^2 e^x dx$

(iii) $\int_0^{\frac{\pi}{2}} \cos x \sin^2 x \, dx$ (iv) $\int_0^1 \tan^{-1} x \, dx$

(v) $\int_1^2 \frac{dx}{(x+3)(x+4)}$ (vi) $\int_0^1 \frac{x^2}{1+x^6} dx$

Solution:

(i) $\int_0^1 \frac{(\tan^{-1} x)^2}{1+x^2} dx$

Put $\tan^{-1} x = t \quad \Rightarrow \quad \frac{1}{1+x^2} dx = dt$

Also when $x = 0, \quad t = 0$

when $x = 1, \; t = \frac{\pi}{4}$

$$\therefore \quad \int_0^1 \frac{(\tan^{-1} x)^2}{1+x^2} dx = \int_0^{\frac{\pi}{4}} t^2 dt$$

$$= \left[\frac{t^3}{3} \right]_0^{\frac{\pi}{4}}$$

$$= \frac{1}{3} \left[\frac{\pi^3}{64} - 0 \right] = \frac{\pi^3}{192}$$

(ii) $\int_0^1 x^2 e^x dx$

Integrating by parts $x^2 \int_0^1 e^x dx - \int_0^1 \left[\frac{d}{dx} x^2 \int e^x dx \right] dx$

$$= \left[x^2e^x\right]_0^1 - \int_0^1 2xe^x dx$$

$$= (e-0) - 2\left\{x\int_0^1 e^x dx - \int_0^1\left[\frac{d}{dx}x\int e^x dx\right]dx\right\}$$

$$= e - 2\left\{(xe^x)_0^1 - \int_0^1 e^x dx\right\}$$

$$= e - 2\left[(e-0) - (e^x)_0^1\right]$$

$$= e - 2[e - (e-1)]$$

$$= e - 2$$

(iii) $\int_0^{\frac{\pi}{2}} \cos x \sin^2 x\, dx$

Put $\sin x = t \Rightarrow \cos x\, dx = dt$

when $x = 0, t = 0$

when $x = \frac{\pi}{2}, t = 1$

$$\therefore \int_0^{\frac{\pi}{2}} \cos x \sin^2 x\, dx = \int_0^1 t^2 dt$$

$$= \left[\frac{t^3}{3}\right]_0^1$$

$$= \frac{1}{3}$$

(iv) $\int_0^1 \tan^{-1} x\, dx$

Integrating by parts,

$$= \tan^{-1} x \int_0^1 dx - \int_0^1 \left[\frac{d}{dx} \tan^{-1} x \int dx \right] dx$$

$$= \left[x \tan^{-1} x \right]_0^1 - \int_0^1 \frac{x}{1+x^2} dx$$

Put $1 + x^2 = t$

$\Rightarrow \quad 2x\, dx = dt$

when $x = 0, t = 1$

when $x = 1, t = 2$

$$\therefore \quad \int_0^1 \tan^{-1} x dx = \left[x \tan^{-1} x \right]_0^1 - \frac{1}{2} \int_1^2 \frac{1}{t} dt$$

$$= \left(\tan^{-1} 1 - 0 \right) - \frac{1}{2} [\log t]_1^2$$

$$= \frac{\pi}{4} - \frac{1}{2} [\log 2 - \log 1]$$

$$= \frac{\pi}{4} - \frac{1}{2} \log 2 \qquad (\because \log 1 = 0)$$

(v) $\displaystyle \int_1^2 \frac{dx}{(x+3)(x+4)}$

Let $\displaystyle \frac{1}{(x+3)(x+4)} = \frac{A}{x+3} + \frac{B}{x+4}$

$1 = A(x + 4) + B(x + 3)$ .....(i)

Putting $x = -3, -4$ in (1), we get

$A = 1, \qquad B = -1$

$$\therefore \quad \int_1^2 \frac{1}{(x+3)(x+4)} = \int_1^2 \frac{1}{x+3} dx - \int_1^2 \frac{1}{x+4} dx$$

$$= [\log(x+3)]_1^2 - [\log(x+4)]_1^2$$

$$= (\log 5 - \log 4) - (\log 6 - \log 5)$$

$$= \log 5 - \log 4 - \log 6 + \log 5$$
$$= 2 \log 5 - \log 4 - \log 6 + \log 5$$
$$= 2 \log 5 - (\log 4 + \log 6)$$
$$= 2 \log 5 - \log (4 \times 6)$$
$$= \log 25 - \log 24 \qquad = \log \frac{25}{24}$$

(vi) $\int_0^1 \frac{x^2}{1+x^6} dx = \int_0^1 \frac{x^2}{1+(x^3)^2} dx$

Put $x^3 = t \quad \Rightarrow \quad 3x^2\, dx = dt$

when $x = 0, t = 0$

when $x = 1, t = 1$

$$\therefore \quad \int_0^1 \frac{x^2}{1+x^6} dx = \frac{1}{3}\int_0^1 \frac{1}{1+t^2} dt$$
$$= \frac{1}{3}\left[\tan^{-1} 0\right]_0^1$$
$$= \frac{1}{3}\left[\tan_1^{-1} - \tan^{-1} 0\right]$$
$$= \frac{1}{3}\left[\frac{\pi}{4} - 0\right] = \frac{\pi}{12}$$

***Example* 21:** Evaluate

(i) $\int_0^1 x\, e^x dx$ (ii) $\int_0^{\frac{\pi}{2}} \sin^2 x\, dx$

(iii) $\int_0^{\pi} x^2 \cos x\, dx$ (iv) $\int_0^{\frac{\pi}{3}} \frac{\cos x}{3 + 4\sin x} dx$

***Solution*:**

(i) $\int_0^1 x\, e^x dx$

Integrating by parts

$$= x\int_0^1 e^x dx - \int_0^1 \left[\frac{d}{dx} x \int e^x dx\right] dx$$

$$= \left[xe^x\right]_0^1 - \left[e^x\right]_0^1$$

$$= (e - o) - (e - 1)$$

$$= e - e + 1 = 1$$

(ii) $$\int_0^{\frac{\pi}{2}} \sin^2 x\, dx = \int_0^{\frac{\pi}{2}} \frac{1 - \cos 2x}{2} dx$$

$$= \frac{1}{2}\left[\int_0^{\frac{\pi}{2}} dx - \int_0^{\frac{\pi}{2}} \cos 2x\, dx\right]$$

$$= \frac{1}{2}\left[x - \frac{\sin 2x}{2}\right]_0^{\frac{\pi}{2}}$$

$$= \frac{1}{2}\left[\frac{\pi}{2} - \frac{1}{2}\sin 2 \times \frac{\pi}{2} - 0 + 0\right]$$

$$= \frac{1}{2}\left[\frac{\pi}{2} - \frac{1}{2} \times 0\right] = \frac{\pi}{4}$$

(iii) $$\int_0^{\pi} x^2 \cos x\, dx$$

Integrating by parts

$$= x^2 \int_0^{\pi} \cos x\, dx - \int_0^{\pi} \left[\frac{d}{dx} x^2 \int \cos x dx\right] dx$$

$$= \left[x^2 \sin x\right]_0^{\pi} - 2\int_0^{\pi} x \sin x\, dx$$

$$= \left[x^2 \sin x\right]_0^{\pi} - 2\left\{x \int_0^{\pi} \sin x\, dx - \int_0^{\pi} \left[\frac{d}{dx} x \int \sin x dx\right] dx\right\}$$

$$= \left[x^2 \sin x\right]_0^{\pi} - 2\left\{x\left[-x\cos x\right]_0^{\pi} + \int_0^{\pi} \cos x\, dx\right\}$$

$$= \left[x^2 \sin x\right]_0^{\pi} - 2\left[-x\cos x\right]_0^{\pi} - 2\left[\sin x\right]_0^{\pi}$$

$$= (\pi^2 \sin \pi - 0) - 2\,[-\pi \cos \pi + 0] - 2\,[\sin \pi - \sin 0]$$

$$= (0 - 0) + 2[\pi\,(-1)\,] - 2\,[\,0 - 0]$$

$$= -2\,\pi$$

(iv) $\displaystyle\int_0^{\frac{\pi}{3}} \frac{\cos x}{3 + 4\sin x}\, dx$

Put $3 + 4 \sin x = t \quad \Rightarrow \quad 4 \cos x\, dx = dt$

when $x = 0,\ t = 3$

when $x = \dfrac{\pi}{3}, \quad t = 3 + 2\sqrt{3}$

$$\therefore \quad \int_0^{\frac{\pi}{3}} \frac{\cos x}{3 + 4\sin x}\, dx = \frac{1}{4}\int_3^{3+2\sqrt{3}} \frac{dt}{t}$$

$$= \frac{1}{4}\left[\log t\right]_3^{3+2\sqrt{3}}$$

$$= \frac{1}{4}\left[\log(3 + 2\sqrt{3}) - \log 3\right]$$

$$= \frac{1}{4}\log\frac{3 + 2\sqrt{3}}{3}$$

***Example* 22:** Evaluate

(i) $\displaystyle\int_3^5 \frac{\sin(\log x)}{x}\, dx$ (ii) $\displaystyle\int_0^{\pi} \frac{dx}{5 + 3\cos x}$ (iii) $\displaystyle\int_0^{\pi} \frac{d\theta}{3 + 2\sin\theta + \cos\theta}$

***Solution*:**

(i) $\displaystyle\int_3^5 \frac{\sin(\log x)}{x}\, dx$

Put $\log x = t \quad \Rightarrow \quad \dfrac{1}{x}\, dx = dt$

when $x = 3, \quad t = \log 3$

when x = 5, t = log 5

$$\therefore \quad \int_3^5 \frac{\sin(\log x)}{x} dx = \int_{\log 3}^{\log 5} \sin t \, dt$$

$$= [-\cos t]_{\log 3}^{\log 5}$$

$$= -\cos(\log 5) + \cos(\log 3)$$

(ii) $$\int_0^{\pi} \frac{dx}{5+3\cos x} = \int_0^{\pi} \frac{dx}{5+3\frac{1-\tan^2\frac{x}{2}}{1+\tan^2\frac{x}{2}}}$$

$$= \int_0^{\pi} \frac{(1+\tan^2\frac{x}{2})dx}{5(1+\tan^2\frac{x}{2})+3(1-\tan^2\frac{x}{2})}$$

$$= \int_0^{\pi} \frac{\sec^2\frac{x}{2}dx}{8+2\tan^2\frac{x}{2}}$$

$$= \frac{1}{2}\int_0^{\pi} \frac{\sec^2\frac{x}{2}dx}{4+\tan^2\frac{x}{2}}$$

Putting $\tan\frac{x}{2} = t \quad \Rightarrow \quad \frac{1}{2}\sec^2\frac{x}{2}dx = dt$

when x = 0, t = 0

when x = π, t = ∞

$$\therefore \quad \int_0^{\pi} \frac{dx}{5+3\cos x} = \int_0^{\infty} \frac{dt}{4+t^2}$$

$$= \frac{1}{2}\left[\tan^{-1}\frac{t}{2}\right]_0^{\infty}$$

$$= \frac{1}{2}\left[\tan^{-1}\infty - \tan^{-1}0\right]$$

$$=\frac{1}{2}\left[\frac{\pi}{2}-0\right]=\frac{\pi}{4}$$

(iii) $$\int_0^{\pi}\frac{dx}{3+2\sin x+\cos x}=\int_0^{\pi}\frac{dx}{3+2\frac{2\tan\frac{x}{2}}{1+\tan^2\frac{x}{2}}+\frac{1-\tan^2\frac{x}{2}}{1+\tan^2\frac{x}{2}}}$$

$$=\int_0^{\pi}\frac{(1+\tan^2\frac{x}{2})dx}{3(1+\tan^2\frac{x}{2})+4\tan\frac{x}{2}+1-\tan^2\frac{x}{2}}$$

$$=\int_0^{\pi}\frac{\sec^2\frac{x}{2}dx}{4+2\tan^2\frac{x}{2}+4\tan\frac{x}{2}}$$

$$=\frac{1}{2}\int_0^{\pi}\frac{\sec^2\frac{x}{2}dx}{\tan^2\frac{x}{2}+2\tan\frac{x}{2}+2}$$

putting $\tan\frac{x}{2}=t \quad\Rightarrow\quad \frac{1}{2}\sec^2\frac{x}{2}dx=dt$

when $x=0,\ t=0$

when $x=\pi,\ t=\infty$

$$\therefore\quad \int_0^{\pi}\frac{dx}{3+2\sin x+\cos x}=\int_0^{\infty}\frac{dt}{t^2+2t+2}$$

$$=\int_0^{\infty}\frac{dt}{(t+1)^2+1}$$

$$=\left[\tan^{-1}(t+1)\right]_0^{\infty}$$

$$=\tan^{-1}\infty-\tan^{-1}1$$

$$=\frac{\pi}{2}-\frac{\pi}{4}=\frac{\pi}{4}$$

***Example* 23:** Evaluate

(i) $\int_0^1 x^5 e^{x^3} dx$ (ii) $\int_0^{\frac{\pi}{2}} \frac{\sin x}{\sin x + \cos x} dx$

(iii) $\int_{\frac{-\pi}{2}}^{\frac{\pi}{2}} \sin^7 x\, dx$ (iv) $\int_0^{\pi} \cos^7 x\, dx$

Solution:

(i) $\int_0^1 x^5 e^{x^3} dx$

Putting $x^3 = t \Rightarrow 3x^2\, dx = dt$

when $x = 0, t = 0$

$x = 1, t = 1$

$$\therefore \int_0^1 x^5 e^{x^3} dx = \frac{1}{3}\int_0^1 t\, e^t dt$$

$$= \frac{1}{3}\left[t\int_0^1 e^t dt - \int_0^1 \left[\frac{d}{dt} t . \int e^t dt\right] dt \right]$$

$$= \frac{1}{3}\left[\left[te^t\right]_0^1 - \left[e^t\right]_0^1 \right]$$

$$= \frac{1}{3}\,[\,(e-0) - (e-1)\,] = \frac{1}{3}$$

(ii) $\int_0^{\frac{\pi}{2}} \frac{\sin x}{\sin x + \cos x} dx$

Let $I = \int_0^{\frac{\pi}{2}} \frac{\sin x}{\sin x + \cos x} dx$ ..... (1)

by property 4, $I = \int_0^{\frac{\pi}{2}} \frac{\sin\left(\frac{\pi}{2} - x\right)}{\sin(\frac{\pi}{2} - x) + \cos(\frac{\pi}{2} - x)} dx$

$$= \int_0^{\frac{\pi}{2}} \frac{\cos x}{\cos x + \sin x} dx \qquad \text{..... (2)}$$

adding (i) and (ii)

$$2I = \int_0^{\frac{\pi}{2}} \frac{\sin x}{\sin x + \cos x} dx + \int_0^{\frac{\pi}{2}} \frac{\cos x}{\sin x + \cos x} dx$$

$$= \int_0^{\frac{\pi}{2}} \frac{\sin x + \cos x}{\sin x + \cos x} dx$$

$$= \int_0^{\frac{\pi}{2}} 1\, dx$$

$$= [x]_0^{\frac{\pi}{2}}$$

$$2I = \frac{\pi}{2}$$

$$\therefore \quad I = \frac{\pi}{4}$$

(iii) $\displaystyle\int_{\frac{-\pi}{2}}^{\frac{\pi}{2}} \sin^7 x\, dx$

$\because \quad f(x) = \sin^7 x$

$$f(-x) = \sin^7(-x) = -\sin^7 x$$

$$= -f(x)$$

$\therefore$ x is an odd function

$$\therefore \quad \int_{\frac{-\pi}{2}}^{\frac{\pi}{2}} \sin^7 x\, dx = 0$$

(iv) $\int_0^{\pi} \cos^7 x\, dx$

$$\because \quad f(x) = \cos^7 x$$

$$f(2a - x) = \cos^7(\pi - x) = \cos^7 x = -f(x)$$

$$\Rightarrow \int_0^{\pi} \cos^7 x\, dx = 0 \qquad \left(\because \int_0^{2a} f(x)\, dx = 0\right)$$

Exercises

Integrate the following w r t x.

1. $4x^3 - 4x^{-5}$
2. $\dfrac{x^3 - 1}{x^2}$
3. $5^x + 3e^x + x^3$
4. $5 \cos x - 3 \sin x + \dfrac{2}{\cos^2 x}$
5. $\int \tan^2 x\, dx$
6. $\int (ax + b)^2 dx$
7. $\dfrac{\sin x}{\sin x - \cos x}$
8. $(3x - 4)^3$
9. $\dfrac{1}{1+x^2} - \dfrac{\cos x}{\sin^2 x}$
10. $\dfrac{1}{\sqrt{1-x^2}} + \dfrac{1}{1+x^2}$
11. $\dfrac{1}{\sqrt{5x+3} + \sqrt{5x+2}}$
12. $e^{a \log x} + e^{x \log a}$
13. $x^2 \cos x^3$
14. $x^2 e^{x^3}$
15. $\dfrac{e^{\tan^{-1} x}}{1+x^2}$
16. $\dfrac{x^5}{\sqrt{1-x^{12}}}$
17. $\dfrac{\cos x}{1+\sin^2 x}$
18. $\dfrac{\sin x \cos x}{1+\sin^4 x}$
19. $\dfrac{(1+\log x)^2}{x}$
20. $\dfrac{1}{x\sqrt{1-(\log x)^2}}$

21. Evaluate $\int \frac{(x+1)(x+\log x)^3}{2x}\,dx$

22. $\int \sin^5 x\,dx$

23. $\int e^x (e^x+1)^3\,dx$

24. $\int \cot x \operatorname{cosec}^3 x\,dx$

25. $\int e^x (a+be^x)^n\,dx$

26. $\int \frac{(\sin^{-1} x)^3}{\sqrt{1-x^2}}\,dx$

27. $\int \frac{(\tan^{-1} x)^4}{1+x^2}\,dx$

28. $\int \sin^4 x \cos x\,dx$

29. $\int \frac{1}{(\tan^{-1} x)^2 (1+x^2)}\,dx$

30. $\int (e^x + e^{-x})(e^x - e^{-x})\,dx$

31. $\int \tan^4 x\,dx$

32. $\int \frac{2x \sin^{-1} x^2}{\sqrt{1-x^4}}\,dx$

33. $\int \frac{1}{\sqrt{1-x^2}\,\sin^{-1} x}\,dx$

34. $\int \frac{\cot x}{(\log \sin x)^3}\,dx$

35. $\int \frac{x^2 \tan^{-1} x^3}{1+x^6}\,dx$

36. $\int \frac{dx}{\sqrt{25+4x^2}}$

37. $\int \frac{x^2}{x^2-4}\,dx$

38. $\int \sqrt{25-9x^2}\,dx$

39. $\int \frac{dx}{\sqrt{(x+1)^2+4}}$

40. $\int \frac{dx}{x\sqrt{x^2-4}}$

41. $\int \cos x \sqrt{4-\sin^2 x}\,dx$

42. $\int \frac{\sec^2 x\,dx}{\sqrt{\tan^2 x+4}}$

43. $\int \frac{\sec^2 x\,dx}{\sqrt{\tan^2 x+4}}$

44. $\int \frac{x\,dx}{a^4+x^4}$

45. Evaluate $\int x \sin 2x\,dx$

46. $\int x \sin^3 x\,dx$

47. $\int x \tan^2 x\,dx$

48. $\int x \sec^2 x\,dx$

49. $\int x \sin nx\,dx$

50. $\int x^3 e^x\,dx$

51. $\int x^2 e^{3x}\, dx$

52. $\int \tan^{-1} x\; dx$

53. $\int \cos^{-1} x\, dx$

54. $\int x^2 \log x\, dx$

55. $\int x \sin^{-1} x\, dx$

56. $\int x^2 \tan^{-1} x\, dx$

57. $\int x^n \log x\, dx$

58. $\int \frac{\log x}{(1+x)^2}\, dx$

59. $\int (\log x)^2\, dx$

60. $\int x^2 \sin^{-1} x\, dx$

61. $\int \sec^3 x\, dx$

62. $\int \frac{\sin^{-1} x}{(1-x^2)^{\frac{3}{2}}}\, dx$

63. $\int \frac{x \sin^{-1} x}{\sqrt{1-x^2}}\, dx$

64. $\int \sqrt{x} \tan^{-1} \sqrt{x}\, dx$

65. $\int \tan^{-1} \sqrt{\frac{1-x}{1+x}}\, dx$

66. $\int \cos^{-1}\left(\frac{1-x^2}{1+x^2}\right) dx$

67. $\int \frac{\log(\sec^{-1} x)}{x\sqrt{x^2-1}}\, dx$

68. Evaluate $\int \frac{x-1}{(x-2)(x-3)}\, dx$

69. Evaluate $\int \frac{x^3+2}{x^3-x}\, dx$

70. Evaluate $\int \frac{dx}{x^2-5x+6}$

71. Evaluate $\int \frac{2x\, dx}{(x^2+1)(x^2+3)}$

72. Evaluate $\int \frac{\cos x\, dx}{(1+\sin x)(2+\sin x)}$

73. Evaluate $\int \frac{1-x^2}{x(1-2x)}\, dx$

74. Evaluate $\int \frac{x\, dx}{(x-1)^2(x+2)}$

75. Evaluate $\int \frac{dx}{1+x+x^2+x^3}$

76. Evaluate $\int \frac{dx}{(x-3)(x^2+4)}$

77. Evaluate $\int \frac{x\, dx}{(x^2-a^2)(x^2-b^2)}$

78. Evaluate $\int \frac{dx}{(x+b)(x^2+a^2)}$

79. $\int \frac{x^3 dx}{x^2-3x+2}$

80. Evaluate $\int \frac{dx}{x^3+1}$

81. Evaluate $\int \frac{x^2dx}{(x+1)(x-2)(x+3)}$

82. Evaluate $\int \frac{e^x dx}{e^{2x}+4e^x+3}$

83. Evaluate $\int \frac{x^2dx}{(x-1)^3(x+1)}$

84. Evaluate the following $\int_0^{\frac{\pi}{2}} (\sin x + \cos x)\, dx$

85. $\int_1^2 (4x^3 - 5x^2 + 6x + 9)dx$

86. $\int_0^{\frac{\pi}{4}} \sec x\, dx$

87. $\int_0^1 \frac{dx}{\sqrt{1-x^2}}$

88. $\int_0^1 \sin^{-1} x\, dx$

89. $\int_0^{\frac{\pi}{2}} \sin^3 x\, dx$

90. $\int_1^2 \frac{dx}{(x+1)(x+2)}$

91. $\int_1^2 \frac{5x^2}{x^2+4x+3} dx$

92. $\int_0^{\frac{\pi}{2}} \frac{\sin\theta}{\sqrt{1+\cos\theta}} d\theta$

93. $\int_0^1 x\, e^{x^2} dx$

94. $\int_0^{\pi} \frac{dx}{5+4\cos x}$

95. $\int_0^{\pi} \frac{\sin x dx}{1+\cos^2 x}$

96. $\int_0^1 \frac{5x^3}{\sqrt{1-x^8}} dx$

97. $\int_0^{\frac{\pi}{2}} \frac{dx}{5+4\sin x}$

98. $\int_0^{\infty} \frac{x^2 dx}{(x^2+a^2)(x^2+b^2)}$

99. $\int_0^1 \frac{x\, dx}{\sqrt{1+x^2}}$

100. $\int_0^{\frac{\pi}{3}} \frac{\cos x}{3+4\sin x} dx$

Answers

1. $x^4 + x^{-4}$ 2. $\frac{1}{2}x^2 + \frac{1}{x}$ 3. $\frac{5^x}{\log 5} + 3e^x + \frac{x^4}{4}$

4. $5 \sin x + 3 \cos x + 2 \tan x$ 5. $\tan x - x$ 6. $\frac{(ax+b)^3}{3a}$

7. $\frac{1}{2}[x + \log(\sin x - \cos x)]$ 8. $\frac{(3x-4)^4}{12}$ 9. $\tan^{-1} x + \text{cosec } x$

10. $\sin^{-1} x + \tan^{-1} x$ 11. $\frac{2}{15}\left[(5x+3)^{\frac{3}{2}} - (5x+2)^{\frac{3}{2}}\right]$

12. $\frac{x^{a+1}}{a+1} + \frac{a^x}{\log a}$ 13. $\frac{1}{3}\sin x^3$ 14. $\frac{1}{3}e^{x^3}$ 15. $e^{\tan^{-1} x}$

16. $\frac{1}{6}\sin^{-1}(x^6)$ 17. $\tan^{-1}(\sin x)$ 18. $\frac{1}{2}\tan^{-1}(\sin^2 x)$

19. $\frac{1}{3}(\log x + 1)^3$ 20. $\sin^{-1}(\log x)$ 21. $\frac{1}{8}(x + \log x)^4$

22. $-\cos x - \frac{\cos^5 x}{5} + \frac{2\cos^2 x}{3}$ 23. $\frac{1}{4}(e^x + 1)^4$ 24. $-\frac{1}{3}\text{cosec}^3 x$

25. $\frac{(a + be^x)^{n+1}}{b(n+1)}$ 26. $\frac{1}{4}(\sin^{-1} x)^4$ 27. $\frac{1}{5}(\tan^{-1} x)^5$

28. $\frac{1}{5}(\sin x)^5$ 29. $-\frac{1}{\tan^{-1} x}$ 30. $\frac{1}{2}(e^x + e^{-x})^2$

31. $\frac{1}{3}\tan^3 x - \tan x + x$ 32. $\frac{1}{2}(\sin^{-1} x^2)^2$ 33. $\log(\sin^{-1} x)$

34. $\frac{-1}{2}(\log \sin x)^{-2}$ 35. $\frac{1}{6}(\tan^{-1} x^3)^2$

36. $\frac{1}{2}\log(2x + \sqrt{25 + 4x^2})$ 37. $x + \log\frac{x-2}{x+2}$

38. $\frac{x}{2}\sqrt{25 - 9x^2} + \frac{25}{6}\sin^{-1}\frac{3x}{5}$ 39. $\log\left\{(x+1) + \sqrt{(x+1)^2 + 4}\right\}$

40. $\frac{1}{2}\sec^{-1}\frac{x}{2}$

41. $\frac{1}{2}\sin x\sqrt{4-\sin^2 x}+2\sin^{-1}\left(\frac{1}{2}\sin x\right)$

42. $\log\left[\tan x+\sqrt{\tan^2 x+4}\right]$

43. $-\log\left[\cos x+\sqrt{4+\cos^2 x}\right]$

44. $\frac{1}{2a^2}\tan^{-1}\frac{x^2}{a^2}$

45. $-\frac{1}{2}x\cos 2x+\frac{1}{4}\sin 2x$

46. $\frac{1}{4}[-3x\cos x+3\sin x+\frac{1}{3}x\cos 3x-\frac{1}{9}\sin 3x]$

47. $x\tan x-\log\sec x-\frac{x^2}{2}$

48. $x \tan x + \log \cos x$

49. $\frac{1}{n}\left[\frac{1}{n}\sin nx-x\cos nx\right]$

50. $e^x[x^3-3x^2+6x-6]$

51. $\frac{1}{3}x^2e^{3x}-\frac{2}{3^2}xe^{3x}+\frac{2}{3^3}e^{3x}$

52. $x\tan^{-1}x-\frac{1}{2}\log(1+x^2)$

53. $x\cos^{-1}x-\sqrt{1-x^2}$

54. $\frac{1}{3}x^3\log x-\frac{x^3}{9}$

55. $\frac{1}{2}x^2\sin^{-1}x+\frac{1}{4}x\sqrt{1-x^2}-\frac{1}{4}\sin^{-1}x$

56. $\frac{1}{3}x^3\tan^{-1}x-\frac{x^2}{6}+\frac{1}{6}\log(x^2+1)$

57. $\frac{1}{n+1}x^{n+1}\log x-\frac{x^{n+1}}{(n+1)^2}$

58. $\frac{-\log x}{x+1}+\log x-\log(x+1)$

59. $x(\log x)^2-2x\log x+2x$

60. $\frac{1}{3}x^3\sin^{-1}x-\frac{1}{9}(1-x^2)^{\frac{3}{2}}+\frac{1}{3}\sqrt{1-x^2}$

61. $\frac{1}{2}$ [sec x tan x + log(sec x + tan x]

62. $\frac{x\sin^{-1}x}{\sqrt{1-x^2}}+\frac{1}{2}\log(1-x^2)+c$

63. $x-(\sin^{-1}x)\sqrt{1-x^2}+c$

64. $\frac{2}{3}x^{\frac{3}{2}}\tan^{-1}\sqrt{x}-\frac{x}{3}+\frac{1}{3}\log(x+1)+c$

65. $\frac{1}{2}\left[x\cos^{-1}x-\sqrt{1-x^2}\right]+c$

66. $2x\tan^{-1}x-2\log\sqrt{1+x^2}+c$

67. $\sec^{-1}[\log(\sec^{-1}x)-1]+c$

68. $2\log(x-3)-\log(x-2)+c$

69. $x - 2\log x + \frac{1}{2}\log(x+1) + \frac{3}{2}\log(x-1) + c$

70. $\log\frac{x-3}{x-2}$ 71. $\frac{1}{2}\log\frac{x^2+1}{x^2+3}$ 72. $\log\left(\frac{1+\sin x}{2+\sin x}\right)$

73. $\frac{x}{2} + \log x - \frac{3}{4}\log(1-2x)$ 74. $\frac{2}{9}\log\left(\frac{x-1}{x+2}\right) - \frac{1}{3(x-1)}$

75. $\frac{1}{2}\log(1+x) - \frac{1}{4}\log(1+x^2) + \frac{1}{2}\tan^{-1} x$

76. $\frac{1}{13}\left[\log(x-3) - \frac{1}{2}\log(x^2+4) - \frac{3}{2}\tan^{-1}\frac{x}{2}\right]$

77. $\frac{1}{2(a^2-b^2)}\log\frac{x^2-a^2}{x^2-b^2}$ 78. $\frac{1}{a^2+b^2}[\log\frac{x+b}{\sqrt{x^2+a^2}} + \frac{b}{a}\tan^{-1}\frac{x}{a}]$

79. $\frac{x^2}{2} + 3x + 8\log(x-2) - \log(x-1)$

80. $\frac{1}{3}\log(x+1) - \frac{1}{6}\log(x^2-x+1) + \frac{1}{\sqrt{3}}\tan^{-1}\frac{2x-1}{\sqrt{3}}$

81. $\frac{9}{10}\log(x+3) + \frac{4}{15}\log(x-2) - \frac{1}{6}\log(x+1)$

82. $\frac{1}{2}[\log(1+e^x) - \log(3+e^x)]$ 83. $\frac{1}{8}\log\frac{x-1}{x+1} - \frac{3}{4(x-1)} - \frac{1}{4(x-1)^2}$

84. 2 85. $\frac{64}{3}$ 86. $\log(1+\sqrt{2})$

87. $\frac{\pi}{2}$ 88. $\frac{\pi}{2} - 1$ 89. $\frac{2}{3}$ 90. $\log 9 - \log 8$

91. $\frac{1}{2}[5\log 3 + 85\log 2 - 45\log 5] + 5$ 92. $-2(1-\sqrt{2})$ 93. $\frac{1}{2}(e-1)$

94. $\frac{\pi}{3}$ 95. $\frac{-\pi}{2}$ 96. $\frac{5\pi}{8}$ 97. $\frac{2}{3}\tan^{-1}\frac{1}{3}$

98. $\frac{\pi}{2(a+b)}$ 99. $\sqrt{2} - 1$ 100. $\frac{1}{4}\log(1+\frac{2}{\sqrt{3}})$

CHAPTER 9

DIFFERENTIAL EQUATIONS OF THE FIRST ORDER AND FIRST DEGREE

Definition

A differential equation is defined to be on equation which involves one or more derivatives or differential coefficients besides independent and dependent variables. For examples

1. $\frac{dy}{dx} = \sin x$
2. $\frac{d^2y}{dx^2} = x^5$
3. $xdx + ydy = 0$
4. $\frac{d^2y}{dx^2} + x^2y = 0$

Kinds of differential equations

Differential equations are divided into two types

1. Ordinary differential equations
2. Partial differential equations

1. Ordinary differential equations

A differential equation which have only one independent variable is called ordinary differential equation. For example

Let x be independent variable and y is a dependent variable, then

$$\frac{dy}{dx} = \sin x \qquad \text{i.e. only one independent variable x.}$$

$$\frac{d^2y}{dx^2} + x^2y = 0$$

2. Partial differential equations

A differential equation which have two or more independent variables is called partial differential equation. For example

1. $x\frac{\partial u}{\partial x}+y\frac{\partial u}{\partial y}=nu$ i.e. two independent variables x and y

2. $\frac{\partial^2 u}{\partial x^2}+\frac{\partial^2 u}{\partial y^2}+\frac{\partial^2 u}{\partial z^2}$ i.e. three independent variables x, y and z

Order of differential equations

The order of the differential equation is the order of the highest derivatives occurring in the differential equations. For example

1. $xdx + ydy = 0$ 2. $\left(\frac{dy}{dx}\right)^2+2y=3x$ 3. $y\frac{dy}{dx}=x^2\left(\frac{dy}{dx}\right)^2+1$

is first order differential equations and

1. $\frac{d^2y}{dx^2}=x^5$ 2. $\left(\frac{d^2y}{dx^2}\right)+3\left(\frac{dy}{dx}\right)^2+y=0$

is second order differential equation

while $\frac{d^3y}{dx^3}+3\frac{d^2y}{dx^2}.\frac{dy}{dx}+x^2\left(\frac{dy}{dx}\right)^3=0$ is of order third

Degree of differential equation

The degree of differential equation is the degree of the highest derivatives occurring in it. For example

1. $\frac{d^2y}{dx^2}+y=0$, is of order 2 and degree 1

2. $\left(\frac{dy}{dx}\right)^2+3\frac{dy}{dx}+y=0$, is of order 1 and degree 2

3. $y=x\frac{dy}{dx}+a\sqrt{1+\left(\frac{dy}{dx}\right)^2}$, is of order 1 and degree 2

Linear differential equation

A differential equation is said to be linear if the dependent variable and all of its derivatives occur only in the first degree. If a differential equation is not linear then it is called non-linear.

***Example* 1:** State the order and degree of the following differential equations. State also whether they are linear or not.

1. $x\frac{dy}{dx}+y=2x$
2. $(1-x^2)\frac{d^2y}{dx^2}-2x\frac{dy}{dx}=1$
3. $\frac{d^2y}{dx^2}+\left(\frac{dy}{dx}\right)^2+y^2=0$
4. $\left(\frac{d^3y}{dx^3}\right)^2-2\left(\frac{dy}{dx}\right)+xy=0$
5. $\left[1+\left(\frac{dy}{dx}\right)^2\right]^{3/2}=a\frac{d^2y}{dx^2}$
6. $\cos x\frac{dy}{dx}+y\sin x=1$

***Solution*:**

1. $x\frac{dy}{dx}+y=2x$ i.e. order 1 and degree 1 i.e. linear

2. $(1-x^2)\frac{d^2y}{dx^2}-2x\frac{dy}{dx}=1$

   $\Rightarrow$ Here, highest differential coefficient is $\frac{d^2y}{dx^2}$

   i.e. order 2 degree 1 i.e. linear

3. $\frac{d^2y}{dx^2}+\left(\frac{dy}{dx}\right)^2+y^2=0$

   $\Rightarrow$ highest differential coefficient is $\frac{d^2y}{dx^2}$

   i.e. order 2 degree 1 but second form have degree 2. i.e. non-linear.

4. $\left(\frac{d^3y}{dx^3}\right)^2-2\left(\frac{dy}{dx}\right)+xy=0$

   $\Rightarrow$ highest differential coefficient is $\frac{d^3y}{dx^3}$ i.e. order 3 and degree 2

   i.e. non linear

5. $\left[1+\left(\frac{dy}{dx}\right)^2\right]^{3/2} = a\frac{d^2y}{dx^2}$

$\Rightarrow$ The highest differential coefficient is $\frac{d^2y}{dx^2}$ i.e. order 2. Now

$$\left[1+\left(\frac{dy}{dx}\right)^2\right]^3 = a^2\left(\frac{d^2y}{dx^2}\right)^2$$, degree is 2. i.e. non linear

6. $\cos x\frac{dy}{dx} + y\sin x = 1$ i.e. order 1 and degree 1, hence linear.

Formation of differential equations

To form the differential equation from a given equation in x and y containing arbitrary constants, the following procedure is adopted.

1. Write down the given equation
2. Differentiate it w.r.t x as many times as are the number of arbitrary constants.
3. Eliminate the arbitrary constants from the equations of the above two steps.

   The resulting equation involving x, y, $\frac{dy}{dx}, \frac{d^2y}{dx^2}$ ...is the required differential equation.

***Example* 1:** Form the differential equation corresponding to the equation

$$y = Ae^{3x} + Be^{5x}$$

***Solution*:** Given equation is

$$y = Ae^{3x} + Be^{5x} \quad \text{..... (i)}$$

Differentiating it w.r.t. 'x', we get

$$\frac{dy}{dx} = 3Ae^{3x} + 5\,Be^{5x} \quad \text{.....(ii)}$$

Differentiating again, we get

$$\frac{d^2y}{dx^2} = 9Ae^{3x} + 25\,Be^{5x} \quad \text{..... (iii)}$$

Eliminating Ae^{3x} and Be^{5x} from eq. (i), (ii) and (iii)

$$\begin{vmatrix} y & 1 & 1 \\ dy/dx & 3 & 5 \\ d^2y/dx^2 & 9 & 25 \end{vmatrix} = 0$$

$$\Rightarrow \quad y(75-45) - 1\left(25\frac{dy}{dx} - 5\frac{d^2y}{dx^2}\right) + 1\left(9\frac{dy}{dx} - 3\frac{d^2y}{dx^2}\right) = 0$$

$$\Rightarrow \quad 30y - 16\frac{dy}{dx} + 2\frac{d^2y}{dx^2} = 0$$

$$\Rightarrow \quad \frac{d^2y}{dx^2} - 8\frac{dy}{dx} + 15y = 0$$

Example 2: Form a differential equation by eliminating the arbitrary constants from the following relation between the variables x and y.

$$y = C_1e^x + C_2\,e^{-x}$$

Solution: Given equation is

$$y = C_1e^x + C_2\,e^{-x} \qquad \text{..... (i)}$$

differentiating it w.r.t. x, we get

$$\frac{dy}{dx} = C_1e^x - C_2e^{-x} \qquad \text{..... (ii)}$$

Differentiating again,

$$\Rightarrow \quad \frac{d^2y}{dx^2} = C_1e^x + C_2e^{-x}$$

$$\Rightarrow \quad \frac{d^2y}{dx^2} = y, \text{ from eq. (ii)}$$

$$\Rightarrow \quad \frac{d^2y}{dx^2} - y = 0$$

Example 3: Form a differential equation by eliminating constants a and b from the following equation.

$y = a \cos x + b \sin x$

Solution: Given equation is

$$y = a \cos x + b \sin x \qquad \text{..... (i)}$$

differentiating it w.r.t 'x', we get

$$\frac{dy}{dx} = -a \sin x + b \cos x \qquad \text{..... (ii)}$$

Differentiating again,

$$\Rightarrow \quad \frac{d^2y}{dx^2} = -a \cos x - b \sin x$$

$$\Rightarrow \quad \frac{d^2y}{dx^2} = -(a \cos x + b \sin x)$$

$$\Rightarrow \quad \frac{d^2y}{dx^2} = -y, \text{ from (i)}$$

$$\Rightarrow \quad \frac{d^2y}{dx^2} + y = 0$$

Example 4: Form the differential equation by eliminating the constant A from the relation $y = A(x + y)$

***Solution*:** Given equation is

$$y = A(x + y) \qquad \text{.....(i)}$$

differentiating it w.r.t. x, we get

$$\frac{dy}{dx} = A\left(1 + \frac{dy}{dx}\right) \qquad \text{.....(ii)}$$

Now, from (i) $A = \dfrac{y}{x+y}$

Putting the value of A in equation (ii), we get

$$\frac{dy}{dx} = \frac{y}{x+y}\left(1 + \frac{dy}{dx}\right)$$

$$\Rightarrow \quad (x + y)\frac{dy}{dx} = y + y\frac{dy}{dx}$$

$$\Rightarrow \quad x\frac{dy}{dx} = y$$

$$\Rightarrow \quad x\frac{dy}{dx} - y = 0$$

***Example* 5:** Form a differential equation by eliminating arbitrary constants

$$y = ax + bx^2$$

***Solution*:** Given equation is

$$y = ax + bx^2 \qquad \text{.....(i)}$$

Differentiating it w.r.t x, we get

$$\frac{dy}{dx} = a + 2b\,x \qquad \text{.....(ii)}$$

differentiating again,

$$\frac{d^2y}{dx^2} = 2b$$

$$\Rightarrow \qquad b = \frac{1}{2}\frac{d^2y}{dx^2}$$

Putting value of b in eq. (ii), we have

$$\frac{dy}{dx} = a + x\frac{d^2y}{dx^2} \Rightarrow a = \frac{dy}{dx} - x\frac{d^2y}{dx^2}$$

∴ Putting the value of a and b in eq. (i), we get

$$y = x\left(\frac{dy}{dx} - x\frac{d^2y}{dx^2}\right) + \frac{x^2}{2}\left(\frac{d^2y}{dx^2}\right)$$

$$2y = 2x\frac{dy}{dx} - 2x^2\frac{d^2y}{dx^2} + x^2\frac{d^2y}{dx^2} \Rightarrow x^2\frac{d^2y}{dx^2} - 2x\frac{dy}{dx} + 2y = 0$$

***Example* 6:** Form a differential equation by eliminating arbitrary constants

$$y = C_1e^{2x} + C_2e^{-2x}$$

***Solution*:** Given equation is

$$y = C_1e^{2x} + C_2\,e^{-2x} \qquad \text{.....(i)}$$

differentiating it w.r.t x, we get

$$\frac{dy}{dx} = 2C_1e^{2x} - 2C_2e^{-2x} \qquad \text{.....(ii)}$$

Differentiating again,

$$\frac{d^2y}{dx^2} = 4C_1e^{2x} + 4C_2e^{-2x}$$

$$\Rightarrow \quad \frac{d^2y}{dx^2} = 4(C_1e^{2x} + C_2e^{-2x})$$

$$\Rightarrow \quad \frac{d^2y}{dx^2} = 4y, \text{ from eq (i)}$$

$$\Rightarrow \quad \frac{d^2y}{dx^2} - 4y = 0$$

***Example* 7:** Form the differential equation by eliminating the constants

$$y = Ae^{3x} + Be^{2x}$$

***Solution*:** Given equation is

$$y = Ae^{3x} + Be^{2x} \qquad \text{..... (i)}$$

differentiating it w.r.t x, we get

$$\frac{dy}{dx} = 3Ae^{3x} + 2Be^{2x} \qquad \text{.....(ii)}$$

Differentiating again,

$$\frac{d^2y}{dx^2} = 9Ae^{3x} + 4Be^{2x} \qquad \text{.....(iii)}$$

Eliminating Ae^{3x} and $B\,e^{2x}$ from (i), (ii) and (iii), we get

$$\begin{vmatrix} y & 1 & 1 \\ dy/dx & 3 & 2 \\ d^2y/dx^2 & 9 & 4 \end{vmatrix} = 0$$

$$\Rightarrow \quad y(12-18) - 1\left(4\frac{dy}{dx} - 2\frac{d^2y}{dx^2}\right) + 1\left(9\frac{dy}{dx} - 3\frac{d^2y}{dx^2}\right) = 0$$

$$\Rightarrow \quad -6y + 5\frac{dy}{dx} - \frac{d^2y}{dx^2} = 0$$

$$\Rightarrow \quad \frac{d^2y}{dx^2} - 5\frac{dy}{dx} + 6y = 0$$

***Example* 8:** Form a differential equation by eliminating arbitrary constants

$$y = A\cos\left(\frac{x}{k}\right) + B\sin\left(\frac{x}{k}\right)$$

***Solution*:** Given equation is

$$y = A\cos\left(\frac{x}{k}\right) + B\sin\left(\frac{x}{k}\right) \qquad .....(i)$$

Differentiating it w.r.t x, we get

$$\frac{dy}{dx} = -\frac{A}{k}\sin\frac{x}{k} + \frac{B}{k}\cos\frac{x}{k} \qquad .....(ii)$$

Differentiating again,

$$\frac{d^2y}{dx^2} = -\frac{A}{k^2}\cos\frac{x}{k} - \frac{B}{k^2}\sin\frac{x}{k}$$

$$\Rightarrow \quad \frac{d^2y}{dx^2} = -\frac{1}{k^2}\left[A\cos\frac{x}{k} + B\sin\frac{x}{k}\right]$$

$$\Rightarrow \quad \frac{d^2y}{dx^2} = -\frac{1}{k^2}y, \qquad \text{from (i)}$$

$$\Rightarrow \quad k^2\frac{d^2y}{dx^2} + y = 0$$

***Example* 9:** Form the differential equation of all circles having radius a and centre as origin

***Solution*:** Centre of the circle is origin i.e. (0, 0) and radius is a. Hence, equation of circle be

$$x^2 + y^2 = a^2 \qquad ..... (i)$$

Differentiating it w.r.t x, we get

$$2x + 2y\frac{dy}{dx} = 0$$

$$\Rightarrow \quad y\frac{dy}{dx} + x = 0$$

***Example* 10:** Form the differential equation by eliminating the constants

$$y = e^x (A\cos x + B\sin x)$$

***Solution*:** Given equation is

$$y = e^x (A\cos x + B\sin x) \qquad .....(i)$$

differentiating it w.r.t x, we get

$$\frac{dy}{dx} = e^x(A\cos x + B\sin x) + e^x(-A\sin x + B\cos x)$$

$$= y + e^x(-A\sin x + B\cos x)$$

$$\frac{dy}{dx} - y = e^x(-A\sin x + B\cos x) \quad \text{..... (ii)}$$

Differentiating again,

$$\frac{d^2y}{dx^2} - \frac{dy}{dx} = e^x(-A\sin x + B\cos x) + e^x(-A\cos x - B\sin x)$$

$$\Rightarrow \quad \frac{d^2y}{dx^2} - \frac{dy}{dx} = \frac{dy}{dx} - y - e^x(A\cos x + B\sin x), \text{ from eq. (ii)}$$

$$\Rightarrow \quad \frac{d^2y}{dx^2} - 2\frac{dy}{dx} + y = -y, \text{ from (i)}$$

$$\Rightarrow \quad \frac{d^2y}{dx^2} - 2\frac{dy}{dx} + 2y = 0$$

***Example* 11:** Form a differential equation by eliminating arbitrary constants

$$y = Ae^{kx} + Be^{-kx}$$

Solution: Given equation is

$$y = Ae^{kx} + Be^{-kx} \quad \text{.....(i)}$$

Differentiating it w.r.t x, we get

$$\frac{dy}{dx} = Ake^{kx} + Bke^{-kx} \quad \text{.....(ii)}$$

Differentiating again,

$$\frac{d^2y}{dx^2} = Ak^2e^{kx} + Bk^2e^{-kx}$$

$$= k^2(Ae^{kx} + Be^{-kx})$$

$$\Rightarrow \quad \frac{d^2y}{dx^2} = k^2y, \text{ from (i)}$$

$$\Rightarrow \quad \frac{d^2y}{dx^2} - k^2y = 0$$

***Example* 12:** Form a differential equation by eliminating a and b

$$(x-a)^2+(y-b)^2=r^2$$

***Solution*:** Given equation is

$$(x-a)^2+(y-b)^2=r^2 \qquad .....(i)$$

Differentiating it w.r.t x, we get

$$2(x-a)+2(y-b)\frac{dy}{dx}=0 \qquad .....(ii)$$

Differentiating again,

$$2+2(y-b)\frac{d^2y}{dx^2}+2\left(\frac{dy}{dx}\right)^2=0 \qquad .....(iii)$$

From (iii), $(y-b)\dfrac{d^2y}{dx^2}=-\left[1+\left(\dfrac{dy}{dx}\right)^2\right]$

$$\Rightarrow \qquad y-b=\frac{-\left[1+\left(\frac{dy}{dx}\right)^2\right]}{\frac{d^2y}{dx^2}}$$

From (ii), $(x-a)=-(y-b)\dfrac{dy}{dx}$

$$=\frac{1+\left(\frac{dy}{dx}\right)^2}{\frac{d^2y}{dx^2}}.\frac{dy}{dx}$$

Putting the value of (x – a) and (y – b) in eq. (i), we have

$$\left[1+\left(\frac{dy}{dx}\right)^2\right]^2\left(\frac{dy}{dx}\right)^2+\left[1+\left(\frac{dy}{dx}\right)^2\right]^2=r^2\left(\frac{d^2y}{dx^2}\right)^2$$

$$\Rightarrow \qquad \left[1+\left(\frac{dy}{dx}\right)^2\right]^2.\left[1+\left(\frac{dy}{dx}\right)^2\right]=r^2\left(\frac{d^2y}{dx^2}\right)^2$$

$$\Rightarrow \qquad \left[1+\left(\frac{dy}{dx}\right)^2\right]^3=r^2\left(\frac{d^2y}{dx^2}\right)^2$$

Exercise 13.1

Form the differential equation by eliminating arbitrary constant from the following equations.

1. $y = mx + \frac{a}{x}$ (parameter m)
2. $y = a\cos 4x + b\sin 4x$
3. $y = ax^2 + bx$
4. $y^2 = 4a(x + a)$
5. $\frac{x}{a} + \frac{y}{b} = 1$
6. $Ax^2 + By^2 = 1$
7. $y = a\cos(wt + b)$
8. $y = Ae^{2x} + Be^{-3x} + Ce^{x}$
9. Find the differential equation of all straight line passing through the origin.
10. Find the differential equation of the circle passing through the origin and having their centre on the x–axis. [Hint: Equation of circle is $x^2 + y^2 + 2gx = 0$]

Answers

1. $x\left(\frac{dy}{dx}\right)^2 - y\frac{dy}{dx} + a = 0$
2. $\frac{d^2y}{dx^2} + 16y = 0$
3. $\frac{1}{2}x^2\frac{d^2y}{dx^2} - x\frac{dy}{dx} + y = 0$
4. $y\frac{d^2y}{dx^2} + \left(\frac{dy}{dx}\right)^2 = 0$
5. $\frac{d^2y}{dx^2} = 0$
6. $xy\frac{d^2y}{dx^2} + x\left(\frac{dy}{dx}\right)^2 - y\frac{dy}{dx} = 0$
7. $\frac{d^2y}{dt^2} + w^2y = 0$
8. $\frac{d^3y}{dx^3} - 7\frac{dy}{dx} + 6y = 0$
9. $y = x\frac{dy}{dx}$
10. $y^2 - x^2 - 2xy\frac{dy}{dx} = 0$

Solution of differential equation: A solution of a differential equation is a functional relation between the variable involved, which satisfies the differential equation.

General solution: If the solution of a differential equation of nth order contains n arbitrary constants, the solution is called the general solution 'or' complete solution. For example,

$$y = \frac{a}{x} + b \text{ is G.S of } \frac{d^2y}{dx^2} + \frac{2}{x}\frac{dy}{dx} = 0$$

Particular solution: A solution obtained by giving particular values to arbitrary constants in the general solution is called a particular solution. For example, Putting a = 2, b = 3.

A particular solution of $\frac{d^2y}{dx^2} + \frac{2}{x}\frac{dy}{dx} = 0$ is $y = \frac{2}{x} + 3$

***Example* 1:** Show that $y = Cx^4$ is a solution of differential equation

$$x\frac{dy}{dx} - 4y = 0$$

***Solution*:** The given equation is $x\frac{dy}{dx} - 4y = 0$ .....(i)

The given function is $y = Cx^4$ .....(ii)

Differentiating it w.r.t x, we get

$$\frac{dy}{dx} = 4Cx^3$$

$$\Rightarrow \quad x\frac{dy}{dx} = 4y, \text{ from (ii)}$$

$$x\frac{dy}{dx} - 4y = 0$$

Which is the same as (i)

Hence the given function is a solution of (i)

***Example* 2:** Verify that $y^2 = 4ax$ satisfies the differential equation

$$y = x\frac{dy}{dx} + a\frac{dx}{dy}$$

***Solution*:** Given differential equation is $y = x\frac{dy}{dx} + a\frac{dx}{dy}$ .....(i)

The given function is $y^2 = 4ax$

Differentiating it w.r.t x, we get

$$2y\frac{dy}{dx} = 4a$$

$$\Rightarrow \quad \frac{dy}{dx} = \frac{2a}{y} \text{ and } \frac{dx}{dy} = \frac{y}{2a}$$

Substituting the values in (i), we get

$$y = x.\frac{2a}{y} + a.\frac{y}{2a}$$

$$= \frac{2ax}{y} + \frac{y}{2}$$

$$= \frac{4ax + y^2}{2y}$$

$$= \frac{y^2 + y^2}{2y}$$

$$= \frac{2y^2}{2y} = y$$

Since the given function satisfies (i)

Hence it is a solution of (i)

***Example* 3:** Show that the function $y = (A+Bx)e^{3x}$ is a solution of the equation

$$\frac{d^2y}{dx^2} - 6\frac{dy}{dx} + 9y = 0$$

***Solution*:** The given function is

$$y = (A + Bx)e^{3x} \quad .....(i)$$

differentiating it w.r.t x, we get

$$\frac{dy}{dx} = 3(A+Bx)e^{3x} + Be^{3x} \quad .....(ii)$$

Differentiating again,

$$\frac{d^2y}{dx^2} = 9(A + Bx)\, e^{3x} + 3Be^{3x} + 3Be^{3x}$$

$$= 6B\, e^{3x} + 9\, e^{3x}\, (A + Bx) \quad .....(iii)$$

$$\therefore \quad \frac{d^2y}{dx^2} - 6\,\frac{dy}{dx} + 9y = 6Be^{3x} + 9e^{3x}(A + Bx) - 18e^{3x}(A + Bx) - 6Be^{3x}$$

$$+ 9e^{3x}(A + Bx) = 0$$

Hence $y = (A + Bx)e^{3x}$ is a solution of given differential equation.

Exercise 13.2

1. Verify that $y = ax^2+b$ is a solution of $x\frac{d^2y}{dx^2} - \frac{dy}{dx} = 0$
2. Show that $y = a \cos nx + b \sin nx$ is a solution of $\frac{d^2y}{dx^2} + n^2y = 0$
3. Show that $y = A \cos(\log x) + B \sin(\log x)$ is a solution of $x^2\frac{d^2y}{dx^2} + x\frac{dy}{dx} + y = 0$

4. Verify that the solution of equation $\frac{dy}{dx} - 2y = 0$ is $y = e^{2x}$

5. Verify that the solution of $y = \frac{dy}{dx} + a\frac{dx}{dy}$ is $y = cx + \frac{a}{c}$

Differential equation of the first order and first degree

All differential equations of the first order can not be solved. They can be solved by some suitable methods if they belong to any one of the following standard form:

1. Variable – separable form of equations
2. Homogeneous equations
3. Non homogeneous equations
4. Linear equations
5. Reducible to linear equations.

Variable-separable form of equations

***Type* I** when the differential equation is of the form

$$\frac{dy}{dx} = f(x)$$

$$\Rightarrow \quad dy = f(x)dx$$

Integrating both sides, we get

$$\int dy = \int f(x)dx$$

$$y = \int f(x)dx + C$$

where C is an arbitrary constant

***Example* 1:** Solve the equation

$$\frac{dy}{dx} = 3x^2 + 2x + 7$$

***Solution*:** The given equation is

$$\frac{dy}{dx} = 3x^2 + 2x + 7$$

$$\Rightarrow \quad dy = (3x^2+2x+7)\,dx$$

On integration

$$\int dy = \int (3x^2 + 2x + 7)dx$$

$$\Rightarrow \quad y = \frac{3x^3}{3} + \frac{2x^2}{2} + 7x + C$$

$$\Rightarrow \quad y = x^3 + x^2 + 7x + C$$

***Example* 2:** Solve the differential equation

$$\frac{dy}{dx} = \sec y$$

***Solution*:** Here differential equation is

$$\frac{dy}{dx} = \sec y$$

$$\Rightarrow \quad \frac{dy}{dx} = \frac{1}{\cos y}$$

$$\Rightarrow \quad \cos y\, dy = dx$$

on integration, $\int \cos y dy = \int dx$

$$\Rightarrow \quad \sin y + C = x$$

$$\Rightarrow \quad x = \sin y + C$$

Type II

Differential equation is of the form

$$\frac{dy}{dx} = f(x).g(y)$$

Can be expressed in the form

$$\frac{dy}{g(y)} = f(x)dx$$

On integration, $\int \frac{dy}{g(y)} = \int f(x)dx + C$

***Example* 3:** Solve the differential equation

$$x(2y-3)dx + (x^2+1)dy = 0$$

***Solution*:** The given equation is

$$x(2y-3)dx + (x^2+1)dy = 0$$

on separating the variables, we get

$$\frac{x}{x^2+1}dx = -\frac{dy}{2y-3}$$

$$\Rightarrow \quad \frac{x}{x^2+1}dx = \frac{dy}{3-2y}$$

On integration

$$\frac{1}{2}\log(x^2+1) = \frac{1}{2}\log(3-2y) + C$$

$$\Rightarrow \quad \log(x^2+1) = \log(3-2y) + \log a$$

$$\Rightarrow \quad \log(x^2+1) = \log(3-2y)a$$

$$x^2+1 = a(3-2y)$$

***Example* 4:** solve $y - x\dfrac{dy}{dx} = a\left(y^2 + \dfrac{dy}{dx}\right)$

***Solution*:** The given equation is

$$y - x\frac{dy}{dx} = ay^2 + a\frac{dy}{dx}$$

$$y - ay^2 = (a+x)\frac{dy}{dx}$$

$$\Rightarrow \quad \frac{dy}{y-ay^2} = \frac{dx}{a+x}$$

$$\Rightarrow \quad \frac{dy}{y(1-ay)} = \frac{dx}{a+x}$$

$$\Rightarrow \quad \left[\frac{1}{y} + \frac{a}{1-ay}\right]dy = \frac{dx}{a+x}$$

Integrating both sides, we get

$$\log y - \log(1-ay) = \log(a+x) + \log c$$

$$\Rightarrow \quad \log \frac{y}{1-ay} = \log c(a+x)$$

$$\frac{y}{1-ay} = c(a+x)$$

$$y = c(1-ay)(a+x)$$

***Example* 5:** Solve $(x^2 - yx^2)\,dy + (y^2 + xy^2)\,dx = 0$

***Solution*:** The given equation is

$$x^2(1-y)\,dy + y^2(1+x)\,dx = 0$$

$$\Rightarrow \quad \frac{1-y}{y^2}dy = -\frac{1+x}{x^2}dx$$

Integrating, we get

$$\int\left(\frac{1}{y^2} - \frac{1}{y}\right)dy = -\int\left(\frac{1}{x^2} + \frac{1}{x}\right)dx + c$$

$$\Rightarrow \quad -\frac{1}{y} - \log y = -\left[-\frac{1}{x} + \log x\right] + c$$

$$\Rightarrow \quad -\frac{1}{y} - \frac{1}{x} = \log y - \log x + \log c$$

$$\Rightarrow \quad -\frac{x+y}{xy} = \log\frac{cy}{x}$$

$$\Rightarrow \quad \frac{cy}{x} = e^{-\left(\frac{x+y}{xy}\right)}$$

***Example* 6:** Solve $9y.\frac{dy}{dx} + 4x = 0$

***Solution*:** The given equation is

$$9y.\frac{dy}{dx} + 4x = 0$$

$$\Rightarrow \quad 9y\,dy = -4x\,dx$$

Integraing, we get

$$9\frac{y^2}{2} = -4\frac{x^2}{2} + c$$

$$\Rightarrow \quad 9y^2 + 4x^2 = 2c$$

$$\Rightarrow \quad 9y^2 + 4x^2 = c$$

***Example* 7:** Solve $\sec^2 x \tan y\, dx + \sec^2 y \tan x\, dy = 0$

Solution: The given equation is

$\sec^2 x \tan y\, dx + \sec^2 y \tan x\, dy = 0$

$$\Rightarrow \quad \frac{\sec^2 x}{\tan x}dx + \frac{\sec^2 y}{\tan y}dy = 0$$

$$\Rightarrow \quad \frac{\sec^2 x}{\tan x}dx = -\frac{\sec^2 y}{\tan y}dy$$

Integrating, we get

$$\log \tan x = -\log \tan y + \log c$$

$$\Rightarrow \quad \log \tan x = \log \frac{c}{\tan y}$$

$$\Rightarrow \quad \tan x = \frac{c}{\tan y}$$

$$\Rightarrow \quad \tan x \,.\, \tan y = c$$

***Example* 8:** If $\frac{dy}{dx} = e^{x+y}$ and it is given that for $x = 1, y = 1$, find y when $x = -1$.

***Solution*:** The given equation is

$$\frac{dy}{dx} = e^{x+y}$$

$$\Rightarrow \quad \frac{dy}{dx} = e^x \,.\, e^y$$

$$\Rightarrow \quad \frac{dy}{e^y} = e^x dx$$

$$\Rightarrow \quad e^{-y}\, dy = e^x\, dx$$

Integrating, we get $\int e^{-y}dy = \int e^{x}dx + c$

$\Rightarrow \quad \dfrac{e^{-y}}{-1} = e^{x} + c$

$\Rightarrow \quad \dfrac{-1}{e^{y}} = e^{x} + c$

For $\quad x = 1, y = 1$ we have

$$\frac{-1}{e} = e + c \quad \Rightarrow \quad c = \frac{-1}{e} - e$$

hence $\quad \dfrac{-1}{e^{y}} = e^{x} - e - \dfrac{1}{e}$

when $\quad x = -1,$

$$\frac{-1}{e^{y}} = e^{-1} - e - \frac{1}{e}$$

$\Rightarrow \quad -\dfrac{1}{e^{y}} = \dfrac{1}{e} - e - \dfrac{1}{e}$

$\Rightarrow \quad -\dfrac{1}{e^{y}} = -e$

$\Rightarrow \quad \dfrac{1}{e^{y}} = e$

$\Rightarrow \quad e^{-y} = e$

$\Rightarrow \quad -y \log e = \log e$

$\Rightarrow \quad -y = 1 \qquad (\because \log e = 1)$

$\Rightarrow \quad y = -1$

Example 9: Solve $\dfrac{dy}{dx} = \dfrac{2x(\log x + 1)}{\sin y + y\cos y}$

Solution: The given equation is

$$\frac{dy}{dx} = \frac{2x(\log x + 1)}{\sin y + y\cos y}$$

$\Rightarrow \quad (\sin y + y \cos y)\, dy = 2x(\log x + 1)dx$

Integrating, we get

$$\int(\sin y + y\cos y)dy = 2\int x(\log x + 1)\,dx + c$$

$$\Rightarrow \quad \int \sin y\,dy + \int y\cos y\,dy = 2\int x\log x\,dx + 2\int x dx + c$$

$$-\cos y + \left|y\int \cos y\,dy - \int 1.\sin y\,dy\right|$$

$$2\left[\log x\int x dx - \int \frac{1}{x}.\frac{x^2}{2}dx\right] + 2.\frac{x^2}{2} + c$$

$$\Rightarrow \quad -\cos y + y\sin y + \cos y = 2\left[\log x.\frac{x^2}{2} - \frac{1}{2}.\frac{x^2}{2}\right] + x^2 + c$$

$$\Rightarrow \quad y\sin y = x^2\log x - \frac{x^2}{2} + x^2 + c$$

$$\Rightarrow \quad y\sin y = x^2\log x + \frac{x^2}{2} + c$$

***Example* 10:** Solve $2\frac{dy}{dx} = \sec y \tan x$

***Solution*:** The given equation is

$$2\frac{dy}{dx} = \sec y \tan x$$

$$\Rightarrow \quad 2\frac{dy}{\sec y} = \tan x\,dx$$

$$\Rightarrow \quad 2\cos y\,dy = \tan x\,dx$$

Integrating, we get

$$2\int \cos y\,dy = \int \tan x\,dx + c$$

$$\Rightarrow \quad 2\sin y = \log\sec x + c$$

***Example* 11:** Solve $(1 - x^2)(1 - y)\,dx = xy\,(1 + y)\,dy$

***Solution*:** The given equation is

$$(1 - x^2)(1 - y)\,dx = xy(1 + y)dy$$

$$\Rightarrow \quad \frac{1-x^2}{x}dx = \frac{y(1+y)}{1-y}dy$$

$$\Rightarrow \quad \left(\frac{1}{x} - x\right)dx = \left(-1 - y - \frac{1}{y-1}\right)dy$$

Integrating, we get

$$\int\left(\frac{1}{x} - x\right)dx = -\int\left(1 + y + \frac{1}{y-1}\right)dy + c$$

$$\Rightarrow \quad \log x - \frac{x^2}{2} = -\left(y + \frac{y^2}{2} + \log(y-1)\right) + c$$

$$\Rightarrow \quad \frac{x^2}{2} - \log x = y + \frac{y^2}{2} + \log(y-1) + c$$

***Example* 12:** Solve $\frac{dy}{dx} = e^{2x+3y}$

***Solution*:** The given equation is

$$\frac{dy}{dx} = e^{2x+3y}$$

$$\Rightarrow \quad \frac{dy}{dx} = e^{2x} . e^{3y}$$

$$\Rightarrow \quad \frac{dy}{e^{3y}} = e^{2x}\, dx$$

Integrating, we get

$$\int e^{-3y} dy = \int e^{2x} dx + c$$

$$\Rightarrow \quad \frac{e^{-3y}}{-3} = \frac{e^{2x}}{2} + c$$

$$\Rightarrow \quad 2e^{-3y} = -3e^{2x} - 3c$$

$$\Rightarrow \quad 2e^{-3y} = -3e^{2x} - 3c \;\Rightarrow\; 2e^{-3y} = -3e^{2x} + c_1 \qquad (\text{put } -3c = c_1)$$

***Example* 13:** Solve $(e^x + 1)\, y\, dy = (y+1)\, e^x\, dx$

***Solution*:** The given equation is

$$(e^x + 1)y\, dy = (y + 1)\, e^x\, dx$$

$$\Rightarrow \quad \frac{y}{y+1}dy = \frac{e^x}{e^x+1}dx$$

$$\Rightarrow \quad \left(1 - \frac{1}{y+1}\right)dy = \frac{e^x}{e^x+1}dx$$

Integrating, we get

$$\int\left(1 - \frac{1}{y+1}\right)dy = \int\frac{e^x}{e^x+1}dx + c$$

$$\Rightarrow \quad y - \log(y + 1) = \log(e^x + 1) + c$$

***Example* 14:** Solve $y(1 - x^2)\dfrac{dy}{dx} + x(1 - y^2) = 0,\ y = 1,\ x = 0$

***Solution*:** The given equation is

$$y(1 - x^2)\frac{dy}{dx} + x(1 - y^2) = 0$$

$$\frac{y}{1-y^2}dy = -\frac{x}{1-x^2}dx$$

Integrating, we get

$$\frac{-1}{2}\log(1-y^2) = \frac{1}{2}\log(1-x^2) + c$$

$$\frac{-1}{2}\log(1-y^2)(1-x^2) = \log c \qquad \text{(put c = log c)}$$

$$\log(1 - y^2)(1 - x^2) = \log c^{-2}$$

$$(1-y^2)(1 - x^2) = c_1 \qquad (\because c_1 = c^{-2})$$

When $x = 0,\ y = 1,$

we have $(1 - 1)(1 - 0) = c_1 \Rightarrow c_1 = 0$

Hence $(1 - y^2)(1 - x^2) = 0$

Differential Equation Reducible to Variable – Separable

Differential equation of the form $\frac{dy}{dx} = f(ax + by + c)$ can be reduced to variable – separable.

To solve the equation

1. Put $z = ax + by + c$
2. Differentiating w.r.t x, i.e., $\frac{dz}{dx} = a + b\frac{dy}{dx}$
3. Put the value of $\frac{dy}{dx}$ and $ax + by + c$ in the given differential equation to get a differential equation in variable x and z.
4. Separate the variable and integrals.
5. Replace the value of z to get the solution.

***Example* 15:** Solve $\frac{dy}{dx} = (4x + y + 1)^2$

***Solution*:** The given equation is

$$\frac{dy}{dx} = (4x + y + 1)^2 \qquad \text{..... (i)}$$

Put $z = 4x + y + 1$

$$\therefore \quad \frac{dz}{dx} = 4 + \frac{dy}{dx} \quad \Rightarrow \quad \frac{dy}{dx} = \frac{dz}{dx} - 4$$

Reduced differential equation of (i) is

$$\frac{dz}{dx} - 4 = z^2$$

$$\Rightarrow \quad \frac{dz}{dx} = z^2 + 4$$

By separating the variable, we have

$$\frac{dz}{z^2 + 4} = dx$$

Integrating, we get

$$\int \frac{dz}{z^2 + 4} = \int dx + c$$

$$\Rightarrow \quad \frac{1}{2}\tan^{-1}\frac{z}{2} = x + c$$

$$\frac{1}{2}\tan^{-1}\left(\frac{4x+y+1}{2}\right) = x + c$$

$$\tan^{-1}\left(\frac{4x+y+1}{2}\right) = 2(x+c)$$

$$4x + y + 1 = 2\tan 2(x + c)$$

***Example* 16:** Solve $\frac{dy}{dx} = \sin(x + y)$

***Solution*:** The given equation is

$$\frac{dy}{dx} = \sin(x+y) \qquad \text{..... (i)}$$

Put $z = x + y$

$\therefore \quad \frac{dz}{dx} = 1 + \frac{dy}{dx} \quad \Rightarrow \quad \frac{dy}{dx} = \frac{dz}{dx} - 1$

$\therefore$ Reducible equation of (i) is

$$\frac{dz}{dx} - 1 = \sin z$$

$$\Rightarrow \quad \frac{dz}{dx} = \sin z + 1$$

By separating the variables, we get

$$\frac{dz}{\sin z + 1} = dx$$

$$\Rightarrow \quad \frac{1-\sin z}{\cos^2 z}dz = dx$$

$$\Rightarrow \quad (\sec^2 z - \sec z \tan z)\, dz = dx$$

Integrating, we get

$$\int \sec^2 z\, dz - \int \sec z \tan z\, dz = \int dx + c$$

$$\Rightarrow \quad \tan z - \sec z = x + c$$

$$\Rightarrow \quad \tan(x+y) - \sec(x+y) = x + c$$

***Example* 17:** Solve $(x + y + 1)\frac{dy}{dx} = 1$

***Solution*:** The given equation is

$$(x+y+1)\frac{dy}{dx} = 1$$

Put $z = x + y + 1$ .....(i)

$\therefore \quad \frac{dz}{dx} = 1 + \frac{dy}{dx} \quad \Rightarrow \quad \frac{dy}{dx} = -1 + \frac{dz}{dx}$

Reducible equation of (i) is

$$z\left(\frac{dz}{dx} - 1\right) = 1$$

$$\Rightarrow \quad z\frac{dz}{dx} - z = 1$$

$$\Rightarrow \quad z\frac{dz}{dx} = 1 + z$$

$$\Rightarrow \quad \frac{z}{z+1}dz = dx$$

Integrating, we get

$$\int\left(1 - \frac{1}{z+1}\right)dz = \int dx + c$$

$$\Rightarrow \quad z - \log(z + 1) = x + c$$

$$\Rightarrow \quad x + y + 1 - \log(x + y + 2) = x + c$$

$$\Rightarrow \quad y + 1 - \log(x + y + 2) = c$$

Exercise 13.3

Solve the following Differential Equations

1. $\frac{dy}{dx} + \frac{1+\cos 2y}{1-\cos 2x} = 0$
2. $\frac{dy}{dx} = e^{y+x} + e^y . x^2$
3. $\frac{dy}{dx} = 1 + x + y + xy$
4. $(x^2 - yx^2)\, dx + (y^2 + x^2 y^2)\, dx = 0$
5. $y\, dx - x\, dy = xy\, dx$
6. $3e^x \tan y\, dx + (1 - e^x) \sec^2 y\, dy = 0$
7. $(1 + x)(1 + y^2)\, dx + (1 + y)(1 + x^2)\, dy = 0$
8. $\frac{dy}{dx} = y \tan 2x$ if $x = 0, y = 2$
9. $\cos x \cos y \frac{dy}{dx} = -\sin x \sin y$
10. $\sec^2 y\, (1 + x^2)\, dy + 2x \tan y\, dx = 0,\ y = \frac{\pi}{4}$ when $x = 1$

11. $(x+1)\dfrac{dy}{dx} = x^2 y$

12. $\dfrac{1}{x}\dfrac{dy}{dx} = \tan^{-1} x$

13. $\dfrac{dy}{dx} = \dfrac{2}{x+2y-3}$

14. $\dfrac{dy}{dx}\cos(x-y) = 1$

15. $\dfrac{dy}{dx} = \cos(x+y) + \sin(x+y)$

16. $x\sqrt{1+y^2}\,dx + y\sqrt{1+x^2}\,dy = 0$

17. $\dfrac{y}{x}\dfrac{dy}{dx} = \dfrac{1+y^2}{1+x^2}$, $y = 2$ when $x = 1$

18. $\left(y - x\dfrac{dy}{dx}\right)x = y$

19. $(x+1)\dfrac{dy}{dx} + 1 = 2e^{-y}$

20. $(x+y)^2\dfrac{dy}{dx} = a^2$

21. $\dfrac{dy}{dx} = \dfrac{x+y+1}{x+y}$

22. $\dfrac{dy}{dx} = 1 + \tan(y-x)$

23. $(1+x^3)\,dy - x^2 y\,dx = 0$, $y = 2$ when $x = 1$.

Answers

1. $\tan y - \cot x = c$

2. $e^x + e^{-y} + \dfrac{x^3}{3} + c = 0$

3. $\log(1+y) = x + \dfrac{x^2}{2} + c$

4. $x - \dfrac{1}{x} = \dfrac{1}{y} + \log y + c$

5. $\log\dfrac{x}{y} = x + c$

6. $(1-e^x)^3 = c\tan y$

7. $\tan^{-1}\left(\dfrac{x+y}{1-xy}\right) + \dfrac{1}{2}\log(1+x^2)(1+y^2) = c$

8. $y = 2(\sec 2x)^{\frac{1}{2}}$

9. $\sin y = c\cos x$

10. $(1+x^2)\tan y = 2$

11. $\log y = \dfrac{x^2}{2} - x + \log(x+1) + c$

12. $y = \dfrac{1}{2}(x^2+1)\tan^{-1}x - \dfrac{1}{2}x + c$

13. $2y = 4\log(x+2y+1) + k, k = c+3$

14. $\cos\left(\dfrac{x-y}{2}\right) = y + c$

15. $\log\left[1 + \tan\dfrac{x+y}{2}\right] = x + c$

16. $\sqrt{1+x^2} + \sqrt{1+y^2} = c$

17. $2y^2 = 5x^2 + 3$

18. $\log\left(\frac{y}{x}\right) = \frac{1}{x} + c$

19. $e^y = e^x + \frac{x^3}{3} + c$

20. $y = a\tan^{-1}\left(\frac{x+y}{a}\right) + c$

21. $2(y - x) = \log(2x + 2y + 1) + c$

22. $\sin(y - x) = e^{x+c}$

23. $y^3 = 4(x^3 + 1)$

Homogeneous Differential Equations

A differential equation of the form $\frac{dy}{dx} = \frac{f(x,y)}{g(x,y)}$, where f(x, y) and g(x, y) are homogeneous functions of same degree in x and y is called homogeneous differential equations.

Solving a Homogeneous Equations: A homogeneous equation can be reduced to a variable separable form.

Put $\quad y = vx$

So that $\quad \frac{dy}{dx} = v + x\frac{dv}{dx}$

Substituting these values in given equations, we get

$$v + x\frac{dv}{dx} = f(v)$$

$$\Rightarrow \quad x\frac{dv}{dx} = f(v) - v$$

$$\Rightarrow \quad \frac{dv}{f(v) - v} = \frac{dx}{x}$$

On integration, we get a solution and finally we replace v by $\frac{y}{x}$

Example 1: Solve $\quad (x^2 - y^2)\,dx + 3\,xy\,dy = 0$

Solution: The given equation is

$$(x^2 - y^2)\,dx + 3xy\,dy = 0$$

$$\Rightarrow \quad \frac{dy}{dx} = \frac{-(x^2 - y^2)}{3xy}$$

$$\frac{dy}{dx} = \frac{y^2 - x^2}{3xy} \quad \text{.....(i)}$$

which is a homogeneous differential equation

Put y = vx

$$\Rightarrow \quad \frac{dy}{dx} = v + x\frac{dv}{dx}$$

$\therefore$ from (i), we have $v + x\dfrac{dv}{dx} = \dfrac{v^2x^2 - x^2}{3x^2v}$

$$\Rightarrow \quad v + x\frac{dv}{dx} = \frac{v^2 - 1}{3v}$$

$$\Rightarrow \quad x\frac{dv}{dx} = \frac{v^2 - 1}{3v} - v$$

$$\Rightarrow \quad x\frac{dv}{dx} = \frac{-2v^2 - 1}{3v}$$

$$\Rightarrow \quad \frac{3v\,dv}{2v^2 + 1} = -\frac{dx}{x}$$

Integrating, we get

$$3\int \frac{vdv}{2v^2 + 1} = -\int \frac{1}{x}dx + c$$

$$\Rightarrow \quad \frac{3}{4}\log(2v^2 + 1) = -\log x + \log c$$

$$\Rightarrow \quad \frac{3}{4}\log\left(\frac{2y^2}{x^2} + 1\right) = -\log x + \log c$$

$$\Rightarrow \quad \frac{3}{4}\log\left(\frac{2y^2 + x^2}{x^2}\right) = -\log x + \log c$$

$$\Rightarrow \quad \frac{3}{4}\log\left(\frac{2y^2 + x^2}{x^2}\right) = \log\frac{c}{x}$$

***Example 2*:** Solve $x(x - y)\,dy + y^2\,dx = 0$

***Solution*:** The given equation is

$$x(x - y)\,dy + y^2\,dx = 0$$

$$\Rightarrow \quad \frac{dy}{dx} = -\frac{y^2}{x(x - y)} \qquad \text{..... (i)}$$

which is a homogeneous differential equation

Put y = vx $\Rightarrow$ $\frac{dy}{dx} = v + x\frac{dv}{dx}$

$\therefore$ from (i), we get

$$v + x\frac{dv}{dx} = \frac{v^2x^2}{x(vx - x)}$$

$$\Rightarrow \quad v + x\frac{dv}{dx} = \frac{v^2}{v-1}$$

$$\Rightarrow \quad x\frac{dv}{dx} = \frac{v^2}{v-1} - v$$

$$\Rightarrow \quad x\frac{dv}{dx} = \frac{v}{v-1}$$

$$\Rightarrow \quad \frac{v-1}{v}dv = \frac{dx}{x}$$

Integrating, we get

$$\int\left(1 - \frac{1}{v}\right)dv = \int\frac{1}{x}dx + c$$

$$\Rightarrow \quad v - \log v = \log x + \log c$$

$$\Rightarrow \quad v = \log v + \log x = \log c$$

$$\Rightarrow \quad v = \log v\,x\,c$$

$$\Rightarrow \quad e^v = v\,x\,c$$

$$\Rightarrow \quad e^{\frac{y}{x}} = yc$$

***Example* 3:** Solve $\frac{dy}{dx} = \frac{x^2 + xy}{x^2 + y^2}$

***Solution*:** The given equation is

$$\frac{dy}{dx} = \frac{x^2 + xy}{x^2 + y^2} \qquad \text{..... (i)}$$

which is a homogeneous differential equation

$\therefore$ Put y = vx $\Rightarrow$ $\frac{dy}{dx} = v + x\frac{dv}{dx}$

$\therefore$ from (i), we get

$$v + x\frac{dv}{dx} = \frac{x^2 + x^2 v}{x^2 + x^2 v^2}$$

$$\Rightarrow \quad v + x\frac{dv}{dx} = \frac{1+v}{1+v^2}$$

$$\Rightarrow \quad x\frac{dv}{dx} = \frac{1+v}{1+v^2} - v$$

$$\Rightarrow \quad x\frac{dv}{dx} = \frac{1-v^3}{1+v^2}$$

$$\Rightarrow \quad \frac{1+v^2}{v^3-1}dv = -\frac{dx}{x}$$

Integrating, we get,

$$\int \frac{v^2}{v^3-1}dv + \int \frac{1}{v^3-1}dv = -\int \frac{1}{x}dx + c \qquad \text{.....(ii)}$$

Now $\quad \dfrac{1}{v^3-1} = \dfrac{1}{(v-1)(v^2+v+1)} = \dfrac{A}{v-1} + \dfrac{Bv+c}{v^2+v+1}$

$$\Rightarrow \quad 1 = A(v^2 + v + 1) + (Bv + c)(v - 1)$$

Equating the coefficient of power of v, we have

$$A + B = 0$$

$$A - B + C = 0$$

$$A - C = 1$$

Put v = 1, we get $A = \dfrac{1}{3}$

Solving these equations to get $B = \dfrac{-1}{3}$, $C = \dfrac{-2}{3}$

$$\therefore \quad \frac{1}{v^3-1} = \frac{1}{3(v-1)} - \frac{v-2}{3(v^2+v+1)}$$

Integrating, we get

$$\int \frac{1}{v^3-1}\,dv = \frac{1}{3}\int \frac{1}{v-1}\,dv - \frac{1}{3}\int \frac{v+2}{v^2+v+1}\,dv$$

$$= \frac{1}{3}\log\,(v-1) - \frac{1}{6}\int \frac{2v+4}{v^2+v+1}\,dv$$

$$= \frac{1}{3}\log\,(v-1) - \frac{1}{6}\int \frac{2v+1+3}{v^2+v+1}\,dv$$

$$= \frac{1}{3}\log\,(v-1) - \frac{1}{6}\int \frac{2v+1}{v^2+v+1}\,dv + \frac{-3}{6}\int \frac{1}{v^2+v+1}\,dv$$

$$= \frac{1}{3}\log\,(v-1) - \frac{1}{6}\log(v^2+v+1) + \frac{-1}{2}\int \frac{1}{\left(v+\frac{1}{2}\right)^2 + \left(\frac{\sqrt{3}}{2}\right)^2}\,dv$$

$$= \frac{1}{3}\log\,(v-1) - \frac{1}{6}\log\,(v^2+v+1) - \frac{1}{2}.\frac{1}{\frac{\sqrt{3}}{2}}\tan^{-1}\frac{v+\frac{1}{2}}{\frac{\sqrt{3}}{2}}$$

$$= \frac{1}{3}\log\,(v-1) - \frac{1}{6}\log\,(v^2+v+1) - \frac{1}{\sqrt{3}}\tan^{-1}\left(\frac{2v+1}{\sqrt{3}}\right)$$

$\therefore$ from (ii), we get

$$\frac{1}{3}\log\,(v^3-1) + \frac{1}{3}\log\,(v-1) - \frac{1}{6}\log\,(v^2+v+1)$$

$$-\frac{1}{\sqrt{3}}\tan^{-1}\left(\frac{2v+1}{\sqrt{3}}\right) = -\log x + \log c$$

$$\Rightarrow \quad \frac{1}{3}\log\left(\frac{y^3-x^3}{x^3}\right) + \frac{1}{3}\log\left(\frac{y-x}{x}\right) - \frac{1}{6}\log\left(\frac{y^2+xy+x}{x}\right)$$

$$-\frac{1}{\sqrt{3}}\tan^{-1}\left(\frac{2y+x}{x\sqrt{3}}\right) = \log\frac{c}{x}$$

***Example* 4:** Solve $\frac{dy}{dx} = \frac{y+x}{x}$

Solution: Given equation is

$$\frac{dy}{dx} = \frac{y+x}{x} \qquad \text{.....(i)}$$

which is a homogeneous differential equation

$\therefore$ Put $y = vx \quad \Rightarrow \quad \frac{dy}{dx} = v + x\frac{dv}{dx}$

$\therefore$ from (i), we get

$$v + x\frac{dv}{dx} = \frac{vx + x}{x}$$

$$\Rightarrow \quad x\frac{dv}{dx} = v + 1 - v$$

$$\Rightarrow \quad xdv = dx$$

$$\Rightarrow \quad dv = \frac{1}{x}dx$$

Integrating, we get

$$\int dv = \int \frac{1}{x}dx + c$$

$$\Rightarrow \quad v = \log x + \log c$$

$$\Rightarrow \quad v = \log xc$$

$$\Rightarrow \quad e^{v} = xc$$

$$\Rightarrow \quad e^{\frac{y}{x}} = xc$$

***Example* 5:** Solve $x\,dy - y\,dx = \sqrt{x^2 + y^2}\,dx$

***Solution*:** Given equation is

$$x\,dy - y\,dx = \sqrt{x^2 + y^2}\,dx$$

$$x\,dy = \left(\sqrt{x^2 + y^2} + y\right)dx$$

$$\frac{dy}{dx} = \frac{\sqrt{x^2 + y^2} + y}{x} \qquad \text{.....(i)}$$

which is a homogeneous differential equation

$\therefore$ Put $y = vx \quad \Rightarrow \quad \frac{dy}{dx} = v + x\frac{dv}{dx}$

$\therefore$ from (i), we have

$$v + x\frac{dv}{dx} = \frac{\sqrt{x^2 + v^2x^2} + vx}{x}$$

$$\Rightarrow \quad x\frac{dv}{dx} = \sqrt{1+v^2} + v - v$$

$$\Rightarrow \quad x\frac{dv}{dx} = \sqrt{1+v^2}$$

$$\Rightarrow \quad \frac{dv}{\sqrt{1+v^2}} = \frac{dx}{x}$$

Integrating, we get

$$\int \frac{dv}{\sqrt{1+v^2}} = \int \frac{1}{x}\,dx + c$$

$$\Rightarrow \quad \log\left[v + \sqrt{1+v^2}\right] = \log x + \log c$$

$$\Rightarrow \quad v + \sqrt{1+v^2} = cx$$

$$\Rightarrow \quad \frac{y}{x} + \sqrt{1 + \frac{y^2}{x^2}} = cx$$

$$\Rightarrow \quad y + \sqrt{x^2 + y^2} = cx^2$$

***Example* 6:** Solve $(x^3 - 3xy^2)dx = (y^3 - 3x^2 y)\,dy$

***Solution*:** Given equation is

$$(x^3 - 3xy^2)\,dx = (y^3 - 3x^2y)\,dy$$

$$\Rightarrow \quad \frac{dy}{dx} = \frac{x^3 - 3xy^2}{y^3 - 3x^2y} \quad .....(i)$$

which is homogeneous equation

$\therefore$ Put $y = vx \quad \Rightarrow \quad \frac{dy}{dx} = v + x\frac{dv}{dx}$

$\therefore$ from (i), we get

$$v + x\frac{dv}{dx} = \frac{x^3 - 3x.x^2v^2}{x^3v^3 - 3x^2.xv}$$

$$\Rightarrow \quad x\frac{dv}{dx} = \frac{1-3v^2}{v^3-3v} - v$$

$$\Rightarrow \quad x\frac{dv}{dx} = \frac{1-v^4}{v^3-3v}$$

$$\Rightarrow \quad \frac{v^3-3v}{1-v^4}dv = \frac{dx}{x} \qquad .....(ii)$$

Now, $\quad \dfrac{v^3-3v}{1-v^4} = \dfrac{v^3-3v}{(1-v)(1+v)(1+v^2)} = \dfrac{A}{1-v} + \dfrac{B}{1+v} + \dfrac{Cv+D}{1+v^2}$

$$\Rightarrow \quad v^3 - 3v = A(1+v)(1+v^2) + B(1-v)(1+v^2) + (Cv+D)(1-v^2)$$

Putting $\quad v = 1, -1$, we get $A = \dfrac{-1}{2}$, $B = \dfrac{1}{2}$, $C = -2$, $D = 0$

$\therefore$ from (ii), we have

$$\left(\frac{1}{2(v-1)} + \frac{1}{2(v+1)} - \frac{2v}{v^2+1}\right)dv = \frac{dx}{x}$$

Integrating, we get

$$\frac{1}{2}\log(v-1) + \frac{1}{2}\log(v+1) - \log(v^2+1) = \log x + \log c$$

$$\log\left(\frac{y}{x}-1\right)^{\frac{1}{2}} + \log\left(\frac{y}{x}+1\right)^{\frac{1}{2}} - \log\left(\frac{y^2}{x^2}+1\right) = \log xc$$

$$\log\frac{\left(\frac{y-x}{x}\right)^{\frac{1}{2}}.\left(\frac{y+x}{x}\right)^{\frac{1}{2}}}{\left(\frac{y^2+x^2}{x^2}\right)} = \log xc$$

$$\frac{(y^2-x^2)^{\frac{1}{2}}}{y^2+x^2} = c$$

$$\left(y^2-x^2\right)^{\frac{1}{2}} = c(y^2+x^2)$$

$$(y^2-x^2) = c^2(y^2+x^2)^2$$

Differential Equation Reducible to Homogeneous form

The equation of the form $\frac{dy}{dx} = \frac{ax + by + c}{a'x + b'y + c'}$ ..... (i)

can be reduced to homogeneous form

Case I: when $\frac{a}{a'} \neq \frac{b}{b'}$

Putting $x = X + h, y = Y + k$, where h and k are constants

$\therefore$ $dx = dX, dy = dY$

From equation (i), we have

$$\frac{dY}{dX} = \frac{aX + bY + (ah + bk + c)}{a'X + b'Y + (a'h + b'k + c')} \quad \text{..... (ii)}$$

Choose h and k, so that equation (ii) may becomes homogeneous

Putting $ah + bk + c = 0$ and $a'h + b'k + c' = 0$

By cross method, $\frac{h}{bc' - b'c} = \frac{k}{ca' - ac'} = \frac{1}{ab' - a'b}$

$\Rightarrow$ $h = \frac{bc' - b'c}{ab' - a'b}, \quad k = \frac{ca' - ac'}{ab' - a'b}.$ when $ab' - a'b \neq 0$

Then $\frac{dY}{dX} = \frac{aX + bY}{a'X + b'Y}$

is a homogeneous equation in X and Y and can be solved by putting $Y = VX$

Case II: When $\frac{a}{a'} = \frac{b}{b'}$ and $ab' - a'b = 0$

Now, $\frac{a}{a'} = \frac{b}{b'} = \frac{1}{m}$ (say)

So that $a' = ma, b' = mb$

$\therefore$ from equation (i), we get

$$\frac{dy}{dx} = \frac{ax + by + c}{m(ax + by) + c'} \quad \text{..... (iii)}$$

Putting $\quad ax + by = z$

$$\Rightarrow \quad a + b\frac{dy}{dx} = \frac{dz}{dx} \quad \Rightarrow \quad \frac{dy}{dx} = \frac{1}{b}\left(\frac{dz}{dx} - a\right)$$

Equation (iii) becomes $\frac{1}{b}\left(\frac{dz}{dx} - a\right) = \frac{z+c}{mz+c'}$

$$\Rightarrow \quad \frac{dz}{dx} = a + \frac{bz+bc}{mz+c'}$$

$$= \frac{(am+b)z + ac' + bc}{(mz+c')}$$

It is the equation solvable by separating the variables

***Example* 7:** Solve $(2x + y + 3)\, dy = (x + 2y - 3)dx$

***Solution*:** Given equation is

$$\frac{dy}{dx} = \frac{x+2y-3}{2x+y-3} \qquad \text{.....(i)}$$

This equation is reducible to homogeneous form

$\therefore$ Putting $x = X + h, \quad y = Y + k$

$$\Rightarrow \quad dx = dX,\ dy = dY$$

$$\Rightarrow \quad \frac{dy}{dx} = \frac{dY}{dX}$$

$\therefore$ from (i), we get

$$\frac{dY}{dX} = \frac{X+h+2Y+2k-3}{2X+2h+Y+k-3} \qquad \text{.....(ii)}$$

$$\frac{dY}{dX} = \frac{X+2Y+(h+2k-3)}{2X+Y+(2h+k-3)}$$

Choose the value of h and k, so that

$$h + 2k - 3 = 0$$

$$2h + k - 3 = 0$$

$$\frac{h}{-6+3} = \frac{k}{-6+3} = \frac{1}{1-4} \quad \Rightarrow \quad h = 1, k = 1$$

Then equation (ii) becomes

$$\frac{dY}{dX} = \frac{X+2Y}{2X+Y} \qquad \text{.....(iii)}$$

which is homogeneous equation

Putting Y = vX $\Rightarrow$ $$\frac{dY}{dX} = v + X\frac{dv}{dX}$$

$\therefore$ from (iii), we get

$$v + X\frac{dv}{dX} = \frac{X + 2vX}{2X + vX}$$

$$\Rightarrow \quad X\frac{dv}{dX} = \frac{1 + 2v}{2 + v} - v$$

$$\Rightarrow \quad x\frac{dv}{dX} = \frac{1 - v^2}{2 + v}$$

$$\Rightarrow \quad \frac{2 + v}{1 - v^2} dv = \frac{dX}{X}$$

Integrating, we get

$$\int \frac{v + 2}{(v-1)(v+1)} dv = -\int \frac{1}{X} dX + c$$

$$\Rightarrow \quad \int \left(\frac{3}{2(v-1)} - \frac{1}{2(v+1)} \right) dv = -\log X + \log c$$

$$\Rightarrow \quad \frac{3}{2}\log(v-1) - \frac{1}{2}\log(v+1) = \log\frac{c}{X}$$

$$\Rightarrow \quad \log(v-1)^3 - \log(v+1) = \log\frac{c^2}{X^2}$$

$$\Rightarrow \quad \frac{(v-1)^3}{v+1} = \frac{c^2}{X^2}$$

$$\Rightarrow \quad \frac{\left(\frac{Y}{X} - 1\right)^3}{\frac{Y}{X} + 1} = \frac{c^2}{X^2}$$

$$\Rightarrow \quad \frac{(Y - X)^3}{Y + X} = c^2$$

$$\Rightarrow \quad \frac{(Y - X)^3}{Y + X - 2} = c^2 \qquad [\because Y = y - k = y - 1;\ X = x - h = x - 1]$$

$$\Rightarrow \quad (y - x)^3 = c^2 (x + y - 2)$$

***Example* 8:** Solve $(2x + 3y - 5)\dfrac{dy}{dx} + (3x + 2y - 5) = 0$

***Solution*:** Given differential equation is

$$\frac{dy}{dx} = -\frac{3x + 2y - 5}{2x + 3y - 5} \qquad \text{.....(i)}$$

This equation is reducible to homogeneous equation

Put $x = X + h,\ y = Y + k,\ \dfrac{dy}{dx} = \dfrac{dY}{dX}$

$\therefore$ equation (i) becomes

$$\frac{dY}{dX} = -\frac{3X + 3h + 2Y + 2k - 5}{2X + 2h + 3Y + 3k - 5}$$

$$\Rightarrow \quad \frac{dY}{dX} = -\frac{3X + 2Y + (3h + 2k - 5)}{2X + 3Y + (2h + 3k - 5)} \qquad \text{.....(ii)}$$

Choose h and k, we have

$$3h + 2k - 5 = 0$$
$$2h + 3k - 5 = 0$$

$$\Rightarrow \quad \frac{h}{-10 + 15} = \frac{k}{-10 + 15} = \frac{1}{9 - 4} \quad \Rightarrow \quad h = 1, k = 1$$

$\therefore$ from (ii) we have

$$\frac{dY}{dX} = -\frac{3X + 2Y}{2X + 3Y} \qquad \text{.....(iii)}$$

which is homogeneous equation

put $\quad Y = vX \quad \Rightarrow \quad \dfrac{dY}{dX} = v + X\dfrac{dv}{dX}$

equation (iii) becomes

$$v + X\frac{dY}{dX} = -\frac{3X + 2vX}{2X + 3vX}$$

$$\Rightarrow \quad X\frac{dv}{dX} = -\frac{3 + 2v}{2 + 3v} - v$$

$$\Rightarrow \quad X\frac{dv}{dX} = -\frac{3v^2 + 4v + 3}{3v + 2}$$

$$\Rightarrow \quad \frac{3v + 2}{3v^2 + 4v + 3}dv = -\frac{dX}{X}$$

Integrating, we get

$$\int \frac{3v+2}{3v^2+4v+3}\,dv = -\int \frac{1}{X}\,dX + c$$

$$\Rightarrow \quad \frac{1}{2}\log(3v^2+4v+3) = -\log X + \log c$$

$$\Rightarrow \quad \log(3v^2+4v+3) = \log \frac{c^2}{X^2}$$

$$\Rightarrow \quad 3\frac{Y^2}{X^2} + 4\frac{Y}{X} + 3 = \frac{c^2}{X^2}$$

$$\Rightarrow \quad 3Y^2 + 4XY + 3X^2 = c^2 \qquad [\because X = x - h = x - 1,\ Y = y - k = y - 1]$$

$$\Rightarrow \quad 3(y-1)^2 + 4(x-1)(y-1) + 3(x-1)^2 = c^2$$

$$\Rightarrow \quad 3x^2 + 3y^2 + 4xy - 10x - 10y = c^2 - 10$$

***Example* 9:** Solve $(3y + 2x + 4)\,dx - (4x + 6y + 5)\,dy = 0$

***Solution*:** Given equation is

$$\frac{dy}{dx} = \frac{3y+2x+4}{4x+6y+5} \qquad \text{.....(i)}$$

here $\frac{a}{a_1} = \frac{b}{b_1}$

∴ (i) becomes

$$\frac{dy}{dx} = \frac{(2x+3y)+4}{2(2x+3y)+5} \qquad \text{.....(ii)}$$

Putting $2x + 3y = z$

$$\Rightarrow \quad 2 + 3\frac{dy}{dx} = \frac{dz}{dx}$$

$$\Rightarrow \quad \frac{dy}{dx} = \frac{1}{3}\left(\frac{dz}{dx} - 2\right)$$

∴ equation (ii) becomes

$$\frac{1}{3}\left(\frac{dz}{dx}-2\right)=\frac{z+4}{2z+5}$$

$$\Rightarrow \qquad \frac{dz}{dx}=\frac{3z+12}{2z+5}+2$$

$$\Rightarrow \qquad \frac{dz}{dx}=\frac{7z+22}{2z+5}$$

$$\Rightarrow \qquad \frac{2z+5}{7z+22}dz=dx$$

Integrating, we get

$$\int\left(\frac{2}{7}-\frac{9}{7}\frac{1}{7z+22}\right)dz=x+c$$

$$\Rightarrow \qquad \frac{2}{7}z-\frac{9}{49}\log(7z+22)=x+c$$

$$\Rightarrow \qquad 14z-9\log(7z+22)=49x+49c$$

$$\Rightarrow \qquad 14(2x+3y)-9\log[7(2x+3y)+22]=49x+c_1$$

$$\Rightarrow \qquad 28x+42y-9\log[14x+21y+22]=49x+c_1$$

$$\Rightarrow \qquad 21x-42y+9\log(14x+21y+22)=c_1$$

***Example* 10:** Solve $\frac{dy}{dx}=\frac{x+y+1}{2x+2y+3}$

***Solution*:** Given equation is

$$\frac{dy}{dx}=\frac{x+y+1}{2(x+y)+3} \qquad \text{.....(i)}$$

here $\frac{a}{a'}=\frac{b}{b'}=\frac{1}{2}$, Put $x+y=z$

$$\Rightarrow \qquad 1+\frac{dy}{dx}=\frac{dz}{dx} \quad \Rightarrow \quad \frac{dy}{dx}=\frac{dz}{dx}-1$$

equation (i) becomes

$$\frac{dz}{dx} - 1 = \frac{z+1}{2z+3}$$

$$\Rightarrow \quad \frac{dz}{dx} = \frac{z+1}{2z+3} + 1$$

$$\Rightarrow \quad \frac{dz}{dx} = \frac{3z+4}{2z+3}$$

$$\Rightarrow \quad \frac{2z+3}{3z+4} dz = dx$$

Integrating, we get

$$\int \frac{2z+3}{3z+4} dz = \int dx + c$$

$$\Rightarrow \quad \frac{1}{3}\int \left[2 + \frac{1}{3z+4}\right] dz = x + c$$

$$\Rightarrow \quad \frac{1}{3}\left[2z + \frac{1}{3}\log(3z+4)\right] = x + c$$

$$\Rightarrow \quad 6z + \log(3z+4) = 9x + 9c$$

$$\Rightarrow \quad 6(x+y) + \log[3(x+y)+4] = 9x + c_1$$

$$\Rightarrow \quad 6x + 6y + \log(3x+3y+4) = 9x + c_1$$

$$\Rightarrow \quad -3x + 6y + \log(3x+3y+4) = c_1$$

Exercise 13.4

Solve the homogeneous equations

1. $x + y\frac{dy}{dx} = 2y$
2. $x(x-y)\,dy = y(x+y)\,dx$
3. $(x^2+y^2)\,dx + 2xy\,dy = 0$
4. $x^2\frac{dy}{dx} = x^2 + xy + y^2$
5. $\left(1+e^{\frac{x}{y}}\right)dx + e^{\frac{x}{y}}\left(1-\frac{x}{y}\right)dy = 0$
6. $\frac{dy}{dx} + \frac{x-2y}{2x-y} = 0$
7. $(x+y)\,dy + (x-y)\,dx = 0$, given that $y = 1$ when $x = 1$.

8. $(x^2 - y^2)\,dx + 2xy\,dy = 0$, given that $y = 1$ when $x = 1$.

9. $\frac{dy}{dx} = \frac{2x + 9y - 20}{6x + 2y - 10}$

10. $(x + 2y)(dx - dy) = dx + dy$

11. $(4x - 6y - 1)\,dx + (3y - 2x - 2)dy = 0$

12. $\frac{dy}{dx} = \frac{y + x}{x}$

Answers

1. $\log(y - x) = c + \frac{x}{y - x}$

2. $\frac{x}{y} + \log(xy) = c$

3. $x^3 + 3xy^2 = c$

4. $\log x = \tan^{-1}\frac{y}{x} + c$

5. $x + y\,e^{\frac{x}{y}} = c$

6. $(y - x) = c\,(x + y)^3$

7. $\log(x^2 + y^2) + 2\tan^{-1}\left(\frac{y}{x}\right) = \frac{\pi}{2} + \log 2$

8. $x^2 + y^2 = 2x$

9. $(y - 2x)^2 = c(x + 2y - 5)$

10. $\log(x + y + \frac{1}{3}) + \frac{3}{2}(y - x) = c$

11. $x - y + \frac{3}{4}\log(8x - 12y - 5) = c$

12. $c.x = e^{\frac{y}{x}}$

Linear Differential Equation: A linear differential equation of first order with constant coefficients is of the form

$$\frac{dy}{dx} + Py = Q \qquad \text{.....(i)}$$

Here P and Q may be a function of x or a constant. It is also called Leibnitz linear differential equation. This type of equation is solved by multiplying by a factor called integrating factor (I.F).

$$\text{I.F} = e^{\int P dx}$$

Then we get

$$e^{\int P dx}\frac{dy}{dx} + P.y.e^{\int P dx} = Q\,e^{\int P dx}$$

$$\Rightarrow \quad \frac{d}{dx}[y.e^{\int P dx}] = Q\,e^{\int P dx}$$

Integrating both sides, we get

$$y\,e^{\int P dx} = \int Q\,e^{\int P dx}\,dx + c$$

'or'

$$y\ \text{I.F} = \int Q\ \text{I.F}\ dx + c$$

which is the required solution

***Example* 1:** Solve $(1 + x^2)\dfrac{dy}{dx} + 2xy = 4x^2$

***Solution*:** Given equation is

$$(1 + x^2)\frac{dy}{dx} + 2xy = 4x^2$$

$$\Rightarrow \quad \frac{dy}{dx} + \frac{2x}{1+x^2}y = \frac{4x^2}{1+x^2} \quad \text{.....(i)}$$

which is a linear differential equation

Here $\quad P = \dfrac{2x}{1+x^2}, \quad Q = \dfrac{4x^2}{1+x^2}$

Now $\quad \text{I.F} = e^{\int P dx}$

$$= e^{\int \frac{2x}{1+x^2} dx}$$

$$= e^{\log(1+x^2)} = 1 + x^2$$

$\therefore$ solution of (i) is

$$y.\ \text{I.F} = \int Q\ \text{I.F}\ dx + c$$

$$\Rightarrow \quad y(1 + x^2) = \int \frac{4x^2}{1+x^2}(1+x^2)\,dx + c$$

$$\Rightarrow \quad y(1+x^2) = \int 4x^2 dx + c$$

$$\Rightarrow \quad y(1 + x^2) = \frac{4}{3}x^3 + c$$

***Example* 2:** Solve $x\frac{dy}{dx}-2y=x^2$

***Solution*:** Given equation is

$$x\frac{dy}{dx}-2y=x^2$$

$$\Rightarrow \quad \frac{dy}{dx}-\frac{2}{x}y=x \qquad \text{.....(i)}$$

which is a linear differential equation

here $\quad P=\frac{-2}{x}, \qquad Q=x$

Now, $\quad I.F = e^{\int P dx} = e^{-\int \frac{2}{x}dx}$

$$= e^{-2\log x}$$

$$= e^{\log x^{-2}} = \frac{1}{x^2}$$

$\therefore$ Solution of (i) is

$$y.I.F = \int Q\,I.F\,dx + c$$

$$\Rightarrow \quad y.\frac{1}{x^2} = \int x.\frac{1}{x^2}\,dx + c$$

$$\frac{y}{x^2} = \int \frac{1}{x}\,dx + c$$

$$\frac{y}{x^2} = \log x + c$$

$$y = x^2 \log x + x^2 c$$

***Example* 3:** Solve $\frac{dy}{dx}+\frac{2x}{1+x^2}y=\frac{1}{(x^2+1)^2}$, given that y = 0 when x = 1

***Solution*:** Given equation is

$$\frac{dy}{dx}+\frac{2x}{1+x^2}y=\frac{1}{(x^2+1)^2} \qquad \text{.....(i)}$$

which is a linear differential equation

here $\quad P=\frac{2x}{1+x^2}, \quad Q=\frac{1}{(x^2+1)^2}$

$$\therefore \quad I.F = e^{\int P dx} = e^{\int \frac{2x}{1+x^2}dx} = e^{\log(1+x^2)} = 1+x^2$$

∴ solution of (i) is

$\Rightarrow \quad y.I.F = \int Q.\,I.F.\,dx + c$

$\Rightarrow \quad y(1 + x^2) = \int \frac{1}{(x^2+1)^2}\,(1 + x^2)\,dx + c$

$\Rightarrow \quad y(1 + x^2) = \int \frac{1}{x^2+1}\,dx + c$

$\Rightarrow \quad y(1 + x^2) = \tan^{-1} x + c$ .....(ii)

Now, when x = 1, y = 0 then from (ii), we get

$0 = \tan^{-1}(1) + c$

$\Rightarrow \quad c = \frac{-\pi}{4}$

∴ from (ii) solution is

$y(1 + x^2) = \tan^{-1}x - \frac{-\pi}{4}$

***Example* 4:** Solve $(1 + x^2)\frac{dy}{dx} + 2xy = \cos x$

***Solution*:** Given equation is

$(1 + x^2)\frac{dy}{dx} + 2xy = \cos x$

$\Rightarrow \quad \frac{dy}{dx} + \frac{2x}{1+x^2}.y = \frac{\cos x}{1+x^2}$ .....(i)

which is a linear differential equation

here $P = \frac{2x}{1+x^2}, \quad Q = \frac{\cos x}{1+x^2}$

$I.F = e^{\int P dx} = e^{\int \frac{2x}{1+x^2}dx} = e^{\log(1+x^2)} = 1 + x^2$

Solution of (i) is

$y.I.F = \int Q.I.F.dx + c$

$\Rightarrow \quad y(1 + x^2) = \int \frac{\cos x}{1+x^2}(1+x^2)dx + c$

$\Rightarrow \quad y(1 + x^2) = \int \cos x\, dx + c$

$\Rightarrow \quad y(1 + x^2) = \sin x + c$

***Example* 5:** Solve $\frac{dy}{dx} + y \tan x = \sin 2x$ with the condition y(0) = 1.

***Solution*:** Given equation is

$$\frac{dy}{dx} + y \cdot \tan x = \sin 2x \qquad \text{.....(i)}$$

which is a linear differential equation

here $\quad P = \tan x, \qquad Q = \sin 2x$

$$\therefore \quad I.F = e^{\int P dx} = e^{\int \tan x \, dx}$$

$$= e^{\log \sec x} = \sec x$$

$\therefore$ solution of (i) is

$$y.I.F = \int Q.I.F\, dx + c$$

$$\Rightarrow \quad y \sec x = \int \sin 2x \sec x \, dx + c$$

$$\Rightarrow \quad y \sec x = 2\int \sin x \, dx + c$$

$$\Rightarrow \quad y \sec x = -2 \cos x + c \qquad \text{.....(ii)}$$

with the condition y(0) = 1 i.e., x = 0, y = 1, we have from (ii)

$$1 = -2 + c \quad \Rightarrow \quad c = 3$$

$\therefore$ from (ii), $y \sec x = -2 \cos x + 3 \quad \Rightarrow \quad y = -2 \cos^2 x + 3 \cos x$

***Example* 6:** Solve the following differential equation subject to the condition y = 1, when x = 0.

$$\frac{dy}{dx} + \frac{y}{x} = x^2$$

***Solution*:** Given equation is

$$\frac{dy}{dx} + \frac{y}{x} = x^2 \qquad \text{.....(i)}$$

which is a linear differential equation

here $\quad P = \frac{1}{x}, \qquad Q = x^2$

$$\therefore \text{ I.F.} = e^{\int P dx} = e^{\int \frac{1}{x} dx} = e^{\log x} = x$$

Solution of (i) is

$$y.I.F = \int Q\, I.F\, dx + c$$

$$\Rightarrow \quad y.x = \int x^2.x\, dx + c$$

$$\Rightarrow \quad xy = \int x^3 dx + c$$

$$\Rightarrow \quad xy = \frac{x^4}{4} + c \qquad \text{.....(ii)}$$

with the condition x = 0, y = 1, then from (ii) we get

$$c = 0$$

$\therefore$ Required solution is $xy = \dfrac{x^4}{4}$

$$y = \frac{x^3}{4}$$

***Example* 7:** Solve the equation $\dfrac{dy}{dx} + y = e^{-x}$ subject to condition x = 0, y = 1.

***Solution*:** Given equation is $\dfrac{dy}{dx} + y = e^{-x}$ .....(i)

which is a linear differential equation

here $\quad P = 1,\ Q_1 = e^{-x}$

$$\therefore \quad I.F = e^{\int P dx} = e^{\int dx} = e^x$$

Solution of (i) is

$$y.I.F = \int Q\, I.F\, dx + c$$

$$\Rightarrow \quad y\, e^x = \int Q\, I.F\, dx + c$$

$$\Rightarrow \quad y\, .\, e^x = \int e^{-x} e^x dx + c$$

$$\Rightarrow \quad y\, e^x = \int dx + c$$

$$\Rightarrow \quad y\, e^x = x + c \qquad \text{.....(ii)}$$

condition x = 0, y = 1, we get 1 = c

$\therefore$ Required solution is $\quad y\, .\, e^x = x + 1$

***Example* 8:** Solve the equation $\frac{dy}{dx} - y = e^{2x}$ subject to condition y = 1 when x = 0.

***Solution*:** Given equation is

$$\frac{dy}{dx} - y = e^{2x} \quad .....(i)$$

which is a linear differential equation

here $P = -1 \quad Q = e^{2x}$

$\therefore \quad I.F = e^{\int P dx} = e^{-\int dx} = e^{-x}$

Solution of (i) is

$$y.I.F = \int Q\, I.F\, dx + c$$

$$y \,.\, e^{-x} = \int e^{2x}.e^{-x} dx + c$$

$$y\, e^{-x} = \int e^{x} dx + c$$

$$y\, e^{-x} = e^{x} + c \quad .....(ii)$$

Putting $x = 0, y = 1$ we get

$1 = 1 + c \quad \Rightarrow \quad c = 0$

$\therefore$ required solution is

$$y \,.\, e^{-x} = e^{x}$$

$\Rightarrow \quad y = e^{2x}$

***Example* 9:** Solve $\frac{dy}{dx} - 3y = x - 1$

***Solution*:** Given equation is

$$\frac{dy}{dx} - 3y = x - 1 \quad .....(i)$$

which is a linear differential equation

here $P = -3, \quad Q = x - 1$

$$I.F = e^{\int P dx} = e^{-3\int dx} = e^{-3x}$$

Solution of (i) is

$$y.I.F = \int Q\, I.F\, dx + c$$

$$\Rightarrow \quad y\,.\,e^{-3x} = \int (x-1)\, e^{-3x} dx + c$$

$$\Rightarrow \quad y\,.\,e^{-3x} = (x-1) \int e^{-3x} dx - \int \left[\frac{d}{dx}(x-1) \int e^{-3x} dx \right] dx + c$$

$$\Rightarrow \quad y\, e^{-3x} = (x-1)\, \frac{e^{-3x}}{(-3)} - \int \frac{e^{-3x}}{(-3)} dx + c$$

$$\Rightarrow \quad y\,.\,e^{-3x} = \frac{-(x-1)}{3} e^{-3x} + \frac{1}{3} \frac{e^{-3x}}{(-3)} + c$$

$$\Rightarrow \quad y = \frac{-(x-1)}{3} - \frac{1}{9} + c\,.\,e^{3x}$$

***Example* 10:** Solve $x\dfrac{dy}{dx} - 2y = 2x$

***Solution*:** Given equation is $x\dfrac{dy}{dx} - 2y = 2x$

$$\Rightarrow \quad \frac{dy}{dx} - \frac{2}{x} y = 2 \qquad \text{.....(i)}$$

which is a linear differential equation

here $\quad P = \dfrac{-2}{x}, \qquad Q = 2$

$$I.F = e^{\int P dx} \quad = e^{-2\int \frac{1}{x} dx} \quad = e^{-2 \log x} \quad = \frac{1}{x^2}$$

Solution of (i) is

$$y.\, I.F = \int Q\, I.F\, dx + c$$

$$y.\frac{1}{x^2} = \int 2.\frac{1}{x^2} dx + c$$

$$\frac{y}{x^2} = -\frac{2}{x} + c$$

$$y = -2x + cx^2$$

***Example* 11:** Solve $x(1-x^2)\dfrac{dy}{dx}+(2x^2-1)\,y=x^3$

***Solution*:** Given equation is $$x(1-x^2)\frac{dy}{dx}+(2x^2-1)\,y=x^3$$

$$\Rightarrow \quad \frac{dy}{dx}+\frac{2x^2-1}{x(1-x^2)}y=\frac{x^3}{x(1-x^2)} \qquad \text{.....(i)}$$

which is a linear differential equation

here $$P=\frac{2x^2-1}{x(1-x^2)},\; Q=\frac{x^3}{x(1-x^2)}$$

Now $$\frac{2x^2-1}{x(1-x^2)}=\frac{2x^2-1}{x(1-x)(1+x)} = \frac{A}{x}+\frac{B}{1-x}+\frac{C}{1+x}$$

$$\Rightarrow \quad 2x^2-1=A(1-x^2)+Bx(1+x)+Cx(1-x)$$

Put $x=0, 1, -1,$ we get $A=-1,\; B=\dfrac{1}{2},\; C=\dfrac{-1}{2}$

hence $$\frac{2x^2-1}{x(1-x^2)}=\frac{-1}{x}+\frac{1}{2(1-x)}-\frac{1}{2(1+x)}$$

$$\int\frac{2x^2-1}{x(1-x^2)}dx=-\int\frac{1}{x}dx+\frac{1}{2}\int\frac{1}{1-x}dx-\frac{1}{2}\int\frac{1}{1+x}dx$$

$$=-\log x-\frac{1}{2}\log(1-x)-\frac{1}{2}\log(1+x)$$

$$=-[\log x+\log(1-x)^{\frac{1}{2}}+\log(1+x)^{\frac{1}{2}}]$$

$$=-\log x\,(1-x^2)^{\frac{1}{2}}$$

$$\therefore \quad \text{I.F} = e^{\int P dx} = e^{\int\frac{2x^2-1}{x(1-x^2)}dx}$$

$$= e^{-\log x(1-x^2)^{\frac{1}{2}}}=\frac{1}{x(1-x^2)^{\frac{1}{2}}}$$

$\therefore$ Solution of (i) is $y \,.\, IF = \int Q\, I.F\, dx + c$

$$\Rightarrow \quad y.\frac{1}{x(1-x^2)^{\frac{1}{2}}} = \int \frac{x^3}{x(1-x^2)}.\frac{1}{x(1-x^2)^{\frac{1}{2}}} dx + c$$

$$\Rightarrow \quad \frac{y}{x(1-x^2)^{\frac{1}{2}}} = \int \frac{x}{(1-x^2)^{\frac{3}{2}}} dx + c$$

$$\Rightarrow \quad \frac{y}{x(1-x^2)^{\frac{1}{2}}} = \frac{1}{(1-x^2)^{\frac{1}{2}}} + c$$

$$\Rightarrow \quad y = x + cx\left(1-x^2\right)^{\frac{1}{2}}$$

***Example* 12:** Solve $\frac{dy}{dx} + y \tan x = \sec x$

***Solution*:** Given equation is

$$\frac{dy}{dx} + y \tan x = \sec x \qquad .....(i)$$

which is a linear differential equation

here $\quad P = \tan x \qquad Q = \sec x$

$$I.F = e^{\int P dx} = e^{\int \tan x\, dx} = e^{\log \sec x} = \sec x$$

Solution of (i) is

$$y.I.F = \int Q.I.F\, dx + c$$

$$\Rightarrow \quad y \sec x = \int \sec x.\sec x\, dx + c$$

$$\Rightarrow \quad y \sec x = \tan x + c$$

$$\Rightarrow \quad y = \sin x + c \cos x$$

***Example* 13:** Solve $\cos^2 x \frac{dy}{dx} + y = \tan x$

***Solution*:** Given equation is

$$\cos^2 x \frac{dy}{dx} + y = \tan x$$

$$\Rightarrow \quad \frac{dy}{dx} + \frac{1}{\cos^2 x} y = \frac{\tan x}{\cos^2 x}$$

$$\Rightarrow \quad \frac{dy}{dx} + \sec^2 x \,.\, y = \tan x \,.\, \sec^2 x \qquad .....(i)$$

which is a linear differential equation

here $P = \sec^2 x \quad Q = \tan x \sec^2 x$

$$\text{I.F} = e^{\int P dx} = e^{\int \sec^2 x \, dx} = e^{\tan x}$$

Solution of (i) is

$$y.\text{I.F} = \int Q \, \text{I.F} \, dx + c$$

$$y \,.\, e^{\tan x} = \int \tan x \sec^2 x \,.\, e^{\tan x} dx + c$$

$$\Rightarrow \quad y\, e^{\tan x} = \int t e^t dt + c \qquad \text{where } t = \tan x$$

$$\Rightarrow \quad y\, e^{\tan x} = t\, e^t - e^t + c$$

$$\Rightarrow \quad y\, e^{\tan x} = \tan x\, e^{\tan x} - e^{\tan x} + c$$

$$\Rightarrow \quad y = \tan x - 1 + c\, e^{-\tan x}$$

***Example* 14:** Solve $(x + y + 1)\dfrac{dy}{dx} = 1$

***Solution*:** Given equation is

$$(x + y + 1)\frac{dy}{dx} = 1$$

$$\Rightarrow \quad \frac{dy}{dx} = \frac{1}{x + y + 1}$$

$$\Rightarrow \quad \frac{dx}{dy} = x + y + 1$$

$$\Rightarrow \quad \frac{dx}{dy} - x = y + 1 \qquad ..... (i)$$

which is linear differential equation

here $\quad P = -1, \qquad Q = y + 1$

$$\text{I.F.} = e^{\int P dy} = e^{-\int dy} = e^{-y}$$

Solution of (i) is

$$x.I.F = \int Q.I.F\,dy + c$$

$$\Rightarrow \quad x\,e^{-y} = \int (y+1)\,e^{-y}\,dy + c$$

$$x\,e^{-y} = -(y+1)\,e^{-y} - e^{-y} + c$$

$$x = ce^{y} - y - 2$$

***Example* 15:** Solve $(1 + y^2)\,dx = (\tan^{-1} y - x)\,dy$

***Solution*:** Given equation is

$$(1 + y^2)\,dx = (\tan^{-1} y - x)dy$$

$$\Rightarrow \quad \frac{dx}{dy} = \frac{\tan^{-1} y - x}{1+y^2}$$

$$\frac{dx}{dy} + \frac{x}{1+y^2} = \frac{\tan^{-1} y}{1+y^2} \qquad \text{..... (i)}$$

which is a linear differential equation

here $\quad P = \dfrac{1}{1+y^2}, \qquad Q = \dfrac{\tan^{-1} y}{1+y^2}$

$$I.F = e^{\int P dy} = e^{\int \frac{1}{1+y^2} dy} = e^{\tan^{-1} y}$$

Solution of (i) is

$$x\,.I.F = \int Q\,I.F.\,dy + c$$

$$\Rightarrow \quad x\,e^{\tan^{-1} y} = \int \frac{\tan^{-1} y}{1+y^2}.e^{\tan^{-1} y}\,dy + c$$

$$\Rightarrow \quad x\,e^{\tan^{-1} y} = \int t\,e^{t}\,dt + c, \qquad \text{where } t = \tan^{-1} y$$

$$\Rightarrow \quad x\,e^{\tan^{-1} y} = t\,e^{t} - e^{t} + c$$

$$\Rightarrow \quad x\,e^{\tan^{-1} y} = \tan^{-1} y\; e^{\tan^{-1} y} - e^{\tan^{-1} y} + c$$

$$\Rightarrow \quad x = c\,e^{-\tan-1\,y} + \tan^{-1} y - 1$$

Equation Reducible to the linear form (Bernoulli's Equation)

A Differential equation of the form

$$\frac{dy}{dx} + Py = Q\,y^n \qquad \text{.....(i)}$$

which is not linear, but it can be made linear equation. We divide both sides by y^n, we have

$$y^{-n}\frac{dy}{dx} + Py^{1-n} = Q \qquad \text{.....(ii)}$$

Putting $y^{1-n} = v$, so that

$$(1-n)\,y^{-n}\frac{dy}{dx} = \frac{dv}{dx}$$

$$\Rightarrow \qquad y^{-n}\frac{dy}{dx} = \frac{1}{1-n}\frac{dv}{dx}$$

Substituting these values in equation (ii) we have

$$\frac{1}{1-n}\frac{dv}{dx} + Pv = Q$$

$$\Rightarrow \qquad \frac{dv}{dx} + (1-n)\,Pv = (1-n)\,Q$$

which is a linear equation in v and x

***Example* 16:** Solve $\frac{dy}{dx} + \frac{y}{x} = y^2 \sin x$

***Solution*:** Given equation is $\frac{dy}{dx} + \frac{y}{x} = y^2 \sin x$

divided by y^2, we have

$$\Rightarrow \qquad y^{-2}\frac{dv}{dx} + \frac{1}{x}\,y^{-1} = \sin x \qquad \text{.....(i)}$$

Which is Bernoulli's equation

Put $y^{-1} = v$, $\Rightarrow \quad -y^{-2}\frac{dy}{dx} = \frac{dv}{dx}$

∴ from (i), we get $\frac{-dv}{dx}+\frac{1}{x}.v = \sin x$

$$\Rightarrow \quad \frac{dv}{dx}-\frac{1}{x}.v = -\sin x \quad .....(ii)$$

which is a linear differential equation

here $P = \frac{-1}{x}$, $Q = -\sin x$

$$I.F = e^{\int P dx} = e^{-\int \frac{1}{x} dx} = e^{-\log x} = \frac{1}{x}$$

Solution of (i) is

$$v.I.F = \int Q\, I.F\, dx + c$$

$$\Rightarrow \quad v\frac{1}{x} = -\int \sin x.\frac{1}{x} dx + c$$

$$\Rightarrow \quad \frac{y^{-1}}{x} = -\int \frac{\sin x}{x} dx + c$$

Since $\int \frac{\sin x}{x} dx$ is not solvable

***Example* 17:** Solve $y = 2px - x^2$

***Solution*:** The given equation is

$$y = 2px - x^2$$

$$\Rightarrow \quad y = 2x\frac{dy}{dx} - x^2 \qquad \left[\because p = \frac{dy}{dx}\right]$$

$$\Rightarrow \quad 2x\frac{dy}{dx} = y + x^2$$

$$\Rightarrow \quad \frac{dy}{dx} - \frac{1}{2x}.y = \frac{x}{2} \quad .....(i)$$

which is linear differential equation

here $P = \frac{-1}{2x}$, $Q = \frac{x}{2}$

$$I.F = e^{\int P dx} = e^{\frac{-1}{2}\int \frac{1}{x} dx} = e^{\frac{-1}{2}\log x} = x^{\frac{-1}{2}} = \frac{1}{\sqrt{x}}$$

Solution of (i) is

$$y.I\,F = \int Q.IF\,dx + c$$

$$\Rightarrow \quad y.\frac{1}{\sqrt{x}} = \int \frac{x}{2}.\frac{1}{\sqrt{x}}\,dx + c$$

$$\Rightarrow \quad \frac{y}{\sqrt{x}} = \frac{1}{2}\int x^{\frac{1}{2}}\,dx + c$$

$$\Rightarrow \quad \frac{y}{\sqrt{x}} = \frac{1}{2}\frac{x^{\frac{3}{2}}}{\frac{3}{2}} + c$$

$$\Rightarrow \quad \frac{y}{\sqrt{x}} = \frac{1}{3}x^{\frac{3}{2}} + c$$

$$\Rightarrow \quad y = \frac{1}{3}x^2 + cx^{\frac{1}{2}}$$

***Example* 18:** Solve $3\,\frac{dy}{dx} + \frac{2}{x+1}y = \frac{x^3}{y^2}$

***Solution*:** Given equation is

$$3\frac{dy}{dx} + \frac{2}{x+1}y = \frac{x^3}{y^2}$$

$$\Rightarrow \quad 3y^2\,\frac{dy}{dx} + \frac{2}{x+1}y^3 = x^3 \qquad .....(i)$$

which is Bernoullis equation

Put $\quad y^3 = v \Rightarrow \quad 3y^2\,\frac{dy}{dx} = \frac{dv}{dx} \qquad$ (differentiating w.r.t x)

$\therefore$ from (i), we have $\frac{dv}{dx} + \frac{2}{x+1}.v = x^3 \qquad .....(ii)$

which is linear differential equation

here $\quad P = \frac{2}{x+1}, \quad Q = x^3$

$$I.F = e^{\int Pdx} = e^{2\int \frac{1}{x+1}dx} = e^{2\log(x+1)} = (x+1)^2$$

$\therefore$ Solution of (i) is

$$v.IF = \int Q.IF\,dx + c$$

$$\Rightarrow \quad v(x+1)^2 = \int x^3(x+1)^2\,dx + c$$

$$\Rightarrow \quad y^3(x+1)^2 = \int (x^5 + 2x^4 + x^3)\,dx + c$$

$$\Rightarrow \quad y^3(x+1)^2 = \frac{x^6}{6} + \frac{2x^5}{5} + \frac{x^4}{4} + c$$

***Example* 19:** Solve $(x+y)^2 \dfrac{dy}{dx} = 4$

***Solution*:** Given equation is

$$(x+y)^2 \frac{dy}{dx} = 4 \qquad \text{.....(i)}$$

Putting $x + y = v$

$$\Rightarrow \quad 1 + \frac{dy}{dx} = \frac{dv}{dx}$$

$$\Rightarrow \quad \frac{dy}{dx} = \frac{dv}{dx} - 1 \qquad \text{.....(ii)}$$

from (i) and (ii), we have

$$v^2\left(\frac{dv}{dx} - 1\right) = 4$$

$$\Rightarrow \quad \frac{dv}{dx} = \frac{4}{v^2} + 1$$

$$\Rightarrow \quad \frac{dv}{dx} = \frac{4+v^2}{v^2}$$

$$\Rightarrow \quad \frac{v^2}{v^2+4}\,dv = dx$$

$$\Rightarrow \quad \left(1 - \frac{4}{v^2+4}\right)dv = dx$$

Integrating, we get $\int\left(1-\frac{4}{v^2+4}\right)dv = \int dx + c$

$$v - \frac{4}{2}\tan^{-1}\frac{v}{2} = x + c$$

$$x + y - 2\tan^{-1}\left(\frac{x+y}{2}\right) = x + c$$

$$y - 2\tan^{-1}\left(\frac{x+y}{2}\right) = c$$

***Example* 20:** Solve $x\,dy + y\,dx = xy$

***Solution*:** Given equation is

$$x\,dy + y\,dx = xy$$

we dividing xy, we get

$$\frac{1}{y}dy + \frac{1}{x}dx = 1$$

Integrating, we get

$$\int\frac{1}{y}dy + \int\frac{1}{x}dx = \int 1\,dx + c$$

$$\Rightarrow \quad \log y + \log x = x + \log c$$

$$\Rightarrow \quad \log\left(\frac{xy}{c}\right) = x$$

$$\Rightarrow \quad \frac{xy}{c} = e^x$$

$$\Rightarrow \quad xy = ce^x$$

Exercise 13.5

Solve the following differential equations

1. $\frac{dy}{dx} + 2y = \sin x$
2. $\frac{dy}{dx} + y = \sin x$
3. $\frac{dy}{dx} + 2y = e^{-2x}$
4. $\frac{dy}{dx} + \frac{y}{x} = x^2$

5. $\sin x \dfrac{dy}{dx} + y \cos x = 2 \sin^2 x \cos x$

6. $x \dfrac{dy}{dx} = 2y + 2x^4 + x^2$

7. $x \dfrac{dy}{dx} + y = xe^{-x}$

8. $4 \dfrac{dy}{dx} + 8y = 5e^{-3x}$

9. $x \dfrac{dy}{dx} + y = x \log x$

10. $x \dfrac{dy}{dx} - y = (x - 1)e^x$

11. $\dfrac{dy}{dx} = y \tan x - 2 \sin x$

12. $\dfrac{dy}{dx} + y \tan x = x^2 \cos^2 x$

13. $(1 + x^2) \dfrac{dy}{dx} + y = e^{\tan^{-1} x}$

14. $(1 + x^2) \dfrac{dy}{dx} + y = \tan^{-1} x$

15. $(x + \tan y)\, dy = \sin 2y\, dx$

16. $e^y \left(\dfrac{dy}{dx} + 1\right) = e^x$

17. $x \dfrac{dy}{dx} + y = xy^3$

18. $\dfrac{dy}{dx} + \dfrac{y}{x} = y^3$

19. $\dfrac{dy}{dx} + 2y \tan x = \sin x$ given that $y = 0$ when $x = \dfrac{\pi}{3}$

20. $(1 + y^2)\, dx = (\tan^{-1} y - x)dy$

21. $y^2 \dfrac{dy}{dx} = x + y^3$

22. $\dfrac{dy}{dx} = e^{x-y} (e^x - e^y)$

Answers

1. $y = \dfrac{1}{5} (2 \sin x - \cos x) + c\, e^{-2x}$

2. $y = \dfrac{1}{2} (\sin x - \cos x) + c\, e^{-x}$

3. $y = x\, e^{-2x} + c\, e^{-2x}$

4. $y = \dfrac{x^3}{4} + \dfrac{c}{x}$

5. $y \sin x = \dfrac{2\sin^3 x}{3} + c$

6. $y = x^4 + x^2 \log x + cx^2$

7. $xy = -e^{-x} (x + 1) + c$

8. $y \,.\, e^{2x} = \dfrac{-5}{4} e^{-x} + c$

9. $4xy = 2x^2 \log x - x^2 + c$

10. $y = e^x + cx$

11. $2y \cos x = \cos 2x + c$

12. $y \sec x = x^2 \sin x + 2 \cos x - 2 \sin x + c$
13. $2y \cdot e^{\tan^{-1}x} = e^{2\tan^{-1}x} + c$
14. $y \cdot e^{\tan^{-1} x} = e^{\tan^{-1} x}(\tan^{-1} x - 1) + c$
15. $x = \tan y + c\sqrt{\tan y}$
16. $e^{x+y} = \frac{1}{2} e^{2x} + c$
17. $1 = xy^2(1 + cx)$
18. $xy^2 (2 + cx) = 1$
19. $y \sec^2 x = \sec x - 2$
20. $x = c\, e^{-\tan^{-1}y} + (\tan^{-1} y - 1)$
21. $y^3 e^{-3x} = c - xe^{-3x} - \frac{1}{3} e^{-3x}$
22. $e^y = ce^{-x} + e^x + 1$

Differential Equation of the form $\frac{d^2y}{dx^2} = f(x)$

This type of equation can be solved by simply integrating it twice i.e.,

$$\frac{d^2y}{dx^2} = f(x)$$

$$\Rightarrow \quad \frac{d}{dx}\left(\frac{dy}{dx}\right) = f(x)$$

integrating, we get $\int \frac{d}{dx}\left(\frac{dy}{dx}\right)dx = \int f(x)dx + c_1$

$$\Rightarrow \quad \frac{dy}{dx} = \int f(x)dx + c_1$$

$$\Rightarrow \quad \frac{dy}{dx} = \phi(x) + c_1 \qquad \left[\because \int f(x)\,dx = \phi(x)\,(\text{say})\right]$$

Integrating again, we get $\int \frac{dy}{dx}dx = \int \phi(x)\,dx + \int c_1 dx + c_2$

$$\Rightarrow \quad y = \int \phi(x)\,dx + c_1 x + c_2$$

$$\Rightarrow \quad y = g(x) + c_1 x + c_2 \qquad \because \left|\int \phi(x)\,dx = g(x)\text{ say}\right|$$

***Example* 1:** Solve $\frac{d^2y}{dx^2} = x \sin x$

***Solution*:** The given equation is

$$\frac{d^2y}{dx^2} = x \sin x$$

Integrating, we get $\int \frac{d}{dx}\left(\frac{dy}{dx}\right)dx = \int x \sin x dx + c_1$

$$\Rightarrow \quad \frac{dy}{dx} = x(-\cos x) + \sin x + c_1$$

Integrating again, we get

$$\int \frac{dy}{dx}.dx = -\int x \cos x\, dx + \int \sin x dx + \int c_1 dx + c_2$$

$$\Rightarrow \quad y = -x \sin x - 2\cos x + c_1 x + c_2$$

***Example* 2:** Solve $\frac{d^2y}{dx^2} = x^2 + e^{2x}$

***Solution*:** Given equation is $\frac{d^2y}{dx^2} = x^2 + e^{2x}$

Integrating, we get

$$\int \frac{d}{dx}\left(\frac{dy}{dx}\right)dx = \int x^2 dx + \int e^{2x} dx + c_1$$

$$\Rightarrow \quad \frac{dy}{dx} = \frac{x^3}{3} + \frac{e^{2x}}{2} + c_1$$

Integrating again, we get

$$\int \frac{dy}{dx} dx = \frac{1}{3}\int x^3 dx + \frac{1}{2}\int e^{2x} dx + \int c_1 dx + c_2$$

$$\Rightarrow \quad y = \frac{1}{3}.\frac{x^4}{4} + \frac{1}{2}.\frac{e^{2x}}{2} + c_1 x + c_2$$

$$\Rightarrow \quad y = \frac{1}{12}x^4 + \frac{1}{4}e^{2x} + c_1 x + c_2$$

***Example* 3:** Solve $\frac{d^2y}{dx^2} = x + \sin x$, subject to the condition that $y = 0$ and $\frac{dy}{dx} = -1$ when $x = 0$.

***Solution*:** Given equation is $\frac{d^2y}{dx^2} = x + \sin x$

Integrating we get

$$\int \frac{d}{dx}\left(\frac{dy}{dx}\right) dx = \int x\,dx + \int \sin x\,dx + c_1$$

$$\Rightarrow \quad \frac{dy}{dx} = \frac{x^2}{2} - \cos x + c_1 \qquad \text{.....(i)}$$

Integrating again, we get $\int \frac{dy}{dx} dx = \frac{1}{2}\int x^2\,dx - \int \cos x\,dx + \int c_1 dx + c_2$

$$\Rightarrow \quad y = \frac{1}{2}.\frac{x^3}{3} - \sin x + c_1 x + c_2$$

$$\Rightarrow \quad y = \frac{1}{6}x^3 - \sin x + c_1 x + c_2 \qquad \text{.....(ii)}$$

when $x = 0$, $\frac{dy}{dx} = -1$ then from (i), we get

$$-1 = -1 + c_1 \quad \Rightarrow \quad c_1 = 0$$

when $x = 0$, $y = 0$, then from (ii), we get

$$0 = c_2$$

Hence, required solution is $y = \frac{1}{6}x^3 - \sin x$

***Example* 4:** Solve $x\frac{d^2y}{dx^2} = 1$, it is being given that $y = -1$ and $\frac{dy}{dx} = 0$ when $x = 1$.

***Solution*:** Given equation is $x\frac{d^2y}{dx^2} = 1$

$$\Rightarrow \quad \frac{d^2y}{dx^2} = \frac{1}{x}$$

Integrating, we get $\int \frac{d}{dx}\left(\frac{dy}{dx}\right) dx = \int \frac{1}{x} dx + c_1$

$$\Rightarrow \quad \frac{dy}{dx} = \log x + c_1 \qquad \text{..... (i)}$$

Integrating again, we get

$$\int \frac{dy}{dx} = \int \log x\,dx + \int c_1 dx + c_2$$

$$y = x\log x - x + c_1 x + c_2 \qquad \text{..... (ii)}$$

when $x = 1$, $\frac{dy}{dx} = 0$ then from (i), we get

$$0 = \log 1 + c_1 \Rightarrow \quad c_1 = 0 \qquad (\because \log 1 = 0)$$

When $x = 1$, $y = -1$ then from (ii), we get

$$-1 = 1 \log 1 - 1 + 0 + c_2$$

$$-1 = 0 - 1 + c_2 \Rightarrow c_2 = 0$$

Hence, required solution is $y = x \log x - x$

Exercise 13.6

Solve the following differential equations.

1. $\frac{d^2y}{dx^2} = 0$
2. $\frac{d^2y}{dx^2} = x^2 + e^{2x}$
3. $\frac{d^2y}{dx^2} = \cos x - \sin x$
4. Solve $\frac{d^2y}{dx^2} = \frac{1}{x} + 6$ it is being given that $y = 0$ and $\frac{dy}{dx} = 0$ when $x = 1$.
5. Solve $\frac{d^2y}{dx^2} = x^2 \sin x$, it is given that $y = 0$ and $\frac{dy}{dx} = 0$ when $x = 1$.
6. Solve $\frac{d^2y}{dx^2} = \log x$, given that $y = 1$ and $\frac{dy}{dx} = 1$ when $x = 1$.
7. Solve $\frac{d^2y}{dx^2} = e^x (\sin x + \cos x)$, given that $y = 1$ and $\frac{dy}{dx} = 0$ when $x = 0$.

Answers

1. $y = c_1 x + c_2$
2. $y = \frac{1}{12}x^4 + \frac{1}{4}e^{2x} + c_1x + c_2$
3. $y = -\cos x + \sin x + c_1 x + c_2$
4. $y = x \log x + 3x^2 - 7x + 4$
5. $y = -x^2 \sin x - 4x \cos x + 6 \sin x - 2x$
6. $y = \frac{1}{2}x^2 \log x - \frac{3}{4}x^2 + \frac{7}{4}$
7. $y = \frac{1}{2} e^x (\sin x - \cos x) + \frac{3}{2}$

CHAPTER 10

LINEAR DIFFERENTIAL EQUATIONS OF HIGHER ORDER WITH CONSTANT COEFFICIENTS

Introduction

The general linear differential equation of the nth order is of the form

$$\frac{d^n y}{dx^n} + b_1 \frac{d^{n-1} y}{dx^{n-1}} + b_2 \frac{d^{n-2} y}{dx^{n-2}} \ldots\ldots\ldots\ldots\ldots\ldots + b_n y = Q \qquad \ldots..(i)$$

Where $b_1, b_2 \ldots\ldots b_n$ are constants and Q is the function of x only. Then the equation is a linear differential equation with constant coefficients.

The Operator D If we write

$$D = \frac{d}{dx},\ D^2 = \frac{d^2}{dx^2},\ D^3 = \frac{d^3}{dx^3} \ldots\ldots D^n = \frac{d^n}{dx^n}$$

The differential equation (i) can be written as

$$D^n y + b_1 D^{n-1} y + b_2 D^{n-2} y + \ldots\ldots. + b_n y = Q$$

$$\Rightarrow \quad [D^n + b_1 D^{n-1} + b_2 D^{n-2} + \ldots.. + b_n] y = Q \qquad \ldots..(ii)$$

$$\Rightarrow \quad f(D) y = Q$$

where $\quad f(D) = D^n + b_1 D^{n-1} + b_2 D^{n-2} + \ldots\ldots.. + b_n$

The general solution of (i) consists of two parts

(i) Complementary function (C.F)

(ii) Particular integral (P.I)

$\Rightarrow \qquad y = \text{C.F} + \text{P.I}$ is the general solution

Complementary function: The general solution of the equation

$$[D^n + b_1 D^{n-1} + \dots\dots\dots + b_n]\, y = 0$$

is called complementary function. It must contains as many arbitrary constants.

Note : If Q = 0, we do not find the particular integral and then general solution is y = C.F

Particular Integral: The general solution of the equation

$$[D^n + b_1 D^{n-1} + \dots\dots + b_n]\, y = Q$$

which does not contain an arbitrary constant called particular integral. Thus

$$\text{P.I.} = \frac{1}{f(D)} Q.$$

and general solution is $\quad y = \text{C.F} + \text{P.I}$

Auxiliary Equation: Let us consider the differential equation

$$(D^n + b_1 D^{n-1} + \dots\dots\dots + b_n)\, y = 0 \qquad \dots\dots(i)$$

Let $\quad y = e^{mx}$ be a solution of this equation, then

$$D_y = me^{mx},\ D^2y = m^2 e^{mx},\ \dots\dots\dots,\ D^n y = m^n \cdot e^{mx}$$

Substituting these values in equation (i), we have

$$(m^n + b_1 m^{n-1} + \dots\dots + b_n)\, e^{mx} = 0$$

$$\text{'or'}\ m^n + b_1 m^{n-1} + \dots\dots\dots + b_n = 0$$

This equation is called auxiliary equation (A.E) of equation (i)

Working Rule: To find the A.E replace D by m in f(D) and take Q = 0 and we find m_1, m_2 ….. be the roots of A.E.

Working Rule for Finding Complementary Function: When Q = 0 and according to the nature of roots of auxiliary equation the following cases arise.

***Case* I:** If all the roots of the auxiliary equation are real and distinct i.e., $m = m_1, m_2, \dots\dots m_n$.

Then $\qquad \text{C.F} = c_1 e^{m_1 x} + c_2 e^{m_2 x} + \dots\dots + c_n^{m_n x}$

***Example* 1:** Solve $\frac{d^2y}{dx^2}+5\frac{dy}{dx}+6y=0$

***Solution*:** Given equation is

$$\frac{d^2y}{dx^2}+5\frac{dy}{dx}+6y=0$$

$\Rightarrow$ $(D^2 + 5D + 6)\, y = 0$ where $\frac{d}{dx} = D$

The auxiliary equation is $m^2 + 5m + 6 = 0$, put $D = m$

$\Rightarrow$ $m^2 + 3m + 2m + 6 = 0$

$\Rightarrow$ $m(m + 3) + 2(m + 3) = 0$

$\Rightarrow$ $(m + 3)\,(m + 2) = 0$

$\Rightarrow$ $m = -2, -3$ (roots are real and distinct)

$\therefore$ Required solution is $y = c_1\, e^{-2x} + c_2\, e^{-3x}$

***Example* 2:** Solve $\frac{d^2y}{dx^2}-7\frac{dy}{dx}+12y=0$

***Solution*:** Given equation is

$$\frac{d^2y}{dx^2}-7\frac{dy}{dx}+12y=0$$

$\Rightarrow$ $(D^2 - 7D + 12)\, y = 0$

The A.E is

$m^2 - 7m + 12 = 0$

$\Rightarrow$ $(m - 4)\,(m - 3) = 0$

$\Rightarrow$ $m = 3, 4$

$\therefore$ The required solution is $y = c_1\, e^{3x} + c_2\, e^{4x}$

***Example* 3:** Solve $\frac{d^2y}{dx^2}-3\frac{dy}{dx}-4y=0$

***Solution*:** Given equation is

$$\frac{d^2y}{dx^2}-3\frac{dy}{dx}-4y=0$$

$\Rightarrow$ $(D^2 - 3D - 4)\, y = 0$

The A.E is

$$m^2 - 3m - 4 = 0$$

$\Rightarrow$ $(m-4)(m+1) = 0$

$\Rightarrow$ $m = -1, 4$

$\therefore$ Required solution is $y = c_1 e^{-x} + c_2 e^{4x}$

***Example* 4:** Solve $\frac{d^3y}{dx^3} - 7\frac{dy}{dx} - 6y = 0$

***Solution*:** Given equation is

$$\frac{d^3y}{dx^3} - 7\frac{dy}{dx} - 6y = 0$$

$\Rightarrow$ $(D^3 - 7D - 6)\,y = 0$

The A.E is

$$m^3 - 7m - 6 = 0$$

$\Rightarrow$ $m^3 + m^2 - m^2 - m - 6m - 6 = 0$

$\Rightarrow$ $m^2(m+1) - m(m+1) - 6(m+1) = 0$

$\Rightarrow$ $(m+1)(m^2 - m - 6) = 0$

$\Rightarrow$ $(m+1)(m+2)(m-3) = 0$

$\Rightarrow$ $m = -1, -2, 3$

$\therefore$ Required solution is $y = c_1 e^{-x} + c_2 e^{-2x} + c_3 e^{3x}$

Case II If the roots of auxiliary equations are equal

(i) i.e., $m_1 = m_2$

then $C.F = (c_1 + c_2 x)e^{m_1 x}$

If (ii) $m_1 = m_2 = m_3$

Then $C.F = (c_1 + c_2 x + c_3 x^2)e^{m_1 x}$

***Example* 5:** Solve $\frac{d^3y}{dx^3} - 4\frac{d^2y}{dx^2} + 4\frac{dy}{dx} = 0$

***Solution*:** Given equation is

$$\frac{d^3y}{dx^3} - 4\frac{d^2y}{dx^2} + 4\frac{dy}{dx} = 0$$

$\Rightarrow$ $(D^3 - 4D^2 + 4D)y = 0$

The A.E is

$$m^3 - 4m^2 + 4m = 0$$

$\Rightarrow$ $m(m^2 - 4m + 4) = 0$

$\Rightarrow$ $m(m - 2)^2 = 0$

$\Rightarrow$ $m = 0, 2, 2$

$\therefore$ The required solution is $y = c_1 e^{ox} + (c_2 + c_3 x) e^{2x}$

$\Rightarrow$ $y = c_1 + (c_2 + c_3 x)e^{2x}$

***Example* 6:** Solve $(D - 1)^2 (D - 2)^3 y = 0$

***Solution*:** Given equation is $(D-1)^2 (D - 2)^3 y = 0$

The A.E is

$$(m - 1)^2 (m - 2)^3 = 0$$

$\Rightarrow$ $m = 1, 1, 2, 2, 2$

$\therefore$ Required solution is

$$y = (c_1 + c_2x)e^x + (c_3 + c_4 x + c_5x^2) e^{2x}$$

***Example* 7:** Solve $\frac{d^4y}{dx^4} - \frac{d^3y}{dx^3} - 9\frac{d^2y}{dx^2} - 11\frac{dy}{dx} - 4y = 0$

***Solution*:** Given equation is $\frac{d^4y}{dx^4} - \frac{d^3y}{dx^3} - 9\frac{d^2y}{dx^2} - 11\frac{dy}{dx} - 4y = 0$

$\Rightarrow$ $(D^4 - D^3 - 9D^2 - 11D - 4) y = 0$

The A.E is

$$m^4 - m^3 - 9m^2 - 11m - 4 = 0$$

$\Rightarrow$ $m^4 + m^3 - 2m^3 - 2m^2 - 7m^2 - 7m - 4m - 4 = 0$

$\Rightarrow$ $m^3(m + 1) - 2m^2(m + 1) - 7m (m + 1) - 4(m + 1) = 0$

$\Rightarrow$ $(m + 1) (m^3 - 2m^2 - 7m - 4) = 0$

$\Rightarrow$ $(m + 1) (m^3 + m^2 - 3m^2 - 3m - 4m - 4) = 0$

$\Rightarrow$ $(m + 1) [m^2(m + 1) - 3m(m + 1) - 4(m + 1)] = 0$

$\Rightarrow$ $(m + 1) (m + 1) (m^2 - 3m - 4) = 0$

$\Rightarrow$ $(m + 1) (m + 1) (m + 1) (m - 4) = 0$

$\Rightarrow$ $m = -1, -1, -1, 4$

$\therefore$ Required solution is $y = (c_1 + c_2x + c_3x^2) e^{-x} + c_4 e^{4x}$

Case III

1. If roots of the auxiliary equation are imaginary

$$\text{i.e., } m = \alpha \pm i\beta$$

Then $C.F = e^{\alpha x}[c_1 \cos \beta x + c_2 \sin \beta x]$

'or' $C.F = e^{\alpha x} c_1 \cos(\beta x + c_2)$

'or' $C.F = e^{\alpha x} c_1 \sin(\beta x + c_2)$

2. If imaginary roots are repeated, i.e., two equal pair i.e.,

$$m_1 = m_2 = \alpha + i\beta,\ m_3 = m_4 = \alpha - i\beta.$$

Then $C.F = e^{\alpha x}[(c_1 + c_2 x)\cos \beta x + (c_3 + c_4 x)\sin \beta x]$

3. When roots are irrational i.e.,

$$m = \alpha \pm \sqrt{\beta}$$

Then $C.F = e^{\alpha x} c_1 \cos h(\sqrt{\beta}\, x + c_2)$

'or' $C.F = e^{\alpha x} c_1 \sin h(\sqrt{\beta}\, x + c_2)$

***Example* 8:** Solve $(D^2 + 1)(D - 1)y = 0$

***Solution*:** Given equation is

$$(D^2 + 1)(D - 1)y = 0$$

The A.E is

$$(m^2 + 1)(m - 1) = 0$$

$\Rightarrow$ $m = \pm i, 1$

$\therefore$ The required solution is

$e^{ox}[c_1 \cos x + c_2 \sin x] + c_3 e^x$

$\Rightarrow$ $c_1 \cos x + c_2 \sin x + c_3 e^x$

***Example* 9:** Solve $\frac{d^3y}{dx^3} - 2\frac{d^2y}{dx^2} + 4\frac{dy}{dx} - 8y = 0$

***Solution*:** Given equation is

$$\frac{d^3y}{dx^3} - 2\frac{d^2y}{dx^2} + 4\frac{dy}{dx} - 8y = 0$$

$\Rightarrow$ $(D^3 - 2D^2 + 4D - 8)y = 0$

The A.E is

$$m^3 - 2m^2 + 4m - 8 = 0$$

$\Rightarrow$ $m^2(m - 2) + 4(m - 2) = 0$

$\Rightarrow$ $(m - 2)(m^2 + 4) = 0$

$\Rightarrow$ $m = 2, \pm 2i$

$\therefore$ The required solution is

$$y = c_1 e^{2x} + e^{ox}[c_2 \cos 2x + c_3 \sin 2x]$$

$\Rightarrow$ $y = c_1 e^{2x} + c_2 \cos 2x + c_3 \sin 2x$

***Example* 10:** Solve $(D + 1)^2 (D^2 + 4) y = 0$

***Solution*:** Given equation is

$$(D + 1)^2 (D^2 + 4) = 0$$

The A.E. is

$$(m + 1)^2 (m^2 + 4) = 0$$

$\Rightarrow$ $m = -1, -1, \pm 2i$

$\therefore$ The required solution is

$$y = (c_1 + c_2 x)e^{-x} + c_3 \cos 2x + c_4 \sin 2x$$

***Example* 11:** Solve $\dfrac{d^4y}{dx^4} - a^4 y = 0$

***Solution*:** The given equation is $\dfrac{d^4y}{dx^4} - a^4 y = 0$

$\Rightarrow$ $(D^4 - a^4) y = 0$

The A.E is

$$m^4 - a^4 = 0$$

$\Rightarrow$ $(m^2 - a^2)(m^2 + a^2) = 0$

$\Rightarrow$ $(m - a)(m + a)(m^2 + a^2) = 0$

$\Rightarrow$ $m = a, -a, \pm ai$

$\therefore$ The required solution is

$$y = c_1 e^{ax} + c_2 e^{-ax} + c_3 \cos ax + c_4 \sin ax$$

***Example* 12:** Solve $\frac{d^4y}{dx^4} + a^4y = 0$

Solution: Given equation is $\frac{d^4y}{dx^4} + a^4y = 0$

$$\Rightarrow \quad (D^4 + a^4)\, y = 0$$

The A.E is

$$m^4 + a^4 = 0$$

$$\Rightarrow \quad m^4 + 2m^2 a^2 + a^4 - 2m^2 a^2 = 0$$

$$\Rightarrow \quad (m^2 + a^2)^2 - 2m^2 a^2 = 0$$

$$\Rightarrow \quad (m^2 + a^2)^2 - (\sqrt{2}\, ma)^2 = 0$$

$$\Rightarrow \quad (m^2 + a^2 + \sqrt{2}\, ma)(m^2 + a^2 - \sqrt{2}\, ma) = 0$$

when $\quad m^2 + \sqrt{2}ma + a^2 = 0 \Rightarrow m = \frac{-\sqrt{2}a \pm \sqrt{2a^2 - 4a^2}}{2}$

$$= \frac{-a}{\sqrt{2}} \pm i\frac{a}{\sqrt{2}}$$

when $\quad m^2 - \sqrt{2ma} + a^2 = 0 \Rightarrow m = \frac{a}{\sqrt{2}} \pm i\frac{a}{\sqrt{2}}$

Hence, the required solution is

$$y = e^{\frac{-a}{\sqrt{2}}x}\left(c_1 \cos\frac{a}{\sqrt{2}}x + c_2 \sin\frac{a}{\sqrt{2}}x\right) + e^{\frac{a}{\sqrt{2}}x}\left(c_3 \cos\frac{a}{\sqrt{2}}x + c_4 \sin\frac{a}{\sqrt{2}}x\right)$$

Exercise 14.1

Solve the Following Differential Equations

1. $2\frac{d^2y}{dx^2} + 3\frac{dy}{dx} - 2y = 0$
2. $\frac{d^2y}{dx^2} - 9y = 0$
3. $\frac{d^3y}{dx^3} - 9\frac{d^2y}{dx^2} + 23\frac{dy}{dx} - 15y = 0$
4. $\frac{d^2y}{dx^2} - m^2y = 0$
5. $(D^3 - 3D^2 + 4)\, y = 0$
6. $(D^3 - D^2 - D + 1)\, y = 0$
7. $(D^4 + 4D^3 - 5D^2 - 36\, D - 36)\, y = 0$
8. $(D - 1)^2 (D - 2)^2\, y = 0$

9. $(D^2 + 4)\, y = 0$

10. $(D^3 + 1)\, y = 0$

11. $\frac{d^4y}{dx^4} + ay = 0$

12. $(D^2 + 1)^2 (D^2 + D + 1)\, y = 0$

13. $(D^3 - D^2 + D - 1)^2\, y = 0$

14. $\frac{d^2y}{dx^2} + y = 0$

15. $(D^3 - 8)\, y = 0$

16. $\frac{d^4y}{dx^4} + 8\frac{d^2y}{dx^2} - 16y = 0$

17. $(D - 1)^2 (D - 2)^3\, y = 0$

18. $\frac{d^3y}{dx^3} - \frac{d^2y}{dx^2} - \frac{dy}{dx} + y = 0$

19. $\frac{d^3y}{dx^3} + y = 0$

20. $\frac{d^2y}{dx^2} + 4\frac{dy}{dx} + 5y = 0$

Answers

1. $y = c_1 e^{\frac{x}{2}} + c_2 e^{-2x}$

2. $y = c_1 e^{3x} + c_2 e^{-3x}$

3. $y = c_1 e^{x} + c_2 e^{3x} + c_3 e^{5x}$

4. $c_1 e^{mx} + c_2 e^{-mx}$

5. $y = (c_1 + c_2 x)\, e^{3x} + c_3 e^{-x}$

6. $y = (c_1 + c_2 x)\, e^{x} + c_3 e^{-x}$

7. $y = c_1 e^{-3x} + c_2 e^{2x} + (c_3 + c_4 x)e^{-2x}$

8. $y = (c_1 + c_2 x)\, e^{x} + (c_3 + c_4 x)e^{2x}$

9. $y = c_1 \cos 2x + c_2 \sin 2x$

10. $y = c_1 e^{-x} + e^{\frac{x}{2}} \left(c_2 \cos \frac{\sqrt{3}}{2} x + c_3 \sin \frac{\sqrt{3}}{2} x\right)$

11. $y = e^{-a^{\frac{1}{4}}\frac{x}{\sqrt{2}}} \left\{c_1 \cos\left(a^{\frac{1}{4}} \frac{x}{\sqrt{2}}\right) + c_2 \sin\left(a^{\frac{1}{4}} \frac{x}{\sqrt{2}}\right)\right\} +$

$e^{a^{\frac{1}{4}}\frac{x}{2}} \left\{c_3 \cos\left(a^{\frac{1}{4}} \frac{x}{\sqrt{2}}\right) + c_4 \sin\left(a^{\frac{1}{4}} \frac{x}{\sqrt{2}}\right)\right\}$

12. $y = (c_1 + c_2 x) \cos x + (c_3 + c_4 x) \sin x + e^{\frac{-x}{2}} \left[c_5 \cos \frac{\sqrt{3}}{2} x + c_6 \sin \frac{\sqrt{3}}{2} x\right]$

13. $y = (c_1 + c_2 x) e^x + (c_3 + c_4 x) \cos x + (c_5 + c_6 x) \sin x$

14. $y = c_1 \cos x + c_2 \sin x$

15. $y = c_1 e^{2x} + e^{-x} [c_2 \cos(x\sqrt{3}) + c_3 \sin (x\sqrt{3})]$

16. $y = (c_1 + c_2 x) \cos 2x + (c_3 + c_4 x) \sin 2x$

17. $y = (c_1 + c_2 x) e^x + (c_3 + c_4 x + c_5 x^2) e^{2x}$

18. $y = (c_1 + c_2 x) e^x + c_3 e^{-x}$

19. $y = c_1 e^{-x} + e^{\frac{x}{2}} \left[c_2 \cos\left(\frac{\sqrt{3}}{2}x\right) + c_3 \sin\left(\frac{\sqrt{3}}{2}x\right) \right]$

20. $y = e^{-2x} (c_1 \cos x + c_2 \sin x)$

Rules for Finding Particular Integral (P.I) when $Q \neq 0$

Rule I : $P.I = \frac{1}{f(D)}.Q$

1. $D.Q = \frac{d}{dx}Q$

2. $\frac{1}{D}.Q = \int Q\,dx$

3. $\frac{1}{D-\alpha} Q = e^{\alpha x} \int e^{-\alpha x} Q\,dx$

4. $\frac{1}{D+\alpha} Q = e^{-\alpha x} \int e^{\alpha x} .Q\,dx$

***Example* 1:** Solve $\frac{d^2y}{dx^2} - 5\frac{dy}{dx} + 6y = e^{4x}$

***Solution*:** Given equation is

$$\frac{d^2y}{dx^2} - 5\frac{dy}{dx} + 6y = e^{4x}$$

$$\Rightarrow \quad (D^2 - 5D + 6)\,y = e^{4x}$$

The A.E is

$$m^2 - 5\,m + 6 = 0$$

$$\Rightarrow \quad m^2 - 3m - 2m + 6 = 0$$

$$\Rightarrow \quad m(m - 3) - 2(m - 3) = 0$$

$$\Rightarrow \quad (m - 3)\,(m - 2) = 0$$

$$\Rightarrow \quad m = 2,\ 3$$

Complementary function is

$$C.F = c_1\, e^{2x} + c_2\, e^{3x}$$

Particular integral is

$$P.I = \frac{1}{(D-3)(D-2)} e^{4x}$$

On breaking into partial fraction

$$P.I = \left[\frac{1}{D-3} - \frac{1}{D-2}\right] e^{4x}$$

$$= \frac{1}{D-3} e^{4x} - \frac{1}{D-2} e^{4x}$$

$$= e^{3x} \int e^{-3x} . e^{4x} dx - e^{2x} \int e^{-2x} . e^{4x} dx$$

$$= e^{3x} \int e^{x}\, dx - e^{2x} \int e^{2x} dx$$

$$= e^{3x} . e^{x} - e^{2x} . \frac{e^{2x}}{2}$$

$$= e^{4x} - \frac{1}{2} e^{4x}$$

Hence, the general solution of the given equation is

$$y = C.F + P.I$$

$$= c_1\, e^{2x} + c_2\, e^{3x} + e^{4x} - \frac{1}{2} e^{4x}$$

***Example* 2:** Solve $\frac{d^2y}{dx^2} + a^2y = \sec ax$

***Solution*:** Given equation is

$$\frac{d^2y}{dx^2} + a^2y = \sec ax$$

$$\Rightarrow \quad (D^2 + a^2)\, y = \sec ax$$

The auxiliary equation is

$$m^2 + a^2 = 0$$

$$\Rightarrow \quad m = \pm\, ia$$

Complementary function is C.F = $c_1 \cos ax + c_2 \sin ax$

Particular integral is

$$\frac{1}{D^2 + a^2} \sec ax$$

$$= \frac{1}{(D + ia)\,(D - ia)} \sec ax$$

On breaking into partial fraction, we get

$$\text{P.I} = \frac{1}{2ai}\left[\frac{1}{D - ia} - \frac{1}{D + ia}\right] \sec ax$$

$$= \frac{1}{2ai}\frac{1}{D - ia}\sec ax - \frac{1}{2ai}\frac{1}{D + ia}\sec ax$$

$$= \frac{1}{2ai}e^{iax}\int e^{-iax} \sec ax\, dx - \frac{1}{2ai}e^{-iax}\int e^{iax} \sec ax\, dx$$

$$= \frac{1}{2ai}e^{iax}\int \frac{\cos ax - i \sin ax}{\cos ax}\, dx - \frac{1}{2ai}e^{-iax}\int \frac{\cos ax + i \sin ax}{\cos ax}\, dx$$

$$[\because e^{ix} = \cos x + i \sin x]$$

$$= \frac{1}{2ai}e^{iax}\int (1 - i \tan ax)\, dx - \frac{1}{2ai}e^{-iax}\int (1 + i \tan ax)\, dx$$

$$= \frac{1}{2ai}\left\{e^{iax}\,(x + \frac{i}{a} \log \cos ax) - e^{-iax}\,(x - \frac{i}{a} \log \cos ax)\right\}$$

$$= \frac{1}{2ai}\left\{x\,(e^{iax} - e^{-iax}) + \frac{i}{a}\log\cos ax(e^{iax} + e^{-iax})\right\}$$

$$= \frac{1}{a}\left\{x\sin ax + \frac{1}{a}(\log\cos ax)\cos ax\right\}$$

$$\left[\because \sin x = \frac{e^{ix} - e^{-ix}}{2i} \quad \cos x = \frac{e^{ix} + e^{-ix}}{2}\right]$$

Hence, the general solution is y = C.F + P.I

$$y = c_1\cos ax + c_2\sin ax + \frac{1}{a}\left\{x\sin ax + \frac{1}{a}\cos ax(\log\cos ax)\right\}$$

Rule II: When $Q = e^{ax}$ then particular integral

$$P.I = \frac{1}{f(D)}e^{ax} = e^{ax}\frac{1}{f(a)}, \qquad f(a) \neq 0$$

If $\quad f(a) = 0$ then $\dfrac{1}{f(D)}e^{ax} = e^{ax}.\dfrac{1}{f(D+a)}.1$

***Example* 3:** Solve $\quad \dfrac{d^2y}{dx^2} - 3\dfrac{dy}{dx} + 2y = e^{5x}$

Solution: Given equation is

$$\frac{d^2y}{dx^2} - 3\frac{dy}{dx} + 2y = e^{5x}$$

$\Rightarrow \quad (D^2 - 3D + 2)\,y = e^{5x}$

Auxiliary equation is $m^2 - 3m + 2 = 0$

$\Rightarrow \quad (m - 1)(m - 2) = 0$

$\Rightarrow \quad m = 1, 2$

Complementary function is

$$C.F = c_1 e^x + c_2 e^{2x}$$

Particular integral is

$$\frac{1}{D^2 - 3D + 2} e^{5x}$$

$$\text{P.I.} = \frac{1}{(5)^2 - 3 \times 5 + 2} e^{5x} \qquad \left[\because \frac{1}{f(D)} e^{ax} = \frac{1}{f(a)} e^{ax} \right]$$

$$= \frac{1}{12} e^{5x}$$

Hence, the general solution is y = C.F + P.I

$$y = c_1 e^x + c_2 e^{2x} + \frac{1}{12} e^{5x}$$

***Example* 4:** Solve $\dfrac{d^2y}{dx^2} + 5\dfrac{dy}{dx} + 6y = e^{2x}$

***Solution*:** Given equation is

$$\frac{d^2y}{dx^2} + 5\frac{dy}{dx} + 6y = e^{2x}$$

$$(D^2 + 5D + 6)\, y = e^{2x}$$

Auxiliary equation is $m^2 + 5m + 6 = 0$

$$(m + 2)(m + 3) = 0$$

$\Rightarrow$ $m = -2, -3$

Complementary function is

$$\text{C.F} = c_1 e^{-2x} + c_2 e^{-3x}$$

Particular Integration is

$$\text{P.I} = \frac{1}{D^2 + 5D + 6} e^{2x}$$

$$= \frac{1}{(2)^2 + 5 \times 2 + 6} e^{2x}$$

$$= \frac{1}{20} e^{2x}$$

Hence, the general solution is y = C.F + P.I

$$y = c_1 e^{-2x} + c_2 e^{-3x} + \frac{1}{20} e^{2x}$$

Example 5: Solve $(D^2 + D - 2)\, y = e^x$

Solution: Given equation is

$$(D^2 + D - 2)y = e^x$$

Auxiliary equation is $m^2 + m - 2 = 0$

$\Rightarrow \quad (m + 2)(m - 1) = 0$

$\Rightarrow \quad m = 1, -2$

Complementary function is C.F $= c_1 e^x + c_2 e^{-2x}$

Particular integral is

$$\text{P.I} = \frac{1}{D^2 + D - 2} e^x$$

$$= \frac{1}{(D+2)(D-1)} e^x, \qquad \text{Here } f(1) = 0$$

$$= \frac{1}{D-1}\left[\frac{1}{D+2} e^x\right]$$

$$= \frac{1}{D-1}\left[\frac{1}{1+2} e^x\right]$$

$$= \frac{1}{3}\frac{1}{D-1} e^x$$

$$= \frac{1}{3} e^x \frac{1}{D+1-1}.1 \qquad \left[\because \frac{1}{f(D)} e^x = e^x \frac{1}{f(D+1)}.1\right]$$

$$= \frac{1}{3} e^x \frac{1}{D}.1$$

$$= \frac{1}{3} e^x . x$$

Hence, the general solution is y = C.F + P.I

$$y = c_1 e^x + c_2 e^{-2x} + \frac{1}{3} x e^x$$

Example 6: Find the general solution of $(D - 1)(D + 2)\, y = e^{3x}$

Solution: Given equation is $(D - 1)(D + 2)\, y = e^{3x}$

Auxiliary equation is $(m - 1)(m + 2) = 0$

$\Rightarrow \quad m = 1, -2$

Complementary function is C.F $= C_1 e^x + c_2 e^{-2x}$

Particular integral is

$$P.I = \frac{1}{(D-1)(D+2)} e^{3x}$$

$$= \frac{1}{(3-1)(3+2)} e^{3x}$$

$$= \frac{1}{10} e^{3x}$$

Hence, the general solution is y = C.F + P.I

$$y = C_1 e^x + c_2 e^{-2x} + \frac{1}{10} e^{3x}$$

***Example* 7:** Find the solution of $\frac{d^2y}{dx^2} - 7\frac{dy}{dx} + y = e^{2x}$

***Solution*:** Given equation is

$$\Rightarrow \quad \frac{d^2y}{dx^2} - 7\frac{dy}{dx} + y = e^{2x}$$

Auxiliary equation is $m^2 - 7m + 1 = 0$

$$\Rightarrow \quad m = \frac{7 \pm \sqrt{49-4}}{2} = \frac{7 \pm \sqrt{45}}{2}$$

$$m = \frac{7}{2} \pm \frac{3\sqrt{5}}{2}$$

Complementary function is

$$C.F = e^{\frac{7}{2}x}\left(c_1 \cos h\left(\frac{3\sqrt{5}}{2}x\right) + c_2 \sinh\left(\frac{3\sqrt{5}}{2}x\right)\right)$$

Particular integral is

$$P.I = \frac{1}{D^2 - 7D + 1} e^{2x}$$

$$= \frac{1}{(2)^2 - 7 \times 2 + 1} e^{2x}$$

$$= -\frac{1}{9} e^{2x}$$

Hence, required general solution is y = C.F + P.I

$$y = e^{\frac{7}{2}x}\left[c_1 \cosh\left(\frac{3\sqrt{5}}{2}x\right) + c_2 \sinh\left(\frac{3\sqrt{5}}{2}x\right)\right] - \frac{1}{9}e^{2x}$$

***Example* 8:** Solve $\frac{d^2y}{dx^2} - 2k\frac{dy}{dx} + k^2y = e^x$

***Solution*:** Given equation is

$$\frac{d^2y}{dx^2} - 2k\frac{dy}{dx} + k^2y = e^x$$

$$(D^2 - 2kD + k^2)\, y = e^x$$

Auxiliary equation is $m^2 - 2km + k^2 = 0$

$\Rightarrow \quad (m - k)^2 = 0$

$\Rightarrow \quad m = k, k$

Complementary function is

$$C.F = (c_1 + c_2 x)e^{kx}$$

Particular integral is

$$P.I = \frac{1}{D^2 - 2kD + k^2}e^x$$

$$= \frac{1}{1 - 2k + k^2}e^x$$

$$= \frac{1}{(1-k)^2}e^x$$

Hence, the general solution is

$$y = C.F + P.I$$

$$y = (c_1 + c_2 x)\, e^{kx} + \frac{1}{(1-k)^2}e^x$$

***Example* 9:** Find the general solution of the differential equation

$$\frac{d^3y}{dx^3} - 3\frac{d^2y}{dx^2} + 4y = e^{2x}$$

***Solution*:** Given equation is

$$\frac{d^3y}{dx^3} - 3\frac{d^2y}{dx^2} + 4y = e^{2x}$$

Auxiliary equation is $m^3 - 3m^2 + 4 = 0$

$\Rightarrow \quad (m + 1)(m^2 - 4m + 4) = 0$

$\Rightarrow \quad (m + 1)(m - 2)^2 = 0$

$\Rightarrow \quad m = -1, 2, 2$

Complementary function is

$$C.F = c_1 e^{-x} + (c_2 + c_3 x) e^{x}$$

Particular integral is

$$P.I = \frac{1}{(D^3 - 3D^2 + 4)} e^{2x}$$

Since, the denominator vanishes when D is replaced by 2. It is failure case i.e., f(2) = 0, then we multiply the numerator by x and differentiate denominator w.r.t D. we get

$$P.I = x.\frac{1}{3D^2 - 6D} e^{2x}$$

It is again failure. We multiply by x and differentiate denominator w.r.t. D. we get

$$P.I = x^2 . \frac{1}{6D - 6} e^{2x}$$

$$= x^2. \frac{1}{6 \times 2 - 6} e^{2x}$$

$$= \frac{1}{6} x^2 e^{2x}$$

Hence, the general solution is

$$y = C.F + P.I$$

$$y = c_1 e^{-x} + (c_2 + c_3 x) e^{2x} + \frac{1}{6} x^2 e^{2x}$$

***Example* 10:** Solve the equation $\frac{d^2y}{dx^2} - 5\frac{dy}{dx} + 6y = 2e^{2x}$

***Solution*:** Given equation is

$$\frac{d^2y}{dx^2} - 5\frac{dy}{dx} + 6y = 2e^{2x}$$

$$(D^2 - 5D + 6)\, y = 2e^{2x}$$

Auxiliary equation is

$$m^2 - 5m + 6 = 0$$

$$\Rightarrow \quad (m-2)(m-3) = 0$$

$$\Rightarrow \quad m = 2, 3$$

Complementary function is

$$C.F = c_1 e^{2x} + c_2 e^{3x}$$

Particular integral is

$$P.I = \frac{1}{D^2 - 5D + 6} 2e^{2x}$$

$$= 2\frac{1}{D^2 - 5D + 6} e^{2x}$$

The denominator vanishes when D replace by 2. It is a case of failure. We multiply numerator by x and differentiate denominator w.r.t D, we get

$$P.I = 2.x\frac{1}{2D - 5} e^{2x}$$

$$= 2.x\frac{1}{2 \times 2 - 5} e^{2x}$$

$$= -2x e^{2x}$$

∴ General solution is

$$y = C.F + P.I$$

$$y = c_1 e^{2x} + c_2 e^{3x} - 2x e^{2x}$$

***Example* 11:** Solve $\frac{d^2y}{dx^2} - 2\frac{dy}{dx} + y = e^x$

***Solution*:** Given equation is

$$\frac{d^2y}{dx^2} - 2\frac{dy}{dx} + y = e^x$$

$$(D^2 - 2D + 1)\, y = e^x$$

Auxiliary equation is

$$m^2 - 2m + 1 = 0$$

$$\Rightarrow \quad (m-1)^2 = 0$$

$$\Rightarrow \quad m = 1, 1$$

Complementary function is

$$C.F = (c_1 + c_2 x)\, e^x$$

Particular integral is

$$P.I = \frac{1}{D^2 - 2D + 1} e^x$$

$$= x \frac{1}{2D - 2} e^x$$

$$= x^2 \cdot \frac{1}{2} e^x$$

$$= \frac{1}{2} x^2 e^x$$

∴ General solution is y = C.F + P.I

$$y = (c_1 + c_2 x)\, e^x + \frac{1}{2} x^2 e^x$$

Exercise 14.2

Solve the Following Differential Equations

1. $\frac{d^2y}{dx^2} + 2p\frac{dy}{dx} + (p^2 + q^2)y = e^{ax}$
2. $\frac{d^2y}{dx^2} - 6\frac{dy}{dx} + 7y = e^x + e^{-x}$
3. $\frac{d^3y}{dx^3} + 3\frac{d^2y}{dx^2} + 3\frac{dy}{dx} + y = e^{-x}$
4. $\frac{d^2y}{dx^2} + 6\frac{dy}{dx} + 9y = 2e^{-3x}$
5. $(D^3 - 1)y = (e^x + 1)^2$
6. $(D - 3)^2\, y = 2e^{3x} + e^{-x}$
7. $\frac{d^3y}{dx^3} + y = 3 + 5e^x$
8. $(D^2 + 2D + 1)\, y = 2e^{3x}$
9. $(D + 3)^2\, y = 50\, e^{2x}$
10. $(D - 1)^2\, (D + 1)^2\, y = e^x$
11. $(3D^2 + D - 14)\, y = 13\, e^{2x}$
12. $(D^3 + 4D^2 + 4D)\, y = 8e^{-2x}$

Answers

1. $y = e^{-px}(c_1 \cos qx + c_2 \sin qx) + \dfrac{e^{ax}}{(a+p)^2 + q^2}$
2. $y = e^{3x}\left(c_1 e^{\sqrt{2}x} + c_2 e^{-\sqrt{2}x}\right) + \dfrac{e^x}{2} + \dfrac{e^{-x}}{14}$
3. $y = (c_1 + c_2 x + c_3 x^2)e^{-x} + \dfrac{1}{6}x^3 e^{-x}$
4. $y = (c_1 + c_2 x)e^{-2x} + x^2 e^{-3x}$
5. $y = e^{\frac{-x}{2}}\left(c_1 \cos\dfrac{\sqrt{3}}{2}x + c_2 \sin\dfrac{\sqrt{3}}{2}x\right) + e^x(c_3 + \dfrac{2}{3}x) + \dfrac{1}{7}e^{2x} - 1$
6. $y = (c_1 + c_2 x)e^{3x} + \dfrac{1}{16}e^{-x} + x^2 e^{3x}$
7. $y = c_1 e^{-x} + e^{\frac{x}{2}}\left(c_2 \cos\dfrac{\sqrt{3}}{2}x + c_3 \sin\dfrac{\sqrt{3}}{2}x\right) + 3 + \dfrac{5}{2}e^x$
8. $y = (c_1 + c_2 x)e^{-x} + \dfrac{1}{8}e^{3x}$
9. $y = (c_1 + c_2 x)e^{-3x} + 2e^{2x}$
10. $y = (c_1 + c_2 x)e^x + (c_3 + c_4 x)e^{-x} + \dfrac{1}{8}x^2 e^{-x}$
11. $y = c_1 e^{\frac{-7}{3}x} + e^{2x}(c_2 + x)$
12. $y = c_1 + (c_2 + c_3 x - 2x^2)e^{-2x}$

Rule III

When $Q = \sin ax$ or $Q = \cos ax$

Particular integral is $\dfrac{1}{f(D^2)}\sin ax = \dfrac{1}{f(-a^2)}\sin ax, \quad f(-a^2) \neq 0$

and $\dfrac{1}{f(D^2)}\cos ax = \dfrac{1}{f(-a^2)}\cos ax, \quad f(-a^2) \neq 0$

If $f(-a^2) = 0$ then $\dfrac{1}{D^2 + a^2}\sin ax = \dfrac{-x}{2a}\cos ax$

and $\dfrac{1}{D^2 + a^2}\cos ax = \dfrac{x}{2a}\sin ax$

***Example* 1:** Solve $\frac{d^2y}{dx^2}+\frac{dy}{dx}+y=\sin 2x$

***Solution*:** Given equation is

$$\frac{d^2y}{dx^2}+\frac{dy}{dx}+y=\sin 2x$$

$$(D^2 + D + 1)y = \sin 2x$$

Auxiliary equation is

$$m^2 + m + 1 = 0$$

$$\Rightarrow \quad m = \frac{-1\pm\sqrt{1-4}}{2} = \frac{-1}{2} \pm \frac{i\sqrt{3}}{2}$$

$\therefore$ Complementary function is

$$C.F = e^{\frac{-x}{2}}\left[c_1 \cos\frac{\sqrt{3}}{2}x + c_2 \sin\frac{\sqrt{3}}{2}x\right]$$

Particular integral is

$$P.I = \frac{1}{D^2+D+1}\sin 2x$$

$$= \frac{1}{-4+D+1}\sin 2x \qquad \left[\because \frac{1}{f(D^2)}\sin ax = \frac{1}{f(-a^2)}\sin ax\right]$$

$$= \frac{1}{D-3}\sin 2x$$

$$= \frac{D+3}{(D-3)(D+3)}\sin 2x$$

$$= \frac{D+3}{D^2-9}\sin 2x$$

$$= \frac{D+3}{-4-9}\sin 2x$$

$$\therefore \quad = \frac{-1}{13}(D+3)\sin 2x$$

$$= \frac{-1}{13}(2\cos 2x + 3\sin 2x)$$

$\therefore$ General solution is

$$y = C.F + P.I$$

$$y = e^{\frac{-x}{2}}[c_1 \cos\frac{\sqrt{3}}{2}x + c_2 \sin\frac{\sqrt{3}}{2}x] - \frac{1}{13}(2\cos 2x + 3\sin 2x)$$

***Example* 2:** Solve $\frac{d^2y}{dx^2} - 3\frac{dy}{dx} + 2y = \cos 3x$

***Solution*:** Given equation is

$$\frac{d^2y}{dx^2} - 3\frac{dy}{dx} + 2y = \cos 3x$$

$$(D^2 - 3D + 2)\,y = \cos 3x$$

Auxiliary equation is $m^2 - 3m + 2 = 0$

$\Rightarrow$ $(m - 1)(m - 2) = 0$

$\Rightarrow$ $m = 1, 2$

Complementary function is

$$C.F = c_1 e^x + c_2 e^{2x}$$

Particular integral is

$$P.I = \frac{1}{D^2 - 3D + 2}\cos 3x$$

$$= \frac{1}{-9 - 3D + 2}\cos 3x$$

$$= \frac{1}{-3D - 7}\cos 3x$$

$$= \frac{-(3D - 7)}{9D^2 - 49}\cos 3x$$

$$= \frac{-(3D - 7)}{9(-9) - 49}\cos 3x$$

$$= \frac{1}{130}(-9 \sin 3x - 7 \cos 3x)$$

$\therefore$ General solution is

$$y = C.F + P.I$$

$$y = c_1 e^x + c_2 e^{2x} - \frac{1}{130}(9 \sin 3x + 7 \cos 3x)$$

***Example* 3:** Solve $(D^2 + 4)\, y = \sin 2x$

***Solution*:** Given equation is

$$(D^2 + 4)\, y = \sin 2x$$

Auxiliary equation is $m^2 + 4 = 0$

$\Rightarrow \quad m = \pm 2i$

Complementary function is $\quad C.F = c_1 \cos 2x + c_2 \sin 2x$

Particular Integral is

$$P.I = \frac{1}{D^2 + 4} \sin 2x$$

$$= \frac{-x}{2.2} \cos 2x \qquad \left[\because \frac{1}{D^2 + a^2} \sin ax = \frac{-x}{2a} \cos ax\right]$$

$$= -\frac{x}{4} \cos 2x$$

$\therefore$ General solution is

$$y = C.F + P.I$$

$$y = c_1 \cos 2x + c_2 \sin 2x - \frac{x}{4} \cos 2x$$

***Example* 4:** Find the general solution of $(D^2 + D + 1)\, y = e^{2x} + \sin 2x$

***Solution*:** Given equation is

$$(D^2 + D + 1)y = e^{2x} + \sin 2x$$

Auxiliary equation is

$$m^2 + m + 1 = 0$$

$$\Rightarrow \quad m = \frac{-1 \pm \sqrt{1-4}}{2} = \frac{-1}{2} \pm i\frac{\sqrt{3}}{2}$$

Complementary function is

$$C.F = e^{\frac{-x}{2}} [c_1 \cos \frac{\sqrt{3}}{2} x + c_2 \sin \frac{\sqrt{3}}{2} x]$$

Particular integral is

$$P.I = \frac{1}{D^2 + D + 1}(e^{2x} + \sin 2x)$$

$$= \frac{1}{D^2 + D + 1} e^{2x} + \frac{1}{D^2 + D + 1} \sin 2x$$

$$= \frac{1}{4 + 2 + 1} e^{2x} + \frac{1}{-4 + D + 1} \sin 2x$$

$$= \frac{1}{7} e^{2x} + \frac{1}{D - 3} \sin 2x$$

$$= \frac{1}{7} e^{2x} + \frac{D + 3}{D^2 - 9} \sin 2x$$

$$= \frac{1}{7} e^{2x} + \frac{D + 3}{-4 - 9} \sin 2x$$

$$= \frac{1}{7} e^{2x} - \frac{1}{13}(D + 3)\sin 2x$$

$$= \frac{1}{7} e^{2x} - \frac{1}{13}(2\cos 2x + 3\sin 2x)$$

$\therefore$ General solution is $y = C.F + P.I$

$$y = e^{\frac{-x}{2}}\left[c_1 \cos\frac{\sqrt{3}}{2}x + c_2 \sin\frac{\sqrt{3}}{2}x\right] + \frac{1}{7} e^{2x} - \frac{1}{13}(2\cos 2x + 3\sin 2x)$$

***Example* 5 :** Find the general solution of $(D^2 + D + 1)y = \sin x$

***Solution*:** Given equation is

$$(D^2 + D + 1)\, y = \sin x$$

Auxiliary equation is

$$m^2 + m + 1 = 0$$

$$\Rightarrow \quad m = \frac{-1 \pm \sqrt{1 - 4}}{2} = \frac{-1}{2} \pm i\frac{\sqrt{3}}{2}$$

Complementary function is

$$C.F = e^{\frac{-x}{2}}[c_1 \cos\frac{\sqrt{3}}{2}x + c_2 \sin\frac{\sqrt{3}}{2}x]$$

Particular integral is

$$P.I = \frac{1}{D^2+D+1}\sin x$$

$$= \frac{1}{-1+D+1}\sin x$$

$$= \frac{1}{D}\sin x$$

$$= -\cos x$$

$\therefore$ General solution is $\quad y = C.F + P.I$

$$y = e^{\frac{-x}{2}}[c_1 \cos\frac{\sqrt{3}}{2}x + c_2 \sin\frac{\sqrt{3}}{2}x] - \cos x$$

***Example* 6:** Solve $\frac{d^2y}{dx^2}+5\frac{dy}{dx}+6y = 2\sin x$

***Solution*:** Given equation is

$$\frac{d^2y}{dx^2}+5\frac{dy}{dx}+6y = 2\sin x$$

$$(D^2+5D+6)\,y = 2\sin x$$

Auxiliary equation is $\quad m^2 + 5m + 6 = 0$

$\Rightarrow \quad (m+2)(m+3) = 0$

$\Rightarrow \quad m = -2, -3$

$\therefore$ Complementary function is

$$C.F = c_1 e^{-2x} + c_2 e^{-3x}$$

Particular Integral is

$$P.I = \frac{1}{D^2+5D+6}(2\sin x)$$

$$= \frac{1}{-1+5D+6}2\sin x$$

$$= \frac{1}{5D+5}2\sin x$$

$$= \frac{5D-5}{25D^2-25} 2\sin x$$

$$= \frac{5D-5}{-25-25} 2\sin x$$

$$= \frac{-2}{50}(5D-5)\sin x$$

$$= -\frac{1}{5}(\cos x - \sin x)$$

$\therefore$ General solution is y = C.F + P.I

$$y = c_1 e^{-2x} + c_2 e^{-3x} - \frac{1}{5}(\cos x - \sin x)$$

***Example* 7:** Solve $\frac{d^2y}{dx^2} - 4y = 2\sin\frac{x}{2}$

***Solution*:** Given equation is

$$\frac{d^2y}{dx^2} - 4y = 2\sin\frac{x}{2}$$

$$(D^2-4)y = 2\sin\frac{x}{2}$$

Auxiliary equation is

$$m^2 - 4 = 0$$

$$\Rightarrow \quad (m-2)(m+2) = 0$$

$$\Rightarrow \quad m = 2, -2$$

Complementary function is C.F = $c_1 e^{2x} + c_2 e^{-2x}$

Particular integral is

$$\text{P.I} = \frac{1}{D^2-4} 2\sin\frac{x}{2}$$

$$= \frac{1}{\frac{-1}{4}-4} 2\sin\frac{x}{2}$$

$$= -\frac{4}{1+16} 2\sin\frac{x}{2}$$

$$= -\frac{8}{17}\sin\frac{x}{2}$$

∴ General solution is y = C.F + P.I

$$y = c_1 e^{2x} + c_2 e^{-2x} - \frac{8}{17}\sin\frac{x}{2}$$

***Example 8*:** Find the General solution of $\frac{d^2y}{dx^2} + 4y = \cos 2x$

***Solution*:** Given equation is

$$\frac{d^2y}{dx^2} + 4y = \cos 2x$$

$$(D^2+4)\,y = \cos 2x$$

Auxiliary equation is

$$m^2 + 4 = 0$$

$$\Rightarrow \quad m = \pm 2i$$

Complementary function is $\quad C.F = c_1 \cos 2x + c_2 \sin 2x$

Particular integral is

$$P.I = \frac{1}{D^2+4}\cos 2x$$

$$= \frac{x}{2.2}\sin 2x \qquad \left[\because \frac{1}{D^2+a^2}\cos ax = \frac{x}{2a}\sin ax\right]$$

$$= \frac{x}{4}\sin 2x$$

∴ General solution is

$$y = C.F + P.I$$

$$y = c_1 \cos 2x + c_2 \sin 2x + \frac{1}{4}x\sin 2x$$

***Example 9*:** Solve $\frac{d^2y}{dx^2} - 4y = \cos^2 x$

***Solution*:** Given equation is

$$\frac{d^2y}{dx^2} - 4y = \cos^2 x$$

$$(D^2 - 4)\,y = \cos^2 x$$

$\therefore$ Auxiliary equation is $m^2 - 4 = 0$

$\Rightarrow \quad (m - 2)(m + 2) = 0$

$\Rightarrow \quad m = 2, -2$

$\therefore$ Complementary function is

$$C.F = c_1 e^{2x} + c_2 e^{-2x}$$

Particular integral is

$$P.I = \frac{1}{D^2 - 4} \cos^2 x$$

$$= \frac{1}{D^2 - 4}\left(\frac{1 + \cos 2x}{2}\right)$$

$$= \frac{1}{2}\left[\frac{1}{D^2 - 4} e^{0.x} + \frac{1}{D^2 - 4} \cos 2x\right]$$

$$= \frac{1}{2}\left[\frac{1}{0 - 4} e^{0.x} + \frac{1}{-4 - 4} \cos 2x\right]$$

$$= \frac{-1}{8} - \frac{1}{16} \cos 2x$$

$\therefore$ General solution is $y = C.F + P.I$

$$y = c_1 e^{2x} + c_2 e^{-2x} - \frac{1}{8} - \frac{1}{16} \cos 2x$$

***Example* 10:** Solve $(D^2 + 1)\, y = \sin x \sin 2x$

***Solution*:** Given equation is

$$(D^2 + 1)\, y = \sin x \sin 2x$$

$\therefore$ Auxiliary equation is

$$m^2 + 1 = 0$$

$\Rightarrow \quad m = \pm i$

$\therefore$ Complementary function is

$$C.F = c_1 \cos x + c_2 \sin x$$

Particular integral is

$$P.I = \frac{1}{D^2+1} \sin x \sin 2x$$

$$= \frac{1}{2} \cdot \frac{1}{D^2+1} (2 \sin x \sin 2x)$$

$$= \frac{1}{2} \frac{1}{D^2+1} (\cos x - \cos 3x)$$

$$= \frac{1}{2} \frac{1}{D^2+1} \cos x - \frac{1}{2} \frac{1}{D^2+1} \cos 3x \backslash$$

$$= \frac{1}{2} \frac{x}{2.1} \sin x - \frac{1}{2} \frac{1}{-9+1} \cos 3x$$

$$= \frac{x}{4} \sin x + \frac{1}{16} \cos 3x$$

∴ General solution is

$$y = C.F + P.I$$

$$y = c_1 \cos x + c_2 \sin x + \frac{x}{4} \sin x + \frac{1}{16} \cos 3x$$

Exercise 14.3

Solve

1. $\frac{d^3y}{dx^3} + y = \cos 2x$
2. $\frac{d^3y}{dx^3} + \frac{d^2y}{dx^2} - \frac{dy}{dx} - y = \cos 2x$
3. $\frac{d^2y}{dx^2} - 4y = e^x + \sin 2x$
4. $\frac{d^2y}{dx^2} + 9y = e^x - \cos 2x$
5. $\frac{d^2y}{dx^2} - 4\frac{dy}{dx} + y = a \cos 2x$
6. $\frac{d^2y}{dt^2} - 3\frac{dy}{dt} + 2y = 10 \sin t$
7. $(2D^2 - 3D + 4)\, y = 3 \cos \frac{x}{2}$
8. $(D^2 + 4)\, y = \sin^2 x$
9. $\frac{d^2y}{dx^2} + 9y = \cos 3x$
10. $(D^2 + 1)\, y = e^x \cos x$
11. $(D^3 - 3D^2 + 9D - 27)\, y = \cos 3x$
12. $(D^2 + 1)^2\, y = \cos 3x$
13. $(D^2 + D + 1)\, y = (1 + \sin x)^2$

Answers

1. $y = c_1 e^{-x} + e^{\frac{x}{2}} (c_2 \cos\frac{\sqrt{3}}{2} x + c_3 \sin\frac{\sqrt{3}}{2} x) + \frac{1}{65} (\cos 2x - 8 \sin 2x)$

2. $y = c_1 e^x + (c_2 + c_3 x) e^{-x} - \frac{2}{25} \sin 2x - \frac{1}{25} \cos 2x$

3. $y = c_1 \cos 2x + c_2 \sin 2x + \frac{1}{5} e^x - \frac{1}{4} x \cos 2x$

4. $y = c_1 \cos 3x + c_2 \sin 3x + \frac{1}{10} e^x - \frac{1}{5} x \cos 2x$

5. $y = c_1 e^{2x} \cosh\left(\sqrt{3}x + c_2\right) - \frac{a}{73} (3 \cos 2x + 8 \sin 2x)$

6. $y = c_1 e^t + c_2 e^{2t} + \sin t + 3 \cos t$

7. $y = (c_1 \cos\frac{\sqrt{23}}{4} x + c_2 \sin\frac{\sqrt{23}}{4} x) e^{\frac{3}{4}x} + \frac{3}{29}(7\cos\frac{x}{2} - 3 \sin\frac{x}{2})$

8. $y = c_1 \cos 2x + c_2 \sin 2x - \frac{1}{8} - \frac{1}{8} x \sin 2x$

9. $y = c_1 \cos 3x + c_2 \sin 3x + \frac{1}{6} x \sin 3x$

10. $y = c_1 \cos x + c_2 \sin x + \frac{1}{2} (e^x + x \sin x)$

11. $y = c_1 e^{3x} + c_2 \cos 3x + c_3 \sin 3x - \frac{1}{36} x (\sin 3x + \cos 3x)$

12. $y = (c_1 + c_2 x) \cos x + (c_3 + c_4 x) \sin x + \frac{1}{64} \cos 3x$

13. $y = \left(c_1 \cos\frac{\sqrt{3}}{2} x + c_2 \sin\frac{\sqrt{3}}{2} x \right) e^{\frac{-x}{2}} + \frac{3}{2} + \frac{1}{26} (3 \cos 2x - 2 \sin 2x) - 2 \cos x$

Rule IV: when $Q = x^m$, m is a positive integer

$$P.I = \frac{1}{f(D)} x^m = [f(D)]^{-1} . x^m$$

First, we bring out common the lowest degree of D from f(D). Then the remaining factor in the denominator is of the form 1 + f(D) or 1–f(D). Take this factor in the numerator. It takes the form $[1 + f(D)]^{-1}$ or $[1 - f(D)]^{-1}$, now expand it by binomial theorem up to D^m and operate on x^m term by term.

Note: Some formulae (by binomial expansions)

1. $(1 + D)^{-1} = 1 - D + D^2 - D^3 + \dots\dots\dots$
2. $(1 - D)^{-1} = 1 + D + D^2 + D^3 + \dots\dots\dots$
3. $(1 + D)^{-2} = 1 - 2D + 3D^2 - 4D^3 + \dots\dots\dots$
4. $(1 - D)^{-2} = 1 + 2D + 3D^2 + 4D^3 + \dots\dots\dots$
5. $(1 + D)^{-3} = 1 - 3D + 6D^2 - 10 D^3 + \dots\dots\dots$
6. $(1 - D)^{-3} = 1 + 3D + 6 D^2 + 10 D^3 + \dots\dots\dots$
7. $(1 + x)^{-n} = 1 - nx + \frac{n(n+1)}{2!} x^2 + \frac{n(n+1)(n+2)}{3!} x^3 + \dots\dots\dots$
8. $(1 - x)^{-n} = 1 + nx + \frac{n(n+1)}{2!} x^2 + \frac{n(n+1)(n+2)}{3!} x^3 + \dots\dots$

***Example* 1:** Solve $\frac{d^2y}{dx^2} - 4\frac{dy}{dx} + 4y = x^2$

Solution: Given equation is

$$\frac{d^2y}{dx^2} - 4\frac{dy}{dx} + 4y = x^2$$

$$(D^2 - 4D + 4)\, y = x^2$$

∴ Auxiliary equation is

$$m^2 - 4m + 4 = 0$$

$$\Rightarrow \quad (m - 2)^2 = 0$$

$$\Rightarrow \quad m = 2, 2$$

∴ Complementary function is

$$C.F = (c_1 + c_2 x)\, e^{2x}$$

Particular integral is

$$P.I = \frac{1}{D^2 - 4D + 4}\, x^2$$

$$= \frac{1}{(D-2)^2}\, x^2$$

$$= \frac{1}{2^2 (1 - \frac{D}{2})^{-2}}\, x^2$$

$$= \frac{1}{4}\left(1 - \frac{D}{2}\right)^{-2} x^2$$

$$= \frac{1}{4}\left(1 + 2.\frac{D}{2} + 3\left(\frac{D}{2}\right)^2 + ...\right) x^2$$

$$= \frac{1}{4}\left(x^2 + Dx^2 + \frac{3}{4}D^2x^2\right)$$

$$= \frac{1}{4}\left(x^2 + 2x + \frac{3}{4} \times 2\right)$$

$$= \frac{1}{4}\left(x^2 + 2x + \frac{3}{2}\right)$$

∴ General solution is

$$y = C.F + P.I$$

$$y = (c_1 + c_2 x)\, e^{2x} + \frac{1}{4}\left(x^2 + 2x + \frac{3}{2}\right)$$

***Example* 2:** Solve $\frac{d^2y}{dx^2} + 2\frac{dy}{dx} + 2y = x$

***Solution*:** Given equation is

$$\frac{d^2y}{dx^2} + 2\frac{dy}{dx} + 2y = x$$

$$(D^2 + 2D + 2)\, y = x$$

∴ Auxiliary equation is

$$m^2 + 2m + 2 = 0$$

$$\Rightarrow \quad m = \frac{-2 \pm \sqrt{4-8}}{2} = -1 \pm i$$

∴ Complementary function is

$$C.F = e^{-x}(c_1 \cos x + c_2 \sin x)$$

Particular integral is

$$P.I = \frac{1}{D^2 + 2D + 2} x$$

$$= \frac{1}{2\left(1 + D + \frac{D^2}{2}\right)} x$$

$$= \frac{1}{2}\left(1 + D + \frac{D^2}{2}\right)^{-1} x$$

$$= \frac{1}{2}\left(1 - D - \frac{D^2}{2} \ldots\ldots\right) x$$

$$= \frac{1}{2}\left(x - Dx - \frac{D^2}{2} x\right)$$

$$= \frac{1}{2}(x - 1)$$

∴ General solution is y = C.F + P.I

$$y = e^{-x}(c_1 \cos x + c_2 \sin x) + \frac{1}{2}(x - 1)$$

Example 3: Solve $(D^2 + 4)\, y = x^2$

Solution: Given equation is

$$(D^2 + 4)\, y = x^2$$

∴ Auxiliary equation is $m^2 + 4 = 0$

$$\Rightarrow \quad m = \pm 2i$$

Complementary function is

$$C.F = c_1 \cos 2x + c_2 \sin 2x$$

Particular integral is

$$P.I = \frac{1}{D^2+4}x^2$$

$$= \frac{1}{4}\frac{1}{1+\frac{D^2}{4}}x^2$$

$$= \frac{1}{4}\left(1+\frac{D^2}{4}\right)^{-1}x^2$$

$$= \frac{1}{4}\left(1-\frac{D^2}{4}+\frac{D^4}{16}....\right)x^2$$

$$= \frac{1}{4}\left(x^2-\frac{1}{4}D^2x^2\right)$$

$$= \frac{1}{4}\left(x^2-\frac{1}{2}\right)$$

$\therefore$ General solution is y = C.F + P.I

$$y = c_1 \cos 2x + c_2 \sin 2x + \frac{1}{4}x^2 - \frac{1}{8}$$

***Example* 4:** Solve $\frac{d^2y}{dx^2} + y = x^2$

***Solution*:** Given equation is

$$\frac{d^2y}{dx^2} + y = x^2$$

$$(D^2 + 1)\, y = x^2$$

Auxiliary equation is

$$m^2 + 1 = 0$$

$\Rightarrow$ $m = \pm i$

$\therefore$ Complementary function is

$$C.F = c_1 \cos x + c_2 \sin x$$

Particular integral is

$$\text{P.I} = \frac{1}{D^2+1}x^2$$
$$= (1+D^2)^{-1}x^2$$
$$= (1 - D^2 + \ldots\ldots)\,x^2$$
$$= (x^2 - D^2x^2)$$
$$= x^2 - 2$$

$\therefore$ General solution is

$$y = \text{C.F} + \text{P.I}$$
$$y = c_1 \cos x + c_2 \sin x + x^2 - 2$$

***Example* 5:** Solve $(D^3 + 3D^2 + 2D)\,y = x^2$

***Solution*:** Given equation is

$$(D^3 + 3D^2 + 2D)\,y = x^2$$

Auxiliary equation is

$$m^3 + 3m^2 + 2m = 0$$
$$\Rightarrow \quad m(m^2 + 3m + 2) = 0$$
$$\Rightarrow \quad m(m+1)(m+2) = 0$$
$$\Rightarrow \quad m = 0, -1, -2$$

$\therefore$ Complementary function is

$$\text{C.F} = c_1 + c_2e^{-x} + c_3e^{-2x}$$

Particular integral is

$$\text{P.I} = \frac{1}{D^3 + 3D^2 + 2D}x^2$$
$$= \frac{1}{2D\left(1 + \frac{3D + D^2}{2}\right)}x^2$$
$$= \frac{1}{2D}\left(1 + \frac{3D + D^2}{2}\right)^{-1}x^2$$
$$= \frac{1}{2D}\left(1 - \frac{3D + D^2}{2} + \left(\frac{3D + D^2}{2}\right)^2 \ldots\ldots\right)x^2$$

$$= \frac{1}{2D}\left(1 - \frac{3D}{2} - \frac{D^2}{2} + \frac{9}{4}D^2.....\right)x^2$$

$$= \frac{1}{2D}\left(1 - \frac{3D}{2} + \frac{7}{4}D^2.....\right)x^2$$

$$= \frac{1}{2D}\left(x^2 - 3x + \frac{7}{2}\right)$$

$$= \frac{1}{2}\left(\frac{x^3}{3} - \frac{3}{2}x^2 + \frac{7}{2}x\right) \qquad \left(\because \frac{1}{D} = \int dx\right)$$

$\therefore$ General solution is

$$y = C.F + P.I$$

$$y = c_1 + c_2\, e^{-x} + c_3\, e^{-2x} + \frac{1}{2}\left(\frac{x^3}{3} - \frac{3}{2}x^2 + \frac{7}{2}x\right)$$

Exercise 14.4

Solve

1. $\dfrac{d^3y}{dx^3} - \dfrac{d^2y}{dx^2} - 6\dfrac{dy}{dx} = 1 + x^2$
2. $(D^2 - 4D + 4)\, y = x^2 + e^x + \cos 2x$
3. $\dfrac{d^2y}{dx^2} - y = 2 + 5x$
4. $(D^2 + 2D + 2)\, y = x^2$
5. $\dfrac{d^3y}{dx^3} + 2\dfrac{d^2y}{dx^2} + 4\dfrac{dy}{dx} + 8y = x$
6. $(D^3 - 8)\, y = x^3$
7. $(D^3 - 4D)\, y = x$
8. $(D^4 + 2D^2 + 1)\, y = x^2$

Answers

1. $y = c_1 + c_2\, e^{-2x} + c_3\, e^{3x} - \frac{1}{18}x^3 + \frac{1}{36}x^2 - \frac{25}{108}x$
2. $y = (c_1 + c_2\, x)\, e^{2x} + \frac{1}{4}(x^2 + 2x + \frac{3}{4}) + e^x - \frac{1}{8}\sin 2x$
3. $y = c_1\, e^x + c_2\, e^{-x} - 2 - 5x$
4. $y = e^{-x}\,(c_1 \cos x + c_2 \sin x) + \frac{1}{2}\,(x^2 - 2x + 1)$

5. $y = c_1 e^{-2x} + c_2 \cos 2x + c_3 \sin 3x + \frac{1}{16}(2x - 1)$
6. $y = c_1 e^{2x} + e^{-x}(c_2 \cos\sqrt{3}x + c_3 \sin\sqrt{3}x) + \frac{1}{8}(x^3 + \frac{3}{4})$
7. $y = c_1 + c_2 e^{2x} + c_3 e^{-2x} - \frac{x^2}{8}$
8. $y = (c_1 + c_2 x)\cos x + (c_3 + c_4 x)\sin x + x^2 - 4$

Rule V

When $Q = e^{ax} . v$, where v is any function of x

$$P.I = \frac{1}{f(D)} e^{ax}.v = e^{ax} \frac{1}{f(D+a)} v$$

***Example* 1:** Solve $\frac{d^2y}{dx^2} - 2\frac{dy}{dx} + y = x^2 e^{3x}$

***Solution*:** Given equation is

$$\frac{d^2y}{dx^2} - 2\frac{dy}{dx} + y = x^2 e^{3x}$$

$$(D^2 - 2D + 1)\, y = x^2 e^{3x}$$

Auxiliary equation is

$$m^2 - 2m + 1 = 0$$

$$\Rightarrow \quad (m - 1)^2 = 0$$

$$\Rightarrow \quad m = 1, 1$$

Complementary function is

$$C.F = (c_1 + c_2 x)\, e^x$$

Particular integral is

$$P.I = \frac{1}{D^2 - 2D + 1} x^2 e^{3x}$$

$$= \frac{1}{(D-1)^2} x^2 e^{3x}$$

$$= e^{3x} \frac{1}{(D+3-1)^2} x^2$$

$$= e^{3x} \frac{1}{(D+2)^2} x^2$$

$$= e^{3x}\,\frac{1}{4}\left(1+\frac{D}{2}\right)^{-2} x^2$$

$$= \frac{1}{4}e^{3x}\left(1-\frac{2D}{2}+3\left(\frac{D}{2}\right)^2 \ldots\ldots\right)x^2$$

$$= \frac{1}{4}e^{3x}\left(x^2-2x+\frac{3}{4}.2\right)$$

$$= \frac{1}{4}e^{3x}\left(x^2-2x+\frac{3}{2}\right)$$

∴ General solution is $y = C.F + P.I$

$$y = (c_1 + c_2\, x)e^x + \frac{1}{4}e^{3x}\left(x^2-2x+\frac{3}{2}\right)$$

***Example* 2:** Solve $(D^2 - 3D + 2)\, y = x\, e^x$

***Solution*:** Given equation is

$$(D^2 - 3D + 2)\, y = xe^x$$

Auxiliary equation is

$$m^2 - 3m + 2 = 0$$

$$\Rightarrow \quad (m - 1)\,(m - 2) = 0$$

$$\Rightarrow \quad m = 1, 2$$

∴ Complementary function is

$$C.F = c_1\, e^x + c_2\, e^{2x}$$

Particular integral is

$$P.I = \frac{1}{D^2-3D+2}\, x\, e^x$$

$$= e^x\, \frac{1}{(D+1)^2-3(D+1)+2}\, x$$

$$= e^x\, \frac{1}{D^2-D}\, x$$

$$= e^x\, \frac{1}{D\,(D-1)}\, x$$

$$= -e^x \frac{1}{D}(1-D)^{-1} x$$

$$= -e^x . \frac{1}{D}(1 + D + ...) x$$

$$= -e^x . \frac{1}{D}(x + 1)$$

$$= -e^x \left(\frac{x^2}{2} + x\right)$$

$\therefore$ General solution is

$$y = C.F + P.I$$

$$y = c_1 e^x + c_2 e^{2x} - e^x \left(\frac{x^2}{2} + x\right)$$

***Example* 3:** Solve $(D^2 - 2D + 5) y = e^{2x} \sin x$

***Solution*:** Given equation is

$$(D^2 - 2D + 5) y = e^{2x} . \sin x$$

Auxiliary equation is

$$m^2 - 2m + 5 = 0$$

$$\Rightarrow \quad m = \frac{2 \pm \sqrt{4-20}}{2} = 1 \pm 2i$$

Complementary function is

$$C.F = e^x (c_1 \cos 2x + c_2 \sin 2x)$$

Particular integral is

$$P.I = \frac{1}{D^2 - 2D + 5} e^{2x} \sin x$$

$$= e^{2x} \frac{1}{(D+2)^2 - 2(D+2) + 5} \sin x$$

$$= e^{2x} \frac{1}{D^2 + 2D + 5} \sin x$$

$$= e^{2x} \frac{1}{-1 + 2D + 5} \sin x$$

$$= e^{2x} \frac{1}{2D+4} \sin x$$

$$= e^{2x} \frac{2D-4}{4D^2-16} \sin x$$

$$= e^{2x} \frac{2D-4}{-4-16} \sin x$$

$$= \frac{-1}{20} e^{2x} (2D \sin x - 4 \sin x)$$

$$= \frac{-1}{10} e^{2x} (\cos x - 2 \sin x)$$

∴ General solution is

$$y = C.F + P.I$$

$$y = e^{x} (c_1 \cos 2x + c_2 \sin 2x) - \frac{1}{10} e^{2x} (\cos x - 2 \sin x)$$

***Example* 4:** Solve $\frac{d^3y}{dx^3} + 3\frac{d^2y}{dx^2} + 3\frac{dy}{dx} + y = e^{-x}$

***Solution*:** Given equation is

$$\frac{d^3y}{dx^3} + 3\frac{d^2y}{dx^2} + 3\frac{dy}{dx} + y = e^{-x}$$

$$(D^3 + 3D^2 + 3D + 1)\, y = e^{-x}$$

Auxiliary equation is

$$m^3 + 3m^2 + 3m + 1 = 0$$

$$\Rightarrow \quad (m + 1)^3 = 0$$

$$\Rightarrow \quad m = -1, -1, -1$$

∴ Complementary function is

$$C.F = (c_1 + c_2\, x + c_3\, x^2)e^{-x}$$

Particular integral is

$$P.I = \frac{1}{(D+1)^3} e^{-x}$$

$$= e^{-x} \frac{1}{(D-1+1)^3}.1$$

$$= e^{-x} \frac{1}{D^3}.1$$

$$= e^{-x} . \frac{x^3}{6}$$

∴ General solution is

$$y = C.F + P.I$$

$$y = (c_1 + c_2 x + c_3 x^2) e^{-x} + \frac{1}{6} x^3 e^{-x}$$

***Example* 5:** Solve $(D + 2)(D - 1)^3 y = e^x$

***Solution*:** Given equation is

$$(D + 2)(D - 1)^3 y = e^x$$

Auxiliary equation is

$$(m + 2)(m - 1)^3 = 0$$

$$m = -2, 1, 1, 1$$

∴ complementary function is $C\text{-}F = C_1 e^{-2x} + (c_2 + c_3 x + c_4 x^2) e^x$

Particular integral is

$$P. I = \frac{1}{(D-1)^3}\left(\frac{1}{D+2} e^x\right)$$

$$= \frac{1}{(D-1)^3}\left(\frac{1}{1+2} e^x\right)$$

$$= \frac{1}{3} \frac{1}{(D-1)^3} e^x$$

$$= \frac{1}{3} e^x \frac{1}{(D+1-1)^3}.1$$

$$= \frac{1}{3} e^x \frac{1}{(D)^3}.1$$

$$= \frac{1}{3} e^x \frac{x^3}{6}$$

$$= \frac{1}{18} x^3 e^x$$

$\therefore$ general solution is y = C.F + P.I.

$$y = c_1 e^{-2x} + (c_2 + c_3 x + c_4 x^2)e^x + \frac{1}{18} x^3 e^x$$

Exercise 14.5

Solve

1. $(D^2 - 2D - 3)y = e^{3x}$
2. $(D^3 + 3D^2 + 3D + 2)y = e^{-x} \sin x$
3. $(D^3 - 3D^2 + 3D - 1)\, y = (x + 1)\, e^x$
4. $(D^2 - 3D + 2)\, y = x\, e^x$
5. $\frac{d^2y}{dx^2} - 2\frac{dy}{dx} + y = x^2 e^{3x}$
6. $\frac{d^2y}{dx^2} - 4\frac{dy}{dx} + y = e^{2x} \sin 2x$
7. $(D^2 - 1)\, y = e^x \cos x$
8. $\frac{d^2y}{dx^2} - 4\frac{dy}{dx} + 12y = (x - 1)\, e^{2x}$
9. $(D + 1)^3\, y = x^2\, e^{-x}$
10. $(D^3 - 3D^2 + 3D - 1)\, y = xe^{-x} + e^x$

Answers

1. $y = c_1\, e^{-x} + c_2\, e^{3x} + \frac{1}{4}\, x\, e^{3x}$
2. $y = c_1\, e^{-2x} + e^{\frac{-x}{2}} \left(c_2 \cos\frac{\sqrt{3}}{2} x + c_3 \sin\frac{\sqrt{3}}{2} x \right) + \frac{1}{2} e^{-x} (\cos x + \sin x)$
3. $y = e^x \left(c_1 + c_2 x + c_3 x^2 + \frac{1}{24} x^4 + \frac{1}{6} x^3 \right)$
4. $y = c_1\, e^x + c_2\, e^{2x} - \frac{1}{2}\, x\, e^x\, (x + 2)$
5. $y = (c_1 + c_2\, x)\, e^x + \frac{1}{8}\, e^{3x}\, (2x^2 - 4x + 3)$

6. $y = e^{2x} (c_1 \cosh \sqrt{3} x + c_2 \sinh \sqrt{3} x) - \frac{1}{7} e^{2x} \sin 2x$

7. $y = c_1 e^x + c_2 e^{-x} + \frac{1}{5} e^x (2 \sin x - \cos x)$

8. $y = c_1 e^{2x} + c_2 e^{-6x} + \frac{1}{16} e^{2x} (x^2 - \frac{9}{4} x)$

9. $y = (c_1 + c_2 x + c_3 x^2) e^{-x} + \frac{1}{60} x^5 e^{-x}$

10. $y = e^x \left(c_1 + c_2 x + c_3 x^2 + \frac{1}{6} x^3 + \frac{1}{24} x^4 \right)$

Rule VI

When Q = x.v, where v is any function of x

$$P.I = \frac{1}{f(D)} x.v = x . \frac{1}{f(D)} v - \frac{f'(D)}{[f(D)]^2} v$$

or

$$\frac{1}{f(D)} x\, v = x \frac{1}{f(D)} v + \left\{ \frac{d}{dD} \frac{1}{f(D)} \right\} v$$

***Example* 1:** Solve $(D^2 + 2D + 1) y = x \cos x$

***Solution*:** Given equation is

$$(D^2 + 2D + 1) y = x \cos x$$

Auxiliary equation is

$$m^2 + 2m + 1 = 0$$

$\Rightarrow$ $(m + 1)^2 = 0$

$\Rightarrow$ $m = -1, -1$

Complementary function is

$$C.F = (c_1 + c_2 x) e^{-x}$$

Particular integral is

$$P.I = \frac{1}{D^2+2D+1} x \cos x$$

$$= x \frac{1}{D^2+2D+1} \cos x + \frac{d}{dD}\left(\frac{1}{D^2+2D+1}\right) \cos x$$

$$= x \frac{1}{-1+2D+1} \cos x + \frac{-2D-2}{(D^2+2D+1)^2} \cos x$$

$$= x \frac{1}{2D} \cos x - \frac{2D+2}{(-1+2D+1)^2} \cos x$$

$$= \frac{x}{2} \sin x - \frac{2D+2}{4D^2} \cos x$$

$$= \frac{x}{2} \sin x - \frac{D+1}{2(-1)} \cos x$$

$$= \frac{x}{2} \sin x + \frac{1}{2}(D+1) \cos x$$

$$= \frac{x}{2} \sin x + \frac{1}{2}(-\sin x + \cos x)$$

∴ General solution is

$$y = C.F + P.I$$

$$y = (c_1 + c_2 x) e^{-x} + \frac{x}{2} \sin x + \frac{1}{2}(\cos x - \sin x)$$

***Example* 2:** Solve $(D^4 + 2D^2 + 1) y = x^2 \cos x$

***Solution*:** Given equation is

$$(D^4 + 2D^2 + 1) y = x^2 \cos x$$

Auxiliary equation is

$$m^4 + 2m^2 + 1 = 0$$

$$\Rightarrow \quad (m^2 + 1)^2 = 0$$

$$\Rightarrow \quad m = \underline{+}i \, {+}i$$

∴ Complementary function is

$$C.F = (c_1 + c_2 x) \cos x + (c_3 + c_4 x) \sin x$$

Particular integral is

$$P.I = \frac{1}{(D^2+1)^2} x^2 \cos x$$

$$= \text{Real part of } \frac{1}{(D^2+1)^2} x^2 . e^{ix} \text{ as } e^{ix} = \cos x + i \sin x$$

$$= \text{Real part of } e^{ix} \frac{1}{[(D+i)^2+1]^2} x^2$$

$$= \text{Real part of } e^{ix} \frac{1}{(D^2+2iD)^2} x^2$$

$$= \text{Real part of } e^{ix} \frac{1}{(2iD)^2 (1+\frac{D^2}{2iD})^2} x^2$$

$$= \text{Real part of } e^{ix} \frac{1}{(-4D^2)} \left(1-\frac{Di}{2}\right)^{-2} x^2$$

$$= \text{Real part of } e^{ix} \left(\frac{-1}{4D^2}\right)\left(1+Di+\frac{2.3}{1.2}\frac{D^2 i^2}{4}....\right) x^2$$

$$= \text{Real part of } e^{ix} \left(\frac{-1}{4D^2}\right)\left(1+Di-\frac{3}{4}D^2+....\right) x^2$$

$$= \text{Real part of } e^{ix} \left(\frac{-1}{4D^2}\right)\left(x^2+2xi-\frac{3}{4}.2\right)$$

$$= \text{Real part of } \frac{e^{ix}}{-4}\left(\frac{x^4}{12}+\frac{ix^3}{3}-\frac{3}{2}.\frac{x^2}{2}\right)$$

$$= \text{Real part of } \left(\frac{\cos x + i\sin x}{-4}\right)\left(\frac{x^4}{12}+i\frac{x^3}{3}-\frac{3}{4}x^2\right)$$

$$= \left(\frac{3x^2}{16}-\frac{x^4}{48}\right)\cos x + \frac{x^3}{12}\sin x$$

$$= \left(\frac{9x^2-x^4}{48}\right) \cos x + \frac{x^3}{12}\sin x$$

$\therefore$ General solution is

$$y = C.F + P.I$$

$$y = (c_1 + c_2 x)\cos x + (c_3 + c_4 x)\sin x + \frac{1}{12}x^3 \sin x + \left(\frac{9x^2 - x^4}{48}\right)\cos x$$

***Example* 3:** Solve $\frac{d^2y}{dx^2} + 4y = x \sin x$

***Solution*:** Given equation is

$$\frac{d^2y}{dx^2} + 4y = x \sin x$$

$$(D^2 + 4)y = x \sin x$$

Auxiliary equation is

$$m^2 + 4 = 0$$

$$\Rightarrow \quad m = \pm 2i$$

$\therefore$ Complementary equation is

$$C.F = c_1 \cos 2x + c_2 \sin 2x$$

Particular integral is $\frac{1}{D^2+4} x \sin x$

$$P.I = x\frac{1}{D^2+4}\sin x + \frac{d}{dD}\frac{1}{D^2+4}\sin x$$

$$= x\frac{1}{-1+4}\sin x + \frac{-2D}{(D^2+4)^2}\sin x$$

$$= \frac{x}{3}\sin x - \frac{2D}{9}\sin x$$

$$= \frac{x}{3}\sin x - \frac{2}{9}\cos x$$

$\therefore$ General solution is

$$y = C.F + P.I$$

$$y = c_1 \cos 2x + c_2 \sin 2x + \frac{x}{3}\sin x - \frac{2}{9}\cos x$$

***Example* 4:** Solve $\frac{d^2y}{dx^2} + y = x^2 \sin 2x$

***Solution*:** Given equation is

$$\frac{d^2y}{dx^2} + y = x^2 \sin 2x$$

$$(D^2 + 1)y = x^2 \sin 2x$$

Auxiliary equation is

$$m^2 + 1 = 0$$

$$\Rightarrow \quad m = \pm i$$

$\therefore$ Complementary function is

$$C.F = c_1 \cos x + c_2 \sin x$$

Particular integral is

$$P.I = \frac{1}{D^2 + 1} x^2 \sin 2x$$

$$= \text{imaginary part of } \frac{1}{D^2 + 1} x^2 e^{2ix}$$

$$= \text{imaginary part of } e^{2ix} \frac{1}{(D + 2i)^2 + 1} x^2$$

$$= \text{imaginary part of } e^{2ix} \frac{1}{D^2 + 4iD - 3} x^2$$

$$= \text{imaginary part of } e^{2ix} \frac{1}{-3\left(1 - \frac{D^2 + 4iD}{3}\right)} x^2$$

$$= \text{imaginary part of } \frac{e^{2ix}}{-3}\left(1 - \frac{D^2 + 4iD}{3}\right)^{-1} x^2$$

$$= \text{imaginary part of}$$

$$\frac{e^{2ix}}{-3}\left(1 + \frac{D^2 + 4iD}{3} + \frac{D^4 + 8iD^3 - 16D^2}{9} + ...\right) x^2$$

$$= \text{imaginary part of } \frac{e^{2ix}}{-3}\left(x^2 + \frac{2+8ix}{3} - \frac{16}{9}.2\right)$$

$$= \text{imaginary part of } \frac{e^{2ix}}{-3}\left(x^2 + \frac{8ix}{3} - \frac{26}{9}\right)$$

$$= \text{imaginary part of } \frac{(\cos 2x + i\sin 2x)}{-3}(x^2 + \frac{8}{3}ix - \frac{26}{9})$$

$$= -\frac{1}{3}\left(\frac{8}{3}x\cos 2x + x^2\sin 2x - \frac{26}{9}\sin 2x\right)$$

$$= \frac{-1}{27}\ (24\,x\cos 2x + 9x^2\sin 2x - 26\sin 2x)$$

$\therefore$ General solution is y = C.F + P.I

$$y = c_1\cos x + c_2\sin x$$

$$-\frac{1}{27}\ (24\,x\cos 2x + 9x^2\sin 2x - 26\sin 2x)$$

Exercise 14.6

Solve the following Differential Equations

1. $(D^2 + 4)\,y = x\sin 2x$
2. $(D^2 - 1)\,y = x^2\cos x$
3. $(D^2 + 3D + 2)\,y = x\sin 2x$
4. $\frac{d^4y}{dx^4} - y = x\sin x$
5. $\frac{d^2y}{dx^2} + a^2y = x\cos ax$

Answers

1. $y = c_1\cos 2x + c_2\sin 2x - \frac{1}{8}x^2\cos 2x$
2. $y = c_1e^x + c_2e^{-x} + x\sin x + \frac{1}{2}(1 - x^2)\cos x$
3. $y = c_1e^{-x} + c_2e^{-2x} - \frac{30x - 7}{200}\cos 2x - \frac{5x - 12}{100}\sin 2x$

4. $y = c_1 e^{-x} + c_2 e^{x} + c_3 \cos x + c_4 \sin x + \frac{1}{8} (x^2 \cos x - 3x \sin x)$

5. $y = c_1 \cos ax + c_2 \sin ax + \frac{1}{4a^2} (x \cos ax + ax^2 \sin ax)$

Miscellaneous Examples

***Example* 1:** Solve $\frac{d^2y}{dx^2} - 2\frac{dy}{dx} + y = x e^x \sin x$

***Solution*:** Given equation is

$$\frac{d^2y}{dx^2} - 2\frac{dy}{dx} + y = x e^x \sin x$$

$$(D^2 - 2D + 1) y = x e^x \sin x$$

Auxiliary equation is

$$m^2 - 2m + 1 = 0$$

$$\Rightarrow \quad (m - 1)^2 = 0$$

$$\Rightarrow \quad m = 1, 1$$

$\therefore$ Complementary function is

$$C.F = (c_1 + c_2 x) e^x$$

Particular integral is

$$P.I = \frac{1}{D^2 - 2D + 1} x e^x \sin x$$

$$= \frac{1}{(D-1)^2} e^x . (x \sin x)$$

$$= e^x \frac{1}{(D+1-1)^2} x \sin x$$

$$= e^x \frac{1}{D^2} x \sin x$$

$$= e^x \frac{1}{D} \int x \sin x \, dx$$

$$= e^x \frac{1}{D} [-x \cos x + \sin x]$$

$$= e^x \int [-x \cos x + \sin x] dx$$

$$= e^x (-x \sin x - \cos x - \cos x)$$

$$= -e^x (x \sin x + 2 \cos x)$$

$\therefore$ General solution is

$$y = C.F + P.I$$

$$y = (c_1 + c_2 x) e^x - e^x(x \sin x + 2 \cos x)$$

***Example* 2:** Solve $\frac{d^2y}{dx^2} - 4\frac{dy}{dx} + 4y = 8x^2 e^{2x} \sin 2x$

***Solution*:** Given equation is

$$\frac{d^2y}{dx^2} - 4\frac{dy}{dx} + 4y = 8 x^2 e^{2x} \sin 2x$$

$$(D^2 - 4D + 4) y = 8 x^2 e^{2x} \sin 2x$$

The auxiliary equation is

$$m^2 - 4m + 4 = 0$$

$$\Rightarrow \quad (m - 2)^2 = 0$$

$$\Rightarrow \quad m = 2, 2$$

$\therefore$ Complementary function is

$$C.F = (c_1 + c_2 x) e^{2x}$$

Particular integral is

$$P.I = \frac{1}{D^2 - 4D + 4} 8 x^2 e^{2x} \sin 2x$$

$$= 8\frac{1}{(D-2)^2} e^{2x} (x^2 \sin 2x)$$

$$= 8 e^{2x} \frac{1}{(D+2-2)^2} x^2 \sin 2x$$

$$= 8 e^{2x} \frac{1}{D^2} x^2 \sin 2x$$

$$= 8\,e^{2x}\frac{1}{D}\int x^2 \sin 2x\,dx$$

$$= 8\,e^{2x}\frac{1}{D}\left(-x^2\frac{\cos 2x}{2}+\frac{x\sin 2x}{2}+\frac{\cos 2x}{4}\right)$$

$$= \frac{8}{2}e^{2x}\int\left(-x^2\cos 2x + x\sin 2x + \frac{\cos 2x}{2}\right)dx$$

$$= 4\,e^{2x}\left[-x^2\frac{\sin 2x}{2} - x\cos 2x + \frac{3}{4}\sin 2x\right]$$

$$= e^{2x}\,[-2x^2 \sin 2x - 4x \cos 2x + 3 \sin 2x]$$

∴ General solution is

$$y = C.F + P.I$$

$$y = (c_1 + c_2\,x)\,e^{2x} + e^{2x}\,[3 \sin 2x - 4x \cos 2x - 2x^2 \sin 2x]$$

***Example* 3:** Solve $(D^2 - 3D + 2)\,y = \cos hx$

***Solution*:** Given equation is

$$(D^2 - 3D + 2)\,y = \cos h\,x$$

Auxiliary equation is

$$m^2 - 3m + 2 = 0$$

$$\Rightarrow \quad (m-1)(m-2) = 0$$

$$\Rightarrow \quad m = 1, 2$$

∴ Complementary function is C.F $= c_1\,e^x + c_2\,e^{2x}$

Particular integral is

$$P.I = \frac{1}{D^2 - 3D + 2}\cos hx$$

$$= \frac{1}{D^2 - 3D + 2}\left(\frac{e^x + e^{-x}}{2}\right) \qquad \left[\because \cosh x = \frac{e^x + e^{-x}}{2}\right]$$

$$= \frac{1}{(D-1)(D-2)}\frac{e^x}{2} + \frac{1}{D^2 - 3D + 2}\frac{e^{-x}}{2}$$

$$= \frac{1}{2} \frac{1}{(D-1)(D-2)} e^{x} + \frac{1}{2} \frac{1}{(-1)^2 - 3(-1) + 2} e^{-x}$$

$$= \frac{-1}{2} \frac{1}{D-1} e^{x} + \frac{1}{12} e^{-x}$$

$$= \frac{-1}{2} e^{x} \frac{1}{D+1-1} .1 + \frac{1}{12} e^{-x}$$

$$= \frac{-1}{2} e^{x} \frac{1}{D} .1 + \frac{1}{12} e^{-x}$$

$$= \frac{-1}{2} x\, e^{x} + \frac{1}{12} e^{-x}$$

∴ General solution is

$$y = C.F + P.I$$

$$y = c_1 e^{x} + c_2 e^{2x} - \frac{1}{2} x\, e^{x} + \frac{1}{12} e^{-x}$$

CHAPTER 11

TRIGONOMETRY

Introduction

Trigonometry is the science of measuring triangles. It is a branch of mathematics which deals with the measurement of angles and the problems applied with angles.

System of Measurement of Angles: There are three system of measurement of angles.

1. **Sexagesimal system (Degree System):** In this system, an angle is measured in degree, minutes and seconds. A complete rotation describes 360°. We represent the degree, minutes and seconds by the symbols $0, ', ''$ respectively.

   $\therefore$ 1 right angle $= 90^\circ$

   1 degree $(1^\circ) = 60'$ (60 minutes)

   1 minute $(1') = 60''$ (60 seconds)

2. **Centesimal System (Grade System):** In this system, right angle is divided into 100 equal parts, each part is called a grade. Each grade is subdivided into 100 equal parts, called minutes and each minutes are divided into 100 equal parts, called seconds.

   1 right angle $= 100^g$

   1 grade $(1^g) = 100'$

   1 minute $(1') = 100''$

3. **Circular system (Radian System):** In this system an angle is measured in radians. A radian is an angle subtended at the centre of a circle by an arc whose length is equal to the radius of the circle. 1 radian $= 1^C$.

Relation between degree and radians: Let r be the radius of circle and l be the length of arc. Then angle $(\theta) = \dfrac{\text{length of arc } (l)}{\text{radius (r)}}$ radians.

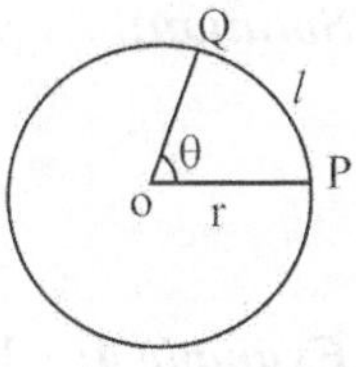

here π radian = 180°

$$\textbf{1 radian} = \frac{180^\circ}{\pi}, \quad 1^\circ = \frac{\pi}{180} \textbf{ radian}, \quad \pi \textbf{ radian} = 200^g = 180^\circ$$

where $\pi = \dfrac{22}{7}$

***Example* 1:** Express the following angles in radian measure: (i) 45° (ii) 18° 20'

***Solution*:** we know that 180° = π radians

(i) $45^\circ = \dfrac{\pi}{180} \times 45 = \dfrac{\pi}{4}$ radians

(ii) $18^\circ 20' = 18\dfrac{20}{60}$ degree $= \left(\dfrac{55}{3}\right)^\circ$

$\because$ 180° = π radians

$$\left(\frac{55}{3}\right)^\circ = \frac{\pi}{180} \times \frac{55}{3} = \frac{11\pi}{108} \text{ radians}$$

***Example* 2:** Find the degree measure corresponding to the following radian measure:

(i) $\left(\dfrac{\pi}{4}\right)^C$ (ii) $(-2)^C$

Solution: we know that $\pi^c = 180^\circ$

(i) $\left(\dfrac{\pi}{4}\right)^c = \dfrac{180}{\pi} \times \dfrac{\pi}{4} = 45^\circ$

(ii) $(-2)^c = \dfrac{180}{\pi} \times (-2)$ degree

$= -\dfrac{180 \times 2 \times 7}{22}$ degree

$= -114.54^\circ$

$= -144^\circ\ 32'\ 24''$ $\quad \because \begin{pmatrix} 0.54^\circ = 54 \times 60' = 32.40' \\ 0.40' = 0.40 \times 60 = 24'' \end{pmatrix}$

Example 3: Change 80^g to degree

Solution: we know that $1^g = \left(\frac{90}{100}\right)^0$

$$80^g = \frac{90}{100} \times 80 = 72^0$$

***Example* 4:** Find the radius of a circle in which a central angle of 45^0 intercepts an arc of 187 cm.

Solution: Let r be the radius of circle. Let $l = 187$

$$\theta = 45^o \quad \Rightarrow \quad \frac{\pi}{180} \times 45 = \frac{\pi}{4} \text{ radian}$$

Now $$r = \frac{l}{\theta} \Rightarrow r = \frac{187 \times 4}{\pi} \Rightarrow r = \frac{187 \times 4 \times 7}{22} = 238 \text{ cm}$$

***Example* 5:** A train is traveling on a curve 700 m radius at 14 km/hr. Through what angle will it return in one minute.

Solution: Given, radius = 700 m
Speed of train = 14 km/h

$$\therefore \text{ distance covered in 1 minute } = \frac{14 \times 1000}{60} \text{ m}$$

$$l = \frac{700}{3} \text{ m}$$

Now, we know that $\theta = \frac{l}{r}$

$$= \frac{700}{3} \times \frac{1}{700} \text{ radians}$$

$$\theta = \frac{1}{3} \text{ radian}$$

Trigonometrical ratio or Trigonometrical Functions

In a ΔABC, BC = Perpendicular (p)

AB = base (B), AC = Hypotenuse (H)

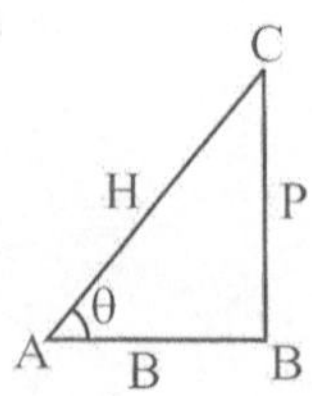

$$\therefore \quad \sin\theta = \frac{P}{H}, \quad \text{cosec}\,\theta = \frac{H}{P}$$

$$\cos\theta = \frac{B}{H}, \quad \sec\theta = \frac{H}{B}$$

$$\tan\theta = \frac{P}{B}, \quad \cot\theta = \frac{B}{P}$$

Fundamental Identities

1. $\sin\theta = \dfrac{1}{\operatorname{cosec}\theta}$
2. $\cos\theta = \dfrac{1}{\sec\theta}$
3. $\tan\theta = \dfrac{1}{\cot\theta}$
4. $\sin^2\theta + \cos^2\theta = 1$
5. $\sin^2\theta = 1 - \cos^2\theta$
6. $\cos^2\theta = 1 - \sin^2\theta$
7. $1 + \tan^2\theta = \sec^2\theta$
8. $\sec^2\theta - \tan^2\theta = 1$
9. $1 + \cot^2\theta = \operatorname{cosec}^2\theta$
10. $\operatorname{cosec}^2\theta - \cot^2\theta = 1$

***Example* 6:**

(i) if $\sin\theta = \dfrac{3}{5}$, find the value of $\cos\theta$ and $\tan\theta$.

(ii) If $\tan A = \dfrac{3}{4}$, find the value of $3\sin A + 4\cos A$

Solution:

(i)

$$\sin\theta = \frac{3}{5},\ \cos^2\theta = 1 - \sin^2\theta$$

$$\cos\theta = \pm\sqrt{1-\sin^2\theta}$$

$$= \pm\sqrt{1-\left(\frac{3}{5}\right)^2}$$

$$= \pm\sqrt{1-\frac{9}{25}}$$

$$= \pm\sqrt{\frac{16}{25}} = \pm\frac{4}{5}$$

$$\tan\theta = \frac{\sin\theta}{\cos\theta} = \frac{\frac{3}{5}}{\frac{4}{5}} = \frac{3}{4}$$

(ii) $\tan A = \frac{3}{4}$,

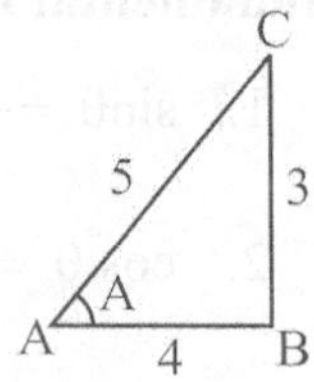

$AC^2 = AB^2 + BC^2$

$= (4)^2 + (3)^2$

$= 16 + 9 = 25$

$AC = 5$

$\therefore \quad \sin A = \frac{3}{5}, \cos A = \frac{4}{5}$

$\therefore \quad 3 \sin A + 4 \cos A = 3 \times \frac{3}{5} + 4 \times \frac{4}{5}$

$= \frac{9}{5} + \frac{16}{5} = 5$

***Example* 7:** Prove that $\frac{1 - \cos A}{\sin A} = \frac{\sin A}{1 + \cos A}$

Solution:

$$\text{L.H.S.} = \frac{1 - \cos A}{\sin A} = \frac{1 - \cos A}{\sin A} \times \frac{1 + \cos A}{1 + \cos A}$$

$$= \frac{1 - \cos^2 A}{\sin A(1 + \cos A)}$$

$$= \frac{\sin^2 A}{\sin A(1 + \cos A)}$$

$$= \frac{\sin A}{1 + \cos A} = \text{R.H.S}$$

Example 8: Prove that $\sqrt{\frac{1 + \cos A}{1 - \cos A}} = \text{cosec } A + \cot A$

Solution:

$$\text{L.H.S} = \sqrt{\frac{1 + \cos A}{1 - \cos A}} = \sqrt{\frac{1 + \cos A}{1 - \cos A} \times \frac{1 + \cos A}{1 + \cos A}}$$

$$= \sqrt{\frac{(1 + \cos A)^2}{1 - \cos^2 A}}$$

$$= \sqrt{\frac{(1 + \cos A)^2}{\sin^2 A}}$$

$$= \frac{1 + \cos A}{\sin A}$$

$$= \frac{1}{\sin A} + \frac{\cos A}{\sin A}$$

$$\mathbf{= \text{cosec } A + \cot A = \text{R.H.S}}$$

***Example* 9:** If $\tan\theta + \sin\theta = m$ and $\tan\theta - \sin\theta = n$, then prove that $m^2 - n^2 = 4\sqrt{mn}$.

Solution: L.H.S $= m^2 - n^2 = (\tan\theta + \sin\theta)^2 - (\tan\theta - \sin\theta)^2$

$$= \tan^2\theta + \sin^2\theta + 2\tan\theta\sin\theta - (\tan^2\theta + \sin^2\theta - 2\tan\theta\sin\theta)$$

$$= 4\tan\theta\sin\theta$$

R.H.S $= 4\sqrt{mn} = 4\sqrt{(\tan\theta + \sin\theta)(\tan\theta - \sin\theta)}$

$$= 4\sqrt{\tan^2\theta - \sin^2\theta}$$

$$= 4\sqrt{\frac{\sin^2\theta}{\cos^2\theta} - \sin^2\theta}$$

$$= 4\sqrt{\sin^2\theta\left(\frac{1}{\cos^2\theta} - 1\right)}$$

$$= 4\sin\theta\sqrt{\sec^2\theta - 1}$$

$$= 4\sin\theta\sqrt{\tan^2\theta}$$

$$= 4\sin\theta\tan\theta$$

$\therefore$ **L.H.S = R.H.S**

***Example* 10:** If $\cos\theta + \sin\theta = \sqrt{2}\cos\theta$, Prove that $\cos\theta - \sin\theta = \sqrt{2}\sin\theta$

Solution: Given $\cos\theta + \sin\theta = \sqrt{2}\cos\theta$

Squaring on both the sides $(\cos\theta + \sin\theta)^2 = (\sqrt{2}\cos\theta)^2$

$$\cos^2\theta + \sin^2\theta + 2\sin\theta\cos\theta = 2\cos^2\theta$$

$$2\sin\theta\cos\theta = 2\cos^2\theta - \cos^2\theta - \sin^2\theta$$

$$2\sin\theta\cos\theta = \cos^2\theta - \sin^2\theta$$

$$2\sin\theta\cos\theta = (\cos\theta - \sin\theta)(\cos\theta + \sin\theta)$$

$$2\sin\theta\cos\theta = (\cos\theta - \sin\theta)\sqrt{2}\cos\theta$$

$$(\because \cos\theta + \sin\theta = \sqrt{2}\cos\theta)$$

$$\cos\theta - \sin\theta = \frac{2\sin\theta\cos\theta}{\sqrt{2}\cos\theta}$$

$$\cos\theta - \sin\theta = \sqrt{2}\sin\theta$$

***Example* 11:** Solve $\frac{\tan\theta+\sec\theta-1}{\tan\theta-\sec\theta+1}=\frac{1+\sin\theta}{\cos\theta}$

Solution:

$$\text{L.H.S} = \frac{\tan\theta+\sec\theta-1}{\tan\theta-\sec\theta+1}$$

$$= \frac{(\tan\theta+\sec\theta)-(\sec^2\theta-\tan^2\theta)}{\tan\theta-\sec\theta+1} \quad (\because \sec^2\theta-\tan^2\theta=1)$$

$$= \frac{(\tan\theta+\sec\theta)-(\sec\theta-\tan\theta)(\sec\theta+\tan\theta)}{\tan\theta-\sec\theta+1}$$

$$= \frac{(\tan\theta+\sec\theta)(1-\sec\theta+\tan\theta)}{(\tan\theta-\sec\theta+1)}$$

$$= \tan\theta + \sec\theta$$

$$= \frac{\sin\theta}{\cos\theta}+\frac{1}{\cos\theta}$$

$$= \frac{\mathbf{1+\sin\theta}}{\mathbf{\cos\theta}} = \mathbf{R.H.S}$$

Trigonometric Table

Angle / T-ratio	0^0	$30^0\left(\frac{\pi}{6}\right)$	$45^0\left(\frac{\pi}{4}\right)$	$60^0\left(\frac{\pi}{3}\right)$	$90^0\left(\frac{\pi}{2}\right)$
sin θ	0	$\frac{1}{2}$	$\frac{1}{\sqrt{2}}$	$\frac{\sqrt{3}}{2}$	1
cos θ	1	$\frac{\sqrt{3}}{2}$	$\frac{1}{\sqrt{2}}$	$\frac{1}{2}$	0
tan θ	0	$\frac{1}{\sqrt{3}}$	1	$\sqrt{3}$	∞
cot θ	∞	$\sqrt{3}$	1	$\frac{1}{\sqrt{3}}$	0
sec θ	1	$\frac{2}{\sqrt{3}}$	$\sqrt{2}$	2	∞
cosec θ	∞	2	$\sqrt{2}$	$\frac{2}{\sqrt{3}}$	1

***Example* 12:** Find the value of $\sin^2 45^0 + \cos^2 45^0 + \tan^2 30^0$

Solution:

$$\sin^2 45^0 + \cos^2 45^0 + \tan^2 30^0 = \left(\frac{1}{\sqrt{2}}\right)^2 + \left(\frac{1}{\sqrt{2}}\right)^2 + \left(\frac{1}{\sqrt{3}}\right)^2$$

$$= \frac{1}{2} + \frac{1}{2} + \frac{1}{3}$$

$$= \frac{3+3+2}{6}$$

$$= \frac{8}{6} = \frac{4}{3}$$

***Example* 13:** If $\tan^2\theta + \sec\theta = 5$. Find $\cos\theta$.

Solution: Given $\tan^2\theta + \sec\theta = 5$

$$\sec^2\theta - 1 + \sec\theta = 5 \qquad (\because \tan^2\theta = \sec^2\theta - 1)$$

$$\sec^2\theta + \sec\theta - 6 = 0$$

$$\sec^2\theta + 3\sec\theta - 2\sec\theta - 6 = 0$$

$$\sec\theta(\sec\theta + 3) - 2(\sec\theta + 3) = 0$$

$$(\sec\theta + 3)(\sec\theta - 2) = 0$$

$$\sec\theta + 3 = 0 \quad \text{or} \quad \sec\theta - 2 = 0$$

$$\sec\theta = -3 \quad \text{or} \quad \sec\theta = 2$$

$$\cos\theta = -\frac{1}{3} \quad \text{or} \quad \cos\theta = \frac{1}{2}$$

***Example* 14:** If $A = 30^0$, verify that $\cos 3A = 4\cos^3 A - 3\cos A$.

Solution: Given $A = 30^0$

$$\text{L.H.S} = \cos 3A = \cos 3 \times 30^0$$

$$= \cos 90^0$$

$$= 0$$

$$\text{R.H.S.} = 4\cos^3 A - 3\cos A = 4\cos^3 30^0 - 3\cos 30^0$$

$$= 4\left(\frac{\sqrt{3}}{2}\right)^3 - 3\times\frac{\sqrt{3}}{2}$$

$$= 4\times\frac{3\sqrt{3}}{8} - \frac{3\sqrt{3}}{2}$$

$$= \frac{3\sqrt{3}}{2} - \frac{3\sqrt{3}}{2}$$

$$= \mathbf{0}$$

$\therefore$ L.H.S = R.H.S

Trigonometric Functions of angles

II quadrant Station i.e. sin and cosec are positive	I quadrant All i.e. all positive
III quadrant Ticket i.e. tan and cot are positive	IV quadrant Collector i.e. cos and sec are positive

Note : To memories the signs of T-ratios in different quadrant. "All stations Ticket Collector".

1. In Ist quadrant, all the trigonometric functions are positive with angle $(90 - \theta)$ and $(360 + \theta)$
2. In IInd quadrant, only $\sin\theta$ and $\operatorname{cosec}\theta$ are positive with angle $(90 + \theta)$ and $(180 - \theta)$.
3. In IIIrd quadrant, only $\tan\theta$ and $\cot\theta$ are positive with angle $(180 + \theta)$ and $(270 - \theta)$.
4. In IVth quadrant, only $\cos\theta$ and $\sec\theta$ are positive with angle $(270 + \theta)$ and $(360 - \theta)$.
   (i) In case of 90^0 and 270^0, the trigonometric functions will be changed i.e., $\sin\theta \to \cos\theta$, $\tan\theta \to \cot\theta$, $\operatorname{cosec}\theta \to \sec\theta$.

   (ii) In case of 180^0 and 360^0, the trigonometric functions will not be changed i.e., $\sin\theta \to \sin\theta$, $\cos\theta \to \cos\theta$, $\tan\theta \to \tan\theta$ etc.

T-ratio / Angle	sin	cos	tan	cot	cosec	sec
$-\theta$	$-\sin\theta$	$\cos\theta$	$-\tan\theta$	$-\cot\theta$	$-\text{cosec}\,\theta$	$\sec\theta$
$90^0-\theta$	$\cos\theta$	$\sin\theta$	$\cot\theta$	$\tan\theta$	$\sec\theta$	$\text{cosec}\,\theta$
$90^0+\theta$	$\cos\theta$	$-\sin\theta$	$-\cot\theta$	$-\tan\theta$	$\sec\theta$	$-\text{cosec}\,\theta$
$180^0-\theta$	$\sin\theta$	$-\cos\theta$	$-\tan\theta$	$-\cot\theta$	$\text{cosec}\,\theta$	$-\sec\theta$
$180^0+\theta$	$-\sin\theta$	$-\cos\theta$	$\tan\theta$	$\cot\theta$	$-\text{cosec}\,\theta$	$-\sec\theta$
$270^0-\theta$	$-\cos\theta$	$-\sin\theta$	$\cot\theta$	$\tan\theta$	$-\sec\theta$	$-\text{cosec}\,\theta$
$270^0+\theta$	$-\cos\theta$	$\sin\theta$	$-\cot\theta$	$-\tan\theta$	$-\sec\theta$	$\text{cosec}\,\theta$
$360^0-\theta$	$-\sin\theta$	$\cos\theta$	$-\tan\theta$	$-\cot\theta$	$-\text{cosec}\,\theta$	$\sec\theta$
$360^0+\theta$	$\sin\theta$	$\cos\theta$	$\tan\theta$	$\cot\theta$	$\text{cosec}\,\theta$	$\sec\theta$

***Example* 15:** Find the following (i) sin (1920^0) (ii) tan (765^0) (iii) cosec (870^0)

Solution: (i)

$$\sin(1920^0) = \sin(360^0 \times 5 + 120^0)$$
$$= \sin 120^0$$
$$= \sin(90^0 + 30^0)$$
$$= \cos 30^0 = \frac{\sqrt{3}}{2}$$

(ii)

$$\tan(765^0) = \tan(360^0 \times 2 + 45^0)$$
$$= \tan 45^0 = 1$$

(iii)

$$\text{cosec}(870^0) = \text{cosec}(360^0 \times 2 + 150^0)$$
$$= \text{cosec}\,150^0$$
$$= \text{cosec}(90^0 + 60^0)$$
$$= \mathbf{\sec 60^0 = 2}$$

***Example* 16:** Prove that $\sin(270^0-\theta)\sin(90^0-\theta) - \cos(270^0-\theta)\cos(90^0+\theta) + 1 = 0$

Solution: L.H.S $= \sin(270^0-\theta)\sin(90^0-\theta) - \cos(270^0-\theta)\cos(90^0+\theta) + 1$

$$= -\cos\theta \, . \, \cos\theta - (-\sin\theta)(-\sin\theta) + 1$$
$$= -\cos^2\theta - \sin^2\theta + 1$$
$$= -(\cos^2\theta + \sin^2\theta) + 1$$
$$= -1 + 1 \qquad (\because \sin^2\theta + \cos^2\theta = 1)$$
$$= \mathbf{0}$$

***Example* 17:** Prove that $\sin 60^0 \cos 30^0 - \cos 150^0 \sin 120^0 = \frac{3}{2}$

Solution: $\text{L.H.S} = \sin 60^0 \cos 30^0 - \cos 150^0 \sin 120^0$

$$= \sin 60^0 \cos 30^0 - \cos (90^0 + 60^0) \sin (90^0 + 30^0)$$

$$= \sin 60^0 \cos 30^0 - (-\sin 60^0) \cos 30^0$$

$$= \frac{\sqrt{3}}{2} \times \frac{\sqrt{3}}{2} + \frac{\sqrt{3}}{2} \times \frac{\sqrt{3}}{2}$$

$$= \mathbf{\frac{3}{4} + \frac{3}{4} = \frac{3}{2}} = \text{R.H.S}$$

***Example* 18:** Prove that $\frac{\tan(360^0 + \theta) + \cot(90^0 - \theta)}{1 + \tan(180^0 - \theta)\cot(270^0 + \theta)} = 2 \sin \theta \cos \theta$

Solution:

$$\text{L.H.S} = \frac{\tan(360^0 + \theta) + \cot(90^0 - \theta)}{1 + \tan(180^0 - \theta)\cot(270^0 + \theta)}$$

$$= \frac{\tan\theta + \tan\theta}{1 + (-\tan\theta)(-\tan\theta)}$$

$$= \frac{2\tan\theta}{1 + \tan^2\theta}$$

$$= \frac{2\tan\theta}{\sec^2\theta}$$

$$= \frac{\frac{2\sin\theta}{\cos\theta}}{\frac{1}{\cos^2\theta}}$$

$$\mathbf{= 2 \sin \theta \cos \theta = R.H.S}$$

Trigonometric ratios of sum and difference of angles

1. $\sin(A + B) = \sin A \cos B + \cos A \sin B$
2. $\sin (A - B) = \sin A \cos B - \cos A \sin B$
3. $\cos (A + B) = \cos A \cos B - \sin A \sin B$
4. $\cos (A - B) = \cos A \cos B + \sin A \sin B$

5. $\tan(A+B) = \dfrac{\tan A + \tan B}{1 - \tan A \tan B}$

6. $\tan(A-B) = \dfrac{\tan A - \tan B}{1 + \tan A \tan B}$

7. $\cot(A+B) = \dfrac{\cot A \cot B - 1}{\cot A + \cot B}$

8. $\cot(A-B) = \dfrac{\cot A \cot B + 1}{\cot A - \cot B}$

9. $\sin(A+B) . \sin(A-B) = \sin^2 A - \sin^2 B = \cos^2 B - \cos^2 A$

10. $\cos(A+B) . \cos(A-B) = \cos^2 A - \sin^2 B = \cos^2 B - \sin^2 A$

11. $\tan(A+B+C) = \dfrac{\tan A + \tan B + \tan C - \tan A \tan B \tan C}{1 - \tan A \tan B - \tan B \tan C - \tan C \tan A}$

***Example* 19:** Obtain the values of $\sin 15^0$, $\cos 15^0$, $\sin 75^0$, $\cos 75^0$.

Solution: (i)

$$\sin 15^0 = \sin(45^0 - 30^0)$$
$$= \sin 45^0 \cos 30^0 - \cos 45^0 \sin 30^0$$
$$= \frac{1}{\sqrt{2}} \times \frac{\sqrt{3}}{2} - \frac{1}{\sqrt{2}} \times \frac{1}{2} = \frac{\sqrt{3}-1}{2\sqrt{2}}$$

(ii)

$$\cos 15^0 = \cos(45^0 - 30^0)$$
$$= \cos 45^0 \cos 30^0 + \sin 45^0 \sin 30^0$$
$$= \frac{1}{\sqrt{2}} \times \frac{\sqrt{3}}{2} + \frac{1}{\sqrt{2}} \times \frac{1}{2} = \frac{\sqrt{3}+1}{2\sqrt{2}}$$

(iii)

$$\sin 75^0 = \sin(45^0 + 30^0)$$
$$= \sin 45^0 \cos 30^0 + \cos 45^0 \sin 30^0$$
$$= \frac{1}{\sqrt{2}} \times \frac{\sqrt{3}}{2} + \frac{1}{\sqrt{2}} \times \frac{1}{2} = \frac{\sqrt{3}+1}{2\sqrt{2}}$$

(iv)

$$\cos 75^0 = \cos(45^0 + 30^0)$$
$$= \cos 45^0 \cos 30^0 - \sin 45^0 \sin 30^0$$
$$= \frac{1}{\sqrt{2}} \times \frac{\sqrt{3}}{2} - \frac{1}{\sqrt{2}} \times \frac{1}{2} = \frac{\sqrt{3}-1}{2\sqrt{2}}$$

***Example* 20:** Prove that $\sin 105^0 + \cos 105^0 = \cos 45^0$

Solution: L.H.S $= \sin 105^0 + \cos 105^0$

$$= \sin(60^0 + 45^0) + \cos(60^0 + 45^0)$$

$$= \sin 60^0 \cos 45^0 + \cos 60^0 \sin 45^0 + \cos 60^0 \cos 45^0 - \sin 60^0 \sin 45^0$$

$$= \frac{\sqrt{3}}{2} \times \frac{1}{\sqrt{2}} + \frac{1}{2} \times \frac{1}{\sqrt{2}} + \frac{1}{2} \times \frac{1}{\sqrt{2}} - \frac{\sqrt{3}}{2} \times \frac{1}{\sqrt{2}}$$

$$= \frac{1}{2\sqrt{2}} + \frac{1}{2\sqrt{2}} = \frac{2}{2\sqrt{2}}$$

$$= \frac{\mathbf{1}}{\sqrt{\mathbf{2}}} = \mathbf{\cos 45^0} = \textbf{R.H.S}$$

***Example* 21:** If $A + B = 45^0$, Prove that $(1 + \tan A)(1 + \tan B) = 2$

Solution: Given $A + B = 45^0$

$$\Rightarrow \quad \tan(A + B) = \tan 45^0$$

$$\frac{\tan A + \tan B}{1 - \tan A \tan B} = 1$$

$$\tan A + \tan B = 1 - \tan A \tan B$$

$$1 + \tan A + \tan B + \tan A \tan B = 2$$

$$(1 + \tan A) + \tan B(1 + \tan A) = 2$$

$$\mathbf{(1 + \tan A)(1 + \tan B) = 2}$$

***Example* 22:** Prove that $\sin(A + B) \sin(A - B) = \sin^2 A - \sin^2 B = \cos^2 B - \cos^2 A$

Solution: L.H.S $= \sin(A + B) \sin(A - B)$

$$= (\sin A \cos B + \cos A \sin B)(\sin A \cos B - \cos A \sin B)$$

$$= \sin^2 A \cos^2 B - \cos^2 A \sin^2 B$$

$$= \sin^2 A(1 - \sin^2 B) - (1 - \sin^2 A) \sin^2 B$$

$$= \sin^2 A - \sin^2 A \sin^2 B - \sin^2 B + \sin^2 A \sin^2 B$$

$$= \sin^2 A - \sin^2 B$$

$$= (1 - \cos^2 A) - (1 - \cos^2 B)$$

$$= \mathbf{\cos^2 B - \cos^2 A} = \textbf{R.H.S}$$

***Example* 23:** Prove that $\tan 5\theta - \tan 3\theta - \tan 2\theta = \tan 5\theta \tan 3\theta \tan 2\theta$.

Solution: Let $\tan 5\theta = \tan(3\theta + 2\theta)$

$$\tan 5\theta = \frac{\tan 3\theta + \tan 2\theta}{1 - \tan 3\theta . \tan 2\theta}$$

$$\tan 5\theta - \tan 5\theta \tan 3\theta \tan 2\theta = \tan 3\theta + \tan 2\theta$$

$$\mathbf{\tan 5\theta - \tan 3\theta - \tan 2\theta = \tan 5\theta \tan 3\theta \tan 2\theta}$$

Formulae to transform the product into sum or difference

1. $2 \sin A \cos B = \sin(A + B) + \sin(A - B)$
2. $2 \cos A \sin B = \sin(A + B) - \sin(A - B)$
3. $2 \cos A \cos B = \cos(A + B) + \cos(A - B)$
4. $2 \sin A \sin B = \cos(A - B) - \cos(A + B)$

Transforming of sums and difference into product

1. $\sin C + \sin D = 2 \sin \frac{C+D}{2} \cos \frac{C-D}{2}$
2. $\sin C - \sin D = 2 \cos \frac{C+D}{2} \sin \frac{C-D}{2}$
3. $\cos C + \cos D = 2 \cos \frac{C+D}{2} \cos \frac{C-D}{2}$
4. $\cos C - \cos D = 2 \sin \frac{C+D}{2} \sin \frac{D-C}{2}$

***Example* 24:** Prove that $\sin 20^0 \sin 40^0 \sin 60^0 \sin 80^0 = \frac{3}{16}$

Solution:

$$L.H.S = \sin 20^0 \sin 40^0 \sin 60^0 \sin 80^0$$

$$= \sin 20^0 \sin 40^0 \times \frac{\sqrt{3}}{2} \times \sin 80^0$$

$$= \frac{\sqrt{3}}{2} \sin 20^0 \times \frac{1}{2} (2 \sin 40^0 \sin 80^0)$$

$$= \frac{\sqrt{3}}{4} \sin 20^0 \; [\cos(80^0 - 40^0) - \cos(80^0 + 40^0)]$$

$$= \frac{\sqrt{3}}{4} \sin 20^0 \ [\cos 40^0 - \cos 120^0]$$

$$= \frac{\sqrt{3}}{4} \sin 20^0 \left(\cos 40^0 + \frac{1}{2} \right)$$

$$= \frac{\sqrt{3}}{8} \left[2 \sin 20^0 \cos 40^0 + \sin 20^0 \right]$$

$$= \frac{\sqrt{3}}{8} \ [\sin(40^0 + 20^0) - \sin(40^0 - 20^0) + \sin 20^0]$$

$$= \frac{\sqrt{3}}{8} \ [\sin 60^0 - \sin 20^0 + \sin 20^0]$$

$$= \frac{\sqrt{3}}{8} \ \sin 60^0 = \frac{\sqrt{3}}{8} \times \frac{\sqrt{3}}{2} = \frac{3}{16} = \text{R.H.S}$$

***Example* 25:** Prove that $\sin \dfrac{5\pi}{18} - \cos \dfrac{4\pi}{9} = \sqrt{3} \sin \dfrac{\pi}{9}$

Solution:

$$\text{L.H.S} = \sin \frac{5\pi}{18} - \cos \frac{4\pi}{9}$$

$$= \sin \frac{5\pi}{18} - \sin \left(\frac{\pi}{2} - \frac{4\pi}{9} \right) \quad \left[\because \cos\theta = \sin\left(\frac{\pi}{2} - \theta\right)\right]$$

$$= \sin \frac{5\pi}{18} - \sin \frac{\pi}{18}$$

$$= 2 \cos \frac{\frac{5\pi}{18} + \frac{\pi}{18}}{2} \sin \frac{\frac{5\pi}{18} - \frac{\pi}{18}}{2}$$

$$= 2 \cos \frac{\pi}{6} \sin \frac{\pi}{9}$$

$$= 2 \times \frac{\sqrt{3}}{2} \sin \frac{\pi}{9} = \sqrt{3} \sin \frac{\pi}{9} = \text{R.H.S}$$

***Example* 26:** Prove that $\dfrac{\sin A + \sin 3A}{\cos A + \cos 3A} = \tan 2A$

Solution:

$$\text{L.H.S.} = \frac{\sin A + \sin 3A}{\cos A + \cos 3A}$$

$$= \frac{2\sin\frac{A+3A}{2}\cos\frac{A-3A}{2}}{2\cos\frac{A+3A}{2}\cos\frac{A-3A}{2}}$$

$$= \mathbf{\frac{\sin 2A}{\cos 2A} = \tan 2A = R.H.S}$$

***Example* 27:** Prove that $= \dfrac{\cos 3\theta + 2\cos 5\theta + \cos 7\theta}{\cos\theta + 2\cos 3\theta + \cos 5\theta} = \dfrac{\cos 5\theta}{\cos 3\theta}$

Solution:

$$\text{L.H.S} = \frac{\cos 3\theta + \cos 7\theta + 2\cos 5\theta}{\cos 5\theta + \cos\theta + 2\cos 3\theta}$$

$$= \frac{2\cos\frac{7\theta+3\theta}{2}.\cos\frac{7\theta-3\theta}{2} + 2\cos 5\theta}{2\cos\frac{5\theta+\theta}{2}\cos\frac{5\theta-\theta}{2} + 2\cos 3\theta}$$

$$= \frac{2\cos 5\theta\cos 2\theta + 2\cos 5\theta}{2\cos 3\theta\cos 2\theta + 2\cos 3\theta}$$

$$= \frac{2\cos 5\theta(\cos 2\theta + 1)}{2\cos 3\theta(\cos 2\theta + 1)}$$

$$= \mathbf{\frac{\cos 5\theta}{\cos 3\theta} = R.H.S}$$

To express the trigonometrical ratios of angle 2A in terms of trigonometrical ratios of angle A.

1. $\sin 2A = 2 \sin A \cos A$
2. $\cos 2A = \cos^2 A - \sin^2 A$

   $= 2 \cos^2 A - 1$

   $= 1 - 2 \sin^2 A$
3. $\tan 2A = \dfrac{2\tan A}{1-\tan^2 A}$
4. $\sin 2A = \dfrac{2\tan A}{1+\tan^2 A}$

5. $\cos 2A = \dfrac{1-\tan^2 A}{1+\tan^2 A}$

6. $\sin^2 x = \dfrac{1-\cos 2x}{2}$

7. $\cos^2 x = \dfrac{1+\cos 2x}{2}$

Proof:

1. We know that $\sin(A + B) = \sin A \cos B + \cos A \sin B$

   Put $B = A$, we get $\sin 2A = 2 \sin A \cos A$

2. Let $\cos(A + B) = \cos A \cos B - \sin A \sin B$

   Put $B = A$, we get $\cos 2A = \cos^2 A - \sin^2 A$

   Now again $\cos 2A = \cos^2 A - (1 - \cos^2 A)$

   $\cos 2A = 2 \cos^2 A - 1$

   again $\cos 2A = \cos^2 A - \sin^2 A$

   $= (1 - \sin^2 A) - \sin^2 A$

   $\cos 2A = 1 - 2 \sin^2 A$

3. Let $\tan (A + B) = \dfrac{\tan A + \tan B}{1-\tan A \tan B}$

   Put $B = A$, we get $\tan 2A = = \dfrac{2\tan A}{1-\tan^2 A}$

4. Let $\sin 2A = 2 \sin A \cos A$

   $= \dfrac{2\sin A \cos A}{\sin^2 A+\cos^2 A}$

   dividing by $\cos^2 A$, we get

   $\sin 2A = = \dfrac{\dfrac{2\sin A}{\cos A}}{\dfrac{\sin^2 A}{\cos^2 A}+1}$

   $\sin 2A = \dfrac{2\tan A}{1+\tan^2 A}$

5. Let $\cos 2A = \cos^2 A - \sin^2 A$

   $= \dfrac{\cos^2 A-\sin^2 A}{\sin^2 A+\cos^2 A}$

dividing by $\cos^2 A$, we have

$$\cos 2A = \frac{1-\tan^2 A}{1+\tan^2 A}$$

Angle 3A in terms of angle A

1. $\sin 3A = 3 \sin A - 4 \sin^3 A$
2. $\cos 3A = 4 \cos^3 A - 3 \cos A$
3. $\tan 3A = \dfrac{3\tan A - \tan^3 A}{1-3\tan^2 A}$

***Example* 28:** Find the value of (i) $\sin 18^0$ (ii) $\cos 18^0$, (iii) $\sin 36^0$ (iv) $\cos 72^0$ (v) $\sin 54^0$

Solution:

(i) $\sin 18^0$

Let $A = 18^0$

$5A = 90^0$

$2A + 3A = 90^0$

$2A = 90^0 - 3A$

$\sin 2A = \sin(90^0 - 3A)$

$\sin 2A = \cos 3A$

$2 \sin A \cos A = 4 \cos^3 A - 3 \cos A$

$2 \sin A = 4 \cos^2 A - 3$

$2 \sin A = 4(1 - \sin^2 A) - 3$

$2 \sin A = 4 - 4\sin^2 A - 3$

$4 \sin^2 A + 2 \sin A - 1 = 0$

$$\sin A = \frac{-2 \pm \sqrt{4+16}}{8}$$

$$= \frac{-2 \pm \sqrt{20}}{8}$$

$$\sin A = \frac{-1 \pm \sqrt{5}}{4}$$

since 18^0 is acute angle, hence $\sin 18^0$ is not negative

$$\therefore \quad \sin 18^0 = = \frac{\sqrt{5}-1}{4}$$

(ii) $\cos 18^0$

Since $\cos\theta = \sqrt{1-\sin^2\theta}$

$$\cos 18^0 = \sqrt{1-\sin^2 18^0}$$

$$= \sqrt{1-\left(\frac{\sqrt{5}-1}{4}\right)^2}$$

$$= \sqrt{1-\frac{5+1-2\sqrt{5}}{16}}$$

$$= \sqrt{1-\frac{6-2\sqrt{5}}{16}}$$

$$= \sqrt{\frac{10+2\sqrt{5}}{16}}$$

$\therefore$ $\cos 18^0 = \frac{1}{4}\sqrt{10+2\sqrt{5}}$

(iii) $\cos 36^0$

Let $\cos 36^0 = \cos 2 \times 18^0 = 1 - 2\sin^2 18^0$

$$= 1-2\left(\frac{\sqrt{5}-1}{4}\right)^2$$

$$= 1-\frac{6-2\sqrt{5}}{8}$$

$$= \frac{2+2\sqrt{5}}{8} = \frac{\sqrt{5}+1}{4}$$

$\therefore$ $\sin 36^0 = \sqrt{1-\cos^2 36^0}$

$$= \sqrt{1-\left(\frac{\sqrt{5}+1}{4}\right)^2}$$

$$= \sqrt{1-\frac{6+2\sqrt{5}}{16}} = \frac{1}{4}\sqrt{10-2\sqrt{5}}$$

(iv) $\cos 72^0 = \cos(90^0 - 18^0)$

$$= \sin 18^0$$

$$= \frac{\sqrt{5}-1}{4}$$

(v) $\sin 54^0 = \sin(90^0 - 36^0)$

$= \cos 36^0$

$= \dfrac{\sqrt{5}+1}{4}$

***Example* 29:** Find the value of (i) $\sin 22\frac{1^0}{2}$ (ii) $\tan 22\frac{1^0}{2}$

Solution: (i) $\cos 45^0 = 1 - 2\sin^2 \dfrac{45^0}{2}$

$$\frac{1}{\sqrt{2}} = 1 - 2\sin^2 \frac{45^0}{2}$$

$$2\sin^2 \frac{45^0}{2} = 1 - \frac{1}{\sqrt{2}}$$

$$\text{Sin}^2 \frac{45^0}{2} = \frac{\sqrt{2}-1}{2\sqrt{2}} \times \frac{\sqrt{2}}{\sqrt{2}}$$

$$\sin^2 \frac{45^0}{2} = \frac{2-\sqrt{2}}{4}$$

$$\sin 22\frac{1^0}{2} = \sqrt{\frac{2-\sqrt{2}}{4}}$$

(ii)
$$\tan 45^0 = \frac{2\tan\frac{45^0}{2}}{1-\tan^2\frac{45^0}{2}} \qquad \left(\because \tan A = \frac{2\tan\frac{A}{2}}{1-\tan^2\frac{A}{2}}\right)$$

$$\Rightarrow \quad 1 = \frac{2\tan 22\frac{1^0}{2}}{1-\tan^2 22\frac{1^0}{2}}$$

$$\Rightarrow \quad 1 - \tan^2 22\frac{1^0}{2} = 2\tan 22\frac{1^0}{2}$$

$$\Rightarrow \quad \tan^2 22\frac{1^0}{2} + 2\tan 22\frac{1^0}{2} - 1 = 0$$

$\Rightarrow \quad \tan^2 22\frac{1}{2}^0 + 2\tan 22\frac{1}{2}^0 + 1 = 2$ (adding 2 on both sides)

$\Rightarrow \quad (\tan 22\frac{1}{2}^0 + 1)^2 = 2$

$\tan 22\frac{1}{2}^0 + 1 = \sqrt{2}$

$\tan 22\frac{1}{2}^0 = \sqrt{2} - 1$

***Example* 30:** Prove that 4 sin A sin(60 – A) sin(60 + A) = sin 3A

Solution:

$$\text{L.H.S} = 4 \sin A \sin (60 - A) \sin (60 + A)$$
$$= 4 \sin A (\sin^2 60 - \sin^2 A)$$
$$= 4 \sin A \left(\frac{3}{4} - \sin^2 A\right)$$
$$= 3 \sin A - 4 \sin^3 A$$
$$\mathbf{= \sin 3A = R.H.S}$$

***Example* 31:** If $0 \le x \le 2\pi$, find $\sin \frac{x}{2}$ and $\cos \frac{x}{2}$. When $\tan x = \frac{-4}{3}$. x lies in II quadrant.

Solution: Given $\tan x = \frac{-4}{3}$

$\cos x = \frac{-3}{5}$ for II quadrant, cos x is negative

Now $\sin\frac{x}{2} = \sqrt{\frac{1-\cos x}{2}}$ $\left(\because \sin^2\frac{x}{2} = \frac{1-\cos x}{2}\right)$

$$= \sqrt{\frac{1+\frac{3}{5}}{2}}$$

$$= \sqrt{\frac{8}{5\times 2}} = \sqrt{\frac{4}{5}} = \frac{2}{\sqrt{5}}$$

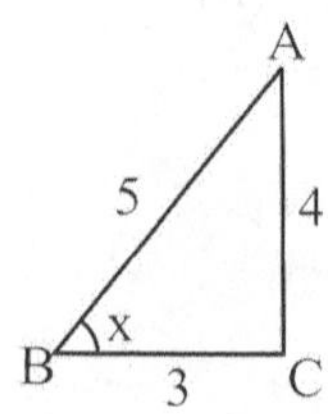

again, $\cos\frac{x}{2} = \sqrt{\frac{1+\cos x}{2}}$

$$= \sqrt{\frac{1-\frac{3}{5}}{2}}$$

$$= \sqrt{\frac{2}{5\times 2}} = \sqrt{\frac{1}{5}} = \frac{1}{\sqrt{5}}$$

***Example* 32:** If $\sin 2A = \frac{4}{5}$. Find the value of tan A

Solution: Given $\sin 2A = \frac{4}{5}$

$$\frac{2\tan A}{1+\tan^2 A} = \frac{4}{5}$$

$$10 \tan A = 4 + 4\tan^2 A$$

$$4\tan^2 A - 10\tan A + 4 = 0$$

$$4\tan^2 A - 8\tan A - 2\tan A + 4 = 0$$

$$4\tan A(\tan A - 2) - 2(\tan A - 2) = 0$$

$$\tan A - 2 = 0 \quad \text{or} \quad 4\tan A - 2 = 0$$

$$\tan A = 2 \quad \text{or} \quad \tan A = \frac{1}{2}$$

***Example* 33:** Prove that $\cos 4x = 1 - 8\sin^2 x\cos^2 x$

Solution: L.H.S = cos 4x

$$= \cos 2(2x)$$

$$= 1 - 2\sin^2 2x$$

$$= 1 - 2(2\sin x\cos x)^2$$

$$= 1 - 2(4\sin^2 x\cos^2 x)$$

$$= 1 - 8\sin^2 x\cos^2 x = \text{R.H.S}$$

***Example* 34:** If $2\cos A = x + \frac{1}{x}$ then prove that $2\cos 3A = x^3 + \frac{1}{x^3}$

Solution: $\cos 3A = 4\cos^3 A - 3\cos A$

$$= 4\left[\frac{1}{2}\left(x + \frac{1}{x}\right)\right]^3 - 3\left[\frac{1}{2}\left(x + \frac{1}{x}\right)\right]$$

$$= 4 \times \frac{1}{8}(x + \frac{1}{x})^3 - \frac{3}{2}(x + \frac{1}{x})$$

$$= \frac{1}{2}\left[x^3 + \frac{1}{x^3} + 3(x + \frac{1}{x}) - 3(x + \frac{1}{x})\right]$$

$$(\because (a+b)^3 = a^3 + b^3 + 3(a+b))$$

$$\cos 3A = \frac{1}{2}\left(x^3 + \frac{1}{x^3}\right)$$

$$2\cos 3A = x^3 + \frac{1}{x^3}$$

***Example* 35:** Prove that $\cos^2 48^0 - \sin^2 12^0 = \dfrac{\sqrt{5}+1}{8}$

Solution:

$$\text{L.H.S} = \cos^2 48^0 - \sin^2 12^0$$

$$= \frac{1}{2}\,[2\cos^2 48^0 - 2\sin^2 12^0] \qquad \left[\begin{array}{r} \because 2\cos^2 A = 1 + \cos 2A \\ 2\sin^2 A = 1 - \cos 2A \end{array}\right]$$

$$= \frac{1}{2}\,[1 + \cos 96^0 - (1 - \cos 24^0)]$$

$$= \frac{1}{2}\,[\cos 96^0 + \cos 24^0]$$

$$= \frac{1}{2}\left[2\cos\frac{96+24}{2}.\cos\frac{96-24}{2}\right]$$

$$= \cos 60^0 \cos 36^0$$

$$= \frac{1}{2}.\frac{\sqrt{5}+1}{4} = \frac{\sqrt{5}+1}{8} = \text{R.H.S}$$

***Example* 36:** Prove that $\sqrt{2+\sqrt{2+2\cos 4\theta}} = 2\cos\theta$

Solution:

$$\text{L.H.S} = \sqrt{2+\sqrt{2+2\cos 4\theta}}$$

$$= \sqrt{2+\sqrt{2+2(2\cos^2 2\theta - 1)}}$$

$$= \sqrt{2+\sqrt{4\cos^2 2\theta}}$$

$$= \sqrt{2+2\cos 2\theta}$$

$$= \sqrt{2+2(2\cos^2\theta - 1)}$$

$$= \sqrt{4\cos^2\theta}$$

$$= 2\cos\theta = \text{R.H.S}$$

***Example* 37:** Prove that $\dfrac{\cos A}{1-\tan A} + \dfrac{\sin A}{1-\cot A} = \sin A + \cos A$

Solution:

$$\text{L.H.S.} = \frac{\cos A}{1-\tan A} + \frac{\sin A}{1-\cot A}$$

$$= \frac{\cos A}{1 - \frac{\sin A}{\cos A}} + \frac{\sin A}{1 - \frac{\cos A}{\sin A}}$$

$$= \frac{\cos^2 A}{\cos A - \sin A} + \frac{\sin^2 A}{\sin A - \cos A}$$

$$= \frac{\cos^2 A}{\cos A - \sin A} - \frac{\sin^2 A}{\cos A - \sin A}$$

$$= \frac{\cos^2 A - \sin^2 A}{\cos A - \sin A}$$

$$= \frac{(\cos A - \sin A)(\cos A + \sin A)}{(\cos A - \sin A)}$$

$$= \sin A + \cos A = \text{R.H.S}$$

Example 38: If $\cot \theta = = \frac{9}{40}$, Find the value of sec θ and cosec θ.

Solution: Given $\cot \theta = \frac{9}{40} = \frac{BC}{AC}$

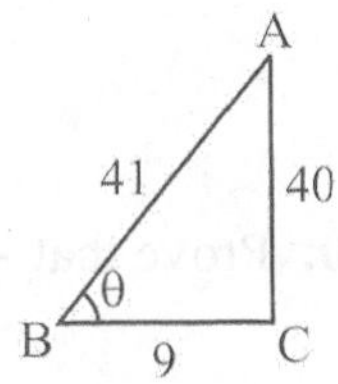

In ΔABC, $AB^2 = AC^2 + BC^2$

$= (40)^2 + (9)^2$

$= 1600 + 81$

$= 1681$

$AB = 41$

∴ $\sec\theta = \frac{AB}{BC} = \frac{41}{9}$

and $\mathbf{cosec\theta = \frac{AB}{AC} = \frac{41}{40}}$

Example 39: If $\tan \theta = \frac{2}{3}$, then find the values of sin θ + cos θ and sec θ + cosec θ.

Solution: Given $\tan \theta = \frac{2}{3} = \frac{AC}{BC}$

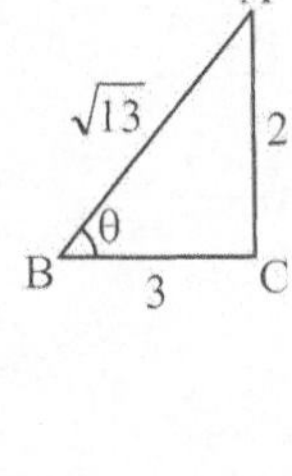

In ΔABC, $AB^2 = AC^2 + BC^2$

$= 4 + 9$

$AB = \sqrt{13}$

$\therefore \quad \sin\theta = \frac{2}{\sqrt{13}}, \qquad \sec\theta = = \frac{\sqrt{13}}{3}$

$\cos\theta = = \frac{3}{\sqrt{13}}, \qquad \operatorname{cosec}\theta = = \frac{\sqrt{13}}{2}$

$$\therefore \quad \sin\theta + \cos\theta = \frac{2}{\sqrt{13}} + \frac{3}{\sqrt{13}}$$

$$= \frac{5}{\sqrt{13}}$$

and $\quad \sec\theta + \operatorname{cosec}\theta = \frac{\sqrt{13}}{3} + \frac{\sqrt{13}}{2}$

$$= \frac{2\sqrt{13} + 3\sqrt{13}}{6}$$

$$= \frac{5}{6}\sqrt{13}$$

***Example* 40:** Prove that $\frac{\cos 2A}{1+\sin 2A} = \tan(45^0 - A)$

$$\text{L.H.S} = \frac{\cos 2A}{1+\sin 2A}$$

$$= \frac{\cos^2 A - \sin^2 A}{\sin^2 A + \cos^2 A + 2\sin A\cos A}$$

$$= \frac{(\cos A - \sin A)(\cos A + \sin A)}{(\cos A + \sin A)^2}$$

$$= \frac{\cos A - \sin A}{\cos A + \sin A}$$

divide by cos A, we have

$$= \frac{1 - \tan A}{1 + \tan A}$$

$$= \frac{\tan 45^0 - \tan A}{1 + \tan 45^0 \tan A} \qquad (\because \tan 45^o = 1)$$

$$= \tan(45^0 - A) = \text{R.H.S}$$

***Example* 41:** Prove that $= \frac{\sin 2A}{1-\cos 2A} = \cot A$

Solution:

$$\text{L.H.S} = \frac{\sin 2A}{1-\cos 2A}$$

$$= \frac{2\sin A\cos A}{1-(1-2\sin^2 A)}$$

$$= \frac{2\sin A\cos A}{2\sin^2 A}$$

$$= \frac{\cos A}{\sin A} = \cot A = \text{R.H.S}$$

***Example* 42:** Prove that $\cos 20^0 \cos 40^0 \cos 60^0 \cos 80^0 = \frac{1}{16}$

Solution:

$$\text{L.H.S} = \cos 20^0 \cos 40^0 \cos 60^0 \cos 80^0$$

$$= \cos 20^0 \cos 40^0 \,.\, \frac{1}{2} \,.\, \cos 80^0$$

$$= \frac{1}{2}\cos 20^0 (\cos 40^0 \cos 80^0)$$

$$= \frac{1}{4} \cos 20^0 (2 \cos 80^0 \cos 40^0)$$

$$= \frac{1}{4} \cos 20^0 [\cos 120^0 + \cos 40^0]$$

$$[\because 2 \cos A \cos B = \cos (A + B) + \cos (A - B)]$$

$$= \frac{1}{4}\cos 20^0 [-\frac{1}{2} + \cos 40^0]$$

$$= \frac{-1}{8}\cos 20^0 + \frac{1}{4}\cos 20^0 \cos 40^0$$

$$= \frac{-1}{8}\cos 20^0 + \frac{1}{8} (2 \cos 40^0 \cos 20^0)$$

$$= \frac{-1}{8}\cos 20^0 + \frac{1}{8}\cos 60^0 + \frac{1}{8}\cos 20^0$$

$$= \frac{1}{8}\cos 60^0 \qquad = \frac{1}{8}\times\frac{1}{2} = \frac{1}{16} = \text{R.H.S}$$

***Example* 43:** Prove that $\cos\frac{10\pi}{13}+\cos\frac{8\pi}{13}+\cos\frac{3\pi}{13}+\cos\frac{5\pi}{13}=0$

Solution:

$$\text{L.H.S} = \cos\frac{10\pi}{13}+\cos\frac{8\pi}{13}+\cos\frac{3\pi}{13}+\cos\frac{5\pi}{13}$$

$$=\left(\cos\frac{10\pi}{13}+\cos\frac{3\pi}{13}\right)+\left(\cos\frac{8\pi}{13}+\cos\frac{5\pi}{13}\right)$$

$$=2\cos\frac{13\pi}{26}\cos\frac{7\pi}{26}+2\cos\frac{13\pi}{26}\cos\frac{3\pi}{26}$$

$$\left[\because \cos c+\cos D = 2\cos\frac{C+D}{2}\cos\frac{C-D}{2}\right]$$

$$=2\cos\frac{\pi}{2}\cos\frac{7\pi}{26}+2\cos\frac{\pi}{2}\cos\frac{3\pi}{26}$$

$$=0+0=0=\text{R.H.S} \qquad \left(\because \cos\frac{\pi}{2}=0\right)$$

***Example* 44:** Prove that $\frac{\sec A-\tan A}{\sec A+\tan A} = 1-2\sec A\tan A+2\tan^2 A$

Solution:

$$\text{L.H.S} = \frac{\sec A-\tan A}{\sec A+\tan A}\times\frac{\sec A-\tan A}{\sec A-\tan A}$$

$$=\frac{(\sec A-\tan A)^2}{\sec^2 A-\tan^2 A}$$

$$=\frac{\sec^2 A+\tan^2 A-2\sec A\tan A}{1}$$

$$=1+\tan^2 A+\tan^2 A-2\sec A\tan A$$

$$=1+2\tan^2 A-2\sec A\tan A = \textbf{R.H.S}$$

***Example* 45:** If $\tan\theta = 1$, find the value of $\sec 2\theta$ and $\text{cosec}\, 2\theta$.

Solution:

since $\cos 2\theta = \frac{1-\tan^2\theta}{1+\tan^2\theta}$

$\therefore$ $\sec 2\theta = \frac{1+\tan^2\theta}{1-\tan^2\theta}$

Since $\tan\theta = 1$ $\therefore$ $\sec 2\theta = \frac{1+1}{1-1}=\frac{2}{0}=\infty$

Now, $\sin 2\theta = \frac{2\tan\theta}{1+\tan^2\theta}$

$\therefore \qquad \text{cosec } 2\theta = \dfrac{1+\tan^2\theta}{2\tan\theta}$

since $\qquad \tan\theta = 1$

$\therefore \qquad \text{cosec } 2\theta = \dfrac{1+1}{2\times 1} = \dfrac{2}{2} = 1$

***Example* 46:** Find the value of each of the following.

(i) $\cos 210^0$ (ii) $\tan \dfrac{11\pi}{5}$ (iii) $\sin \dfrac{5\pi}{3}$

Solution:

(i)
$$\cos 210^0 = \cos(180^0 + 30^0)$$
$$= -\cos 30^0$$
$$= \frac{-\sqrt{3}}{2}$$

(ii)
$$\tan \frac{11\pi}{5} = \tan \frac{11\times 180}{5}$$
$$= \tan 396^0$$
$$= \tan(360^0 + 36^0)$$
$$= \tan 36^0$$
$$= \frac{\sin 36^0}{\cos 36^0}$$
$$= \frac{\dfrac{\sqrt{10-2\sqrt{5}}}{4}}{\dfrac{\sqrt{5}+1}{4}} = \frac{\sqrt{10-2\sqrt{5}}}{\sqrt{5}+1}$$

(iii)
$$\sin \frac{5\pi}{3} = \sin \frac{5\times 180}{3}$$
$$= \sin 300^0$$
$$= \sin(360^0 - 60^0)$$
$$= -\sin 60^0$$
$$= -\frac{\sqrt{3}}{2}$$

***Example* 47:** Prove that $\dfrac{1-\cos 2A}{1+\cos 2A} = \tan^2 A$

Solution:

$$\text{L.H.S} = \frac{1-\cos 2A}{1+\cos 2A}$$

$$= \frac{1-(1-2\sin^2 A)}{1+2\cos^2 A-1}$$

$$= \frac{2\sin^2 A}{2\cos^2 A}$$

$$\mathbf{= \tan^2 A = R.H.S}$$

***Example* 48:** Prove that cotA + tan (180⁰ + A) + tan (90⁰ + A) + tan (360⁰ – A) = 0

Solution:

$$\text{L.H.S} = \cot A + \tan(180^0 + A) + \tan(90^0 + A) + \tan(360^0 - A)$$

$$= \cot A + \tan A - \cot A - \tan A$$

$$\mathbf{= 0 \qquad = R.H.S}$$

***Example* 49:** Prove that $\dfrac{\cos 11^0 + \sin 11^0}{\cos 11^0 - \sin 11^0} = \tan 56^0$

Solution:

$$\text{L.H.S} = \frac{\cos 11^0 + \sin 11^0}{\cos 11^0 - \sin 11^0}$$

dividing by cos 11^0, we have

$$= \frac{1+\tan 11^0}{1-\tan 11^0}$$

$$= \frac{\tan 45^0 + \tan 11^0}{1-\tan 45^0 \tan 11^0} \qquad (\because \tan 45^0 = 1)$$

$$= \tan(45^0 + 11^0)$$

$$\mathbf{= \tan 56^0 \qquad R.H.S}$$

***Example* 50:** Prove that $\dfrac{\cos 3A - \cos 5A}{\sin 5A - \sin 3A} = \tan 4A$

Proof:

$$\text{L.H.S} = \frac{\cos 3A - \cos 5A}{\sin 5A - \sin 3A}$$

$$= \frac{2\sin\frac{3A+5A}{2}\sin\frac{5A-3A}{2}}{2\cos\frac{5A+3A}{2}\sin\frac{5A-3A}{2}}$$

$$= \mathbf{\frac{\sin 4A}{\cos 4A} = \tan 4A} \qquad = \textbf{R.H.S}$$

***Example* 51:** Prove that $\sqrt{\dfrac{1+\sin A}{1-\sin A}} = \sec A + \tan A$

Solution:

$$\text{L.H.S} = \sqrt{\frac{1+\sin A}{1-\sin A}}$$

$$= \sqrt{\frac{(1+\sin A)}{1-\sin A)} \times \frac{(1+\sin A)}{(1+\sin A)}}$$

$$= \sqrt{\frac{(1+\sin A)^2}{1-\sin^2 A}}$$

$$= \frac{1+\sin A}{\cos A}$$

$$= \frac{1}{\cos A} + \frac{\sin A}{\cos A}$$

$$= \mathbf{\sec A + \tan A} \qquad = \textbf{R.H.S}$$

***Example* 52:** Prove that $\dfrac{1-\cos^2 A}{1-\sin^2 A} = \tan^2 A$

Solution:

$$\text{L.H.S} = \frac{1-\cos^2 A}{1-\sin^2 A}$$

$$= \frac{\sin^2 A}{\cos^2 A}$$

$$= \mathbf{\tan^2 A} \qquad = \textbf{R.H.S}$$

***Example* 53:** Prove that $\dfrac{1+\cos\theta+\sin\theta}{1+\cos\theta-\sin\theta}=\dfrac{1+\sin\theta}{\cos\theta}$

Solution: $\text{L.H.S} = \dfrac{1+\cos\theta+\sin\theta}{1+\cos\theta-\sin\theta}$

Dividing by cos θ, we get

$$= \frac{\sec\theta+1+\tan\theta}{\sec\theta+1-\tan\theta}$$

$$= \frac{(\sec\theta+\tan\theta)+(\sec^2\theta-\tan^2\theta)}{\sec\theta-\tan\theta+1}$$

$$= \frac{(\sec\theta+\tan\theta)\times(1+\sec\theta-\tan\theta)}{(1+\sec\theta-\tan\theta)}$$

$$= \sec\theta+\tan\theta$$

$$= \frac{1}{\cos\theta}+\frac{\sin\theta}{\cos\theta}$$

$$= \mathbf{\frac{1+\sin\theta}{\cos\theta}} \qquad \mathbf{= R.H.S}$$

***Example* 54:** Find tan θ and cosec θ if $\cos\theta = \dfrac{15}{17}$

Solution: Let $\sin\theta = \sqrt{1-\cos^2\theta}$

$$= \sqrt{1-\left(\frac{15}{17}\right)^2}$$

$$= \sqrt{1-\frac{225}{289}}$$

$$= \sqrt{\frac{289-225}{289}} = \sqrt{\frac{64}{289}} = \frac{8}{17}$$

∴ $\tan\theta = = \dfrac{\sin\theta}{\cos\theta} = \dfrac{8}{15}$

and $\text{cosec}\,\theta = \dfrac{1}{\sin\theta} = \dfrac{17}{8}$

***Example* 55:** If $0 \leq x \leq 2\pi$, find cos x and tanx when sin x $= \frac{-1}{2}$, x lies in fourth quadrant.

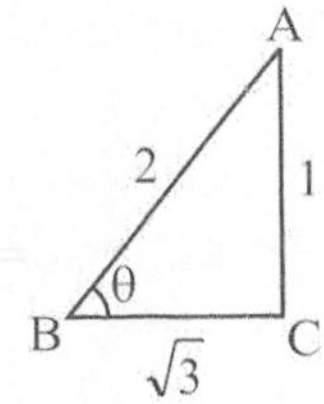

Solution: In fourth quadrant cos x is positive

$$\therefore \quad \cos x = \frac{BC}{AC} = \frac{\sqrt{3}}{2}$$

$$\therefore \quad \tan x = \frac{\sin x}{\cos x} = \frac{-\frac{1}{2}}{\frac{\sqrt{3}}{2}} = \frac{-1}{\sqrt{3}}$$

***Example* 56:** Solve $\sin 75^0 + \cos 75^0 + \sin 15^0 + \cos 15^0$

Solution: $\sin 75^0 + \cos 75^0 + \sin 15^0 + \cos 15^0$

$$= \sin(45^0 + 30^0) + \cos(45^0 + 30^0) + \sin(45^0 - 30^0) + \cos(45^0 - 30^0)$$

$$= \sin 45^0 \cos 30^0 + \cos 45^0 \sin 30^0 + \cos 45^0 \cos 30^0 - \sin 45^0 \sin 30^0 +$$
$$\sin 45^0 \cos 30^0 - \cos 45^0 \sin 30^0 + \cos 45^0 \cos 30^0 + \sin 45^0 \sin 30^0$$

$$= 2 \sin 45^0 \cos 30^0 + 2 \cos 45^0 \cos 30^0$$

$$= 2 \times \frac{1}{\sqrt{2}} \times \frac{\sqrt{3}}{2} + 2 \times \frac{1}{\sqrt{2}} \times \frac{\sqrt{3}}{2}$$

$$= \sqrt{\frac{3}{2}} + \sqrt{\frac{3}{2}} = \frac{2\sqrt{3}}{\sqrt{2}}$$

***Example* 57:** Prove that $\frac{\sin A - \sin 5A + \sin 9A - \sin 13A}{\cos A - \cos 5A - \cos 9A + \cos 13A} = \cot 4A$

Solution: $\text{L.H.S} = \frac{\sin A - \sin 5A + \sin 9A - \sin 13A}{\cos A - \cos 5A - \cos 9A + \cos 13A}$

$$= \frac{2\cos\frac{A+5A}{2}.\sin\frac{A-5A}{2} + 2\cos\frac{9A+13A}{2}.\sin\frac{9A-13A}{2}}{2\sin\frac{A+5A}{2}\sin\frac{5A-A}{2} + 2\sin\frac{13A+9A}{2}\sin\frac{9A-13}{2}}$$

$$= \frac{2[\cos 3A \sin(-2A) + \cos 11A \sin(-2A)]}{2[\sin 3A \sin 2A + \sin 11A \sin(-2A)]}$$

$$= \frac{-\sin 2A\cos 3A - \sin 2A\cos 11A}{\sin 2A\sin 3A - \sin 2A\sin 11A}$$

$$= \frac{\cos 3A + \cos 11A}{\sin 11A - \sin 3A}$$

$$= \frac{2\cos\frac{3A+11A}{2}\cos\frac{3A-11A}{2}}{2\cos\frac{11A+3A}{2}\sin\frac{11A-3A}{2}}$$

$$= \frac{\cos(-4A)}{\sin 4A}$$

$$= \frac{\cos 4A}{\sin 4A} \qquad (\because \cos(-\theta) = \cos\theta)$$

= cot 4A = R.H.S

Exercises

1. Prove that $\frac{\tan A}{1-\cot A} + \frac{\cot A}{1-\tan A} = 1 + \sec A \operatorname{cosec} A$
2. Prove that $\sin\theta\,(1 + \tan\theta) + (1+ \cot\theta)\cos\theta = \sec\theta + \operatorname{cosec}\theta$
3. Prove that $\cos^6 A + \sin^6 A = 1 - 3\sin^2 A\cos^2 A$
4. Find the value of

   (i) $\sin 135^0$ (ii) $\operatorname{cosec} 150^0$ (iii) $\cot 120^0$ (iv) $\tan(-210^0)$ (v) $\cos(-1950^0)$
5. Prove that $\sin(270^0 - \theta)\sin(90^0 - \theta) - \cos(270^0 - \theta)\cos(90^0 + \theta) + 1 = 0$
6. If $\alpha + \beta = \frac{\pi}{2}$ and $\sin\alpha = \frac{1}{2}$ then find value of β.
7. If $\sin\alpha = \frac{3}{5}, \cos\beta = \frac{9}{41}$ then prove that

   $\cos(\alpha + \beta) = \frac{-84}{205}$ and $\sin(\alpha - \beta) = = \frac{-133}{205}$
8. Prove that $\sin(45^0 + A) \,.\, \sin(45^0 - A) = = \frac{1}{2}\cos 2A$

9. Prove that $\sin\theta + \sin\left(\frac{2\pi}{3}+\theta\right) + \sin\left(\frac{4\pi}{3}+\theta\right) = 0$.

10. Prove that $\cos(36^0 - A)\cos(36^0 + A) + \cos(54^0 + A)\cos(54^0 - A) = \cos 2A$

11. Prove that $\frac{\sin A + \sin 3A + \sin 5A}{\cos A + \cos 3A + \cos 5A} = \tan 3A$

12. Prove that $\frac{\sin^2 A - \sin^2 B}{\sin A \cos A - \sin B \cos B} = \tan(A+B)$

13. If $\tan\theta = \frac{3}{4}$ and $\tan\phi = \frac{5}{12}$, then find the value of $\tan(\theta + \phi)$ and $\cot(\theta + \phi)$.

14. Prove that $1 + \tan A \tan\frac{A}{2} = \tan A \cot\frac{A}{2} - 1 = \sec A$.

15. If $\tan\alpha = \frac{n}{n+1}$ and $\tan\beta = \frac{1}{2n+1}$ then prove that $\tan(\alpha+\beta) = 1$.

16. If $\cos\theta = \frac{4}{5}$, then find the value of $\tan 2\theta$.

17. If $\tan A = \frac{1}{7}$ and $\tan B = \frac{1}{3}$ then prove that $\cos 2A = \sin 4B$.

18. Prove that $\frac{\tan(\frac{\pi}{4}+\theta) - \tan(\frac{\pi}{4}-\theta)}{\tan(\frac{\pi}{4}+\theta) + \tan(\frac{\pi}{4}-\theta)} = \sin 2\theta$.

19. Prove that $1 + \tan A \tan 2A = \sec 2A$

20. Prove that $\frac{1-\tan^2(45^0 - A)}{1+\tan^2(45^0 - A)} = \sin 2A$.

21. If $\sin\alpha = \frac{1}{3}$, then find $\sin 3\alpha$.

22. Prove that $\sec\left(\frac{\pi}{4}+\theta\right)\sec\left(\frac{\pi}{4}-\theta\right) = 2\sec 2\theta$

23. Prove that $\sin\frac{\pi}{10} + \sin\frac{13\pi}{10} = \frac{-1}{2}$

24. Prove that $\tan 6^0 \tan 42^0 \tan 66^0 \tan 78^0 = 1$

25. Prove that $\frac{\cot\theta + \text{cosec}\,\theta - 1}{\cot\theta - \text{cosec}\,\theta + 1} = \text{cosec}\,\theta + \cot\theta.$

26. Prove that $\sqrt{\tan\theta + \sin\theta} \times \sqrt{\tan\theta - \sin\theta} = \tan\theta \,.\, \sin\theta$

27. If $8\sin\theta = 4 + \cos\theta$, find $\sin\theta$.

28. If $\tan^2\theta + \sec\theta = 5$, find $\cos\theta$.

29. If $A = 45^0$, verify that

(i) $\sin 2A = 2\sin A\cos A$ (ii) $\cos 2A = 1 - 2\sin^2 A$

30. Find $\sec\theta$ and $\tan\theta$ if $\sin\theta = \frac{12}{13}$ lies in second quadrant.

Find the values of the following:

31. $\sin^2 30^0 + \sin^2 45^0 + \sin^2 60^0$

32. $\tan^2\frac{\pi}{3} + \text{cosec}^2\frac{\pi}{4} + 3\sec^2\frac{\pi}{6}$

33. $4\sin^2 30^0 - \frac{3}{4}\sec^2 45^0 + \cot^2 30^0 - 2\cos^2 60^0.$

34. Find the values of

(i) $\tan\frac{11\pi}{6}$ (ii) $\cos(-480^0)$ (iii) $\text{cosec}\,870^0$ (iv) $\cot\frac{15\pi}{4}$

35. Simplify $\frac{\cos(180^0 + \theta)\cos(-\theta)}{\sin(180^0 - \theta)\cos(90^0 + \theta)}$

36. Prove that $\frac{\tan(360^0 + \theta) + \cot(90^0 - \theta)}{1 + \tan(180^0 - \theta)\cot(270^0 + \theta)} = 2\sin\theta\cos\theta.$

37. Prove that $\sin(A + 45^0) = \frac{1}{\sqrt{2}}(\sin A + \cos A)$

38. Prove that $\tan 3A - \tan 2A - \tan A = \tan 3A \,.\, \tan 2A \,.\, \tan A.$

39. If $A + B = \frac{\pi}{4}$, then show that $(1 + \tan A)(1 + \tan B) = 2$

40. Prove that $\tan\left(\frac{\pi}{4} + \frac{x}{2}\right) = \sec x + \tan x$

41. Prove that $\dfrac{\tan A + \tan B}{\tan A - \tan B} = \dfrac{\sin(A+B)}{\sin(A-B)}$

42. Prove that $\sin\dfrac{5\pi}{18} - \cos\dfrac{4\pi}{9} = \sqrt{3}\sin\dfrac{\pi}{9}$

43. Prove that $\sin(60^0 + A) + \sin(60^0 - A) = \sqrt{3}\cos A$.

44. Prove that $\cos\dfrac{\pi}{8} + \cos\dfrac{3\pi}{8} + \cos\dfrac{5\pi}{8} + \cos\dfrac{7\pi}{8} = 0$

45. Prove that $\dfrac{\sin(A-B) + \sin A + \sin(A+B)}{\cos(A-B) + \cos A + \cos(A+B)} = \tan A$

46. Prove that $\dfrac{\sin\dfrac{5\pi}{12} - \sin\dfrac{\pi}{12}}{\cos\dfrac{5\pi}{12} + \cos\dfrac{\pi}{12}} = \dfrac{1}{\sqrt{3}}$

47. Prove that $\dfrac{\sin 2A + \sin 2B}{\sin 2A - \sin 2B} = \dfrac{\tan(A+B)}{\tan(A-B)}$

48. Prove that $\sin 10^0 \sin 30^0 \sin 50^0 \sin 70^0 = \dfrac{1}{16}$.

49. Prove that $\dfrac{\sin\theta + \sin 2\theta}{1 + \cos\theta + \cos 2\theta} = \tan\theta$.

50. Prove that $\dfrac{\sec 8\theta - 1}{\sec 4\theta - 1} = \dfrac{\tan 8\theta}{\tan 2\theta}$

51. If $\sin 2A = \dfrac{4}{5}$, find the value of $\tan A$ $(0 \le A \le \dfrac{\pi}{2})$

52. Prove that $\dfrac{\sin 3A}{\sin A} - \dfrac{\cos 3A}{\cos A} = 2$

53. Prove that $\tan\left(\dfrac{\pi}{4} + A\right) - \tan\left(\dfrac{\pi}{4} - A\right) = 2\tan 2A$

54. Prove that $(\cos A - \cos B)^2 + (\sin A - \sin B)^2 = 4\sin\dfrac{A-B}{2}$

55. Prove that $\tan 70^0 = \tan 20^0 + 2\tan 50^0$

Answers

4. (i) $\frac{1}{\sqrt{2}}$ (ii) 2 (iii) $\frac{-1}{\sqrt{3}}$ (iv) $\frac{-1}{\sqrt{3}}$ (v) $\frac{-\sqrt{3}}{2}$

6. 60^0

13. $\frac{56}{33}, \frac{63}{16}$

16. $\frac{24}{7}$

21. $\frac{23}{27}$

27. $\frac{3}{5}$ or $\frac{5}{13}$

28. $\frac{1}{2}$ or $\frac{-1}{3}$

30. $\sec\theta = \frac{-13}{5}$, $\tan\theta = \frac{-12}{5}$

31. $\frac{3}{2}$

32. 17

33. $\frac{1}{2}$

34. (i) $\frac{-1}{\sqrt{3}}$ (ii) $\frac{-1}{2}$ (iii) 2 (iv) –1

35. $\cot^2\theta$

51. 2 or $\frac{1}{2}$

CHAPTER 12

LOGARITHM

Introduction

The british mathematician John Napier was the first person to introduce the concept of logarithm in 1550.

The process to find the multiplication, division, indices and roots of large numbers are very time consuming, cumbersome and error prone. Therefore, the technique of logarithm was developed to minimize the complications of calculations and possibilities of any errors.

Definition

The logarithm of any positive number to a given base is equal to the power to which the base should be raised to get the given number.

If there be a positive number a, where $a \neq 1$, then the logarithm of a number N to the base a is the power to which a must be raised in order to be equal to the number i.e. if

$$a^x = N$$

Then $$\log_a N = x$$

***Example* 1:** Write each of the following in the form of logarithms:

(i) $12^3 = 1728$ (ii) $2^{-5} = \frac{1}{32}$ (iii) $10^{-3} = 0.001$

***Solution*:** (i) $12^3 = 1728 \Rightarrow \log_{12} 1728 = 3$

(ii) $2^{-5} = \frac{1}{32} \Rightarrow \log_2 \frac{1}{32} = -5$

(iii) $10^{-3} = 0.001 \Rightarrow \log_{10} 0.001 = -3$

***Example* 2:** Write the following in exponential form:

(i) $\log_2 128 = 7$ (ii) $\log_9 729 = 3$

***Solution*:** (i) $\log_2 128 = 7 \Rightarrow 2^7 = 128$

(ii) $\log_9 729 = 3 \Rightarrow 9^3 = 729$

Properties of Logarithms:

(i) the logarithm of 1 on any base a $(a \neq 0)$ is zero i.e.

$$a^o = 1 \Rightarrow \log_a 1 = 0$$

(ii) The logarithm of any number to the base it self is always 1 i.e.

$$a^1 = a \Rightarrow \log_a a = 1$$

(iii) Base will never be ∞ or 0.

(iv) The logarithm of zero on any base is negatively infinite i.e.

$$a^{-\infty} = 0 \Rightarrow \log_a 0 = -\infty$$

Fundamental laws of logarithm

***First Law*:** The logarithm of the product of two numbers is equal to the sum of the logarithm of these numbers to the same base. i.e.

$$\text{Log}_a (mn) = \log_a m + \log_a n$$

***Proof*:** Let $\log_a m = x$ and $\log_a n = y$

$\Rightarrow$ $m = a^x$ and $n = a^y$

$\therefore$ $mn = a^x . a^y$

$= a^{x+y}$

Again by definition

$$\log_a (mn) = x + y$$

$$\log_a (mn) = \log_a m + \log_a n$$

Second Law: The logarithm of the quotient of two numbers is equal to the difference of the logarithm of the numerator and the logarithm of the denominator to the same base. i.e.

$$\log_a\left(\frac{m}{n}\right) = \log_a m - \log_a n$$

Proof: Let $\log_a m = x$ and $\log_a n = y$

$\Rightarrow$ $m = a^x$ and $n = a^y$

$$\therefore \quad \frac{m}{n} = \frac{a^x}{a^y}$$

$$= a^{x-y}$$

$\therefore$ By definition,

$$\log_a\left(\frac{m}{n}\right) = x - y$$

$$\therefore \quad \log_a\left(\frac{m}{n}\right) = \log_a m - \log_a n$$

Third Law: The logarithm of a number raised to a power is equal to the product of the index of the power and the logarithm of the number to the same base. i.e

$$\log_a (m^n) = n \log_a m$$

Proof: Let $\log_a m = x \Rightarrow m = a^x$

$$\therefore \quad m^n = (a^x)^n$$

$$= a^{xn}$$

$\therefore$ By definition, $\log_a (m^n) = nx$

$$\log_a (m^n) = n \log_a m$$

Fourth Law: The base of logarithm can be changed. If the logarithm of numbers are given to the base a, then the logarithm of those numbers can be obtained to another base b. i.e.

$$\log_b m = \log_a m \times \log_b a$$

Proof: Let $\log_b m = x$, $\log_a m = y$, $\log_b a = z$

$\Rightarrow$ $b^x = m$, $a^y = m$ and $b^z = a$

$$\therefore \quad b^x = a^y$$

$\Rightarrow \quad b^x = (b^z)^y$

$\Rightarrow \quad b^x = b^{yz}$

$\Rightarrow \quad x = yz$

$\Rightarrow \quad \log_b m = \log_a m \times \log_b a$

$$\Rightarrow \quad \log_a m = \frac{\log_b m}{\log_b a}$$

Fifth Law: The product of logarithm of number b to base a with logarithm of number a to base b is equal to 1. i.e.

$$\log_a b \times \log_b a = 1$$

Proof: Let $\log_a b = x \Rightarrow a^x = b$

$\Rightarrow \log_b (a^x) = \log_b b$

$\Rightarrow x \log_b a = 1$

$\Rightarrow \log_a b \times \log_b a = 1$

or $$\log_b a = \frac{1}{\log_a b}$$

Example 3: Prove that $\log_a (1 + 2 + 3) = \log_a 1 + \log_a 2 + \log_a 3$.

Solution: L.H.S. $= \log_a (1 + 2 + 3) = \log_a 6$

R.H.S. $= \log_a 1 + \log_a 2 + \log_a 3 = \log_a (1 \times 2 \times 3)$

$(\because \log_a (m \times n \times p) = \log_a m + \log_a n + \log_a p)$

$= \log_a 6$

$\therefore$ **L.H.S = R.H.S**

Example 4: If $\log (m + n) = \log m + \log n$, show that $m = \dfrac{n}{n-1}$

Solution: Given $\log (m + n) = \log m + \log n$

$\Rightarrow \log (m + n) = \log (mn)$

$\Rightarrow \quad m + n = mn$

$\Rightarrow \quad n = mn - m$

$\Rightarrow \quad n = m (n - 1)$

$$\Rightarrow \quad \mathbf{m = \frac{n}{n-1}}$$

***Example* 5:** Find the value of x, when $\dfrac{\log_2 144}{\log_2 12} = \log_2 x$.

***Solution*:**

$$\text{L.H.S} = \frac{\log_2 144}{\log_2 12} = \frac{\log_2 (12\times 12)}{\log_2 12}$$

$$= \frac{\log_2 (12)^2}{\log_2 12}$$

$$= \frac{2\log_2 12}{\log_2 12} = 2$$

Thus, we have $\log_2 x = 2$

$\Rightarrow x = 2^2$

$\Rightarrow \mathbf{x = 4}$

***Example* 6:** If $\log\dfrac{x+y}{2} = \dfrac{\log x + \log y}{2}$, then prove that x = y

***Solution*:** Given $\log\dfrac{x+y}{2} = \dfrac{\log x + \log y}{2}$

$\Rightarrow \log\dfrac{x+y}{2} = \dfrac{1}{2}\log(x.y)$

$\Rightarrow \log\dfrac{x+y}{2} = \log\ (xy)^{\frac{1}{2}}$

$\Rightarrow \dfrac{x+y}{2} = (xy)^{\frac{1}{2}}$

$\Rightarrow \left(\dfrac{x+y}{2}\right)^2 = xy$

$\Rightarrow x^2 + y^2 + 2xy = 4xy$

$\Rightarrow x^2 + y^2 - 2xy = 0$

$\Rightarrow (x - y)^2 = 0$

$\Rightarrow x - y = 0$

$\Rightarrow x = y$

***Example* 7:** If $a^2 + b^2 = 7ab$, prove that $\log\dfrac{a+b}{3} = \dfrac{\log a + \log b}{2}$

Solution: Given $a^2 + b^2 = 7ab$

$\Rightarrow a^2 + b^2 + 2ab = 7ab + 2ab$

$\Rightarrow (a + b)^2 = 9ab$

$$\Rightarrow \quad \left(\frac{a+b}{3}\right)^2 = ab$$

$$\Rightarrow \quad \log\left(\frac{a+b}{3}\right)^2 = \log(ab)$$

$$\Rightarrow \quad 2\log\left(\frac{a+b}{3}\right) = \log a + \log b$$

$$\Rightarrow \quad \mathbf{\log\left(\frac{a+b}{3}\right) = \frac{\log a + \log b}{2}}$$

***Example* 8:** If $\log(mn) = \log m - \log n$, show that $n = 1$.

***Solution*:** Given $\log(mn) = \log m - \log n$

$\Rightarrow \quad \log m + \log n = \log m - \log n$

$\Rightarrow \quad 2\log n = 0$

$\Rightarrow \quad \log n = 0$

$\Rightarrow \quad \log n \pm \log 1 \qquad (\because \log 1 = 0)$

$\Rightarrow \quad \mathbf{n = 1}$

***Example* 9:** Prove that $x^{\log y - \log z} . y^{\log z - \log x} . z^{\log x - \log y} = 1$

***Solution*:** L.H.S. $A = x^{\log y - \log z} . y^{\log z - \log x} . z^{\log x - \log y}$

$\log A = \log [x^{\log y - \log z} . y^{\log z - \log x} . z^{\log x - \log y}]$

$= \log x^{\log y - \log z} + \log y^{\log z - \log x} + \log z^{\log x - \log y}$

$= (\log y - \log z) \log x + (\log z - \log x)\log y + (\log x - \log y) \log z$

$= \log x \log y - \log x \log z + \log y \log z - \log x \log y + \log x \log z - \log y \log z$

$\log A = 0$

$\Rightarrow \log A = \log 1 \qquad (\because \log 1 = 0)$

$\Rightarrow A = 1$

$\therefore$ L.H.S = R.H.S

Types of Logarithm

1. **Natural Logarithm:** The logarithm of a number to the base e is called natural logarithm. It is also called Napieian logarithm. The value of e = 2.7186 (approximate)

2. **Common Logarithm:** The logarithm of a number to the base 10 is called common logarithm. In numerical calculations common logarithms are used. The common logarithm of a number consists of two parts:

   1. The integral part of logarithm is called characteristic, and

   2. The decimal part is called the mantissa.

To find characteristic of a common logarithm: The following rules are:

Rule 1: Number greater than 1.

The characteristic of a common logarithm is a positive integer and is one less than the number of digits in the integral part of that number.

Number	**Number of digits in the integral part of the number**	**Characteristic**
(i) 2.70827	1	1 – 1 = 0
(ii) 56.787	2	2 – 1 = 1
(iii) 134.82	3	3 – 1 = 2
(iv) 1348.2	4	4 – 1 = 3

Rule 2: Number less than 1.

If the positive number begins with a decimal point, the characteristic of its common logarithm is negative and is numerically one greater than the number of zeros immediately after the decimal point. The negative characteristics are written as a symbol (–) 'bar' over it, i.e. –1, –2, –3 etc., are negative characteristics which are written as $\bar{1}, \bar{2}, \bar{3}$ etc., and read as one bar, two bar, three bar etc.

Number	**Number of zeros after decimal point in this number (– n)**	**Adding 1**	**Charactristic**
(i) 0 .9207	0	0 + 1 = 1	–1 or $\bar{1}$
(ii) 0.00392	2	2 + 1 = 3	–3 or $\bar{3}$
(iii) 0.0342	1	1 + 1 = 2	–2 or $\bar{2}$
(iv) 0.00007027	4	4 + 1 = 5	–5 or $\bar{5}$

To find mantissa of a common logarithm

To find the mantissa of common logarithm of a number, special tables have been formed. These tables are called logarithmic table.

Procedure:

(i) Ignore the decimal point, use the digits irrespective of the decimal point. Also ignore zero if it is not a significant figure. In case of more than four significant figures, round off the number up to four significant figures while using the four figure log tables.

(ii) The first two digits are used for determining the row of logarithm table. If there is only one digit then we add a zero.

(iii) Proceed along that row till that column (in the main body) of third digit is reached. Read off the figure in that row and column.

(iv) Proceed along that row till that common (in the main difference) of fourth digit is reached and read off this figure.

(v) Add the two figures obtained in step (iii) and (iv). This is the required mantissa of the common logarithm.

***Example* 10:** Obtain the logarithm of 90.

***Solution*:** The given is greater than 1 so characteristic is $2 - 1 = 1$

***For mantissa*:** First look down the left column to locate 90, look across the 90 row and locate the entry under column 90 labelled 0. Read off this figure 9542. Hence mantissa of $\log_{10} 90 = 0.9542$

$$\therefore \log_{10} 90 = \text{characteristic} + \text{mantissa}$$
$$= 1 + 0.9542$$
$$= \mathbf{1.9542}$$

***Example* 11:** Find the logarithm of 0.0004291

***Solution*:** ***characteristic*:** The given is less than 1 and the number of zeros after the decimal point is 3 hence, the characteristic $= -(3 + 1)$ or -4 or $\bar{4}$.

***Mantissa*:** Ignore decimal point consider the number 4291

The figure 42 under 9 in main table $= 6325$

The figure 42 under 1 in main difference $= \quad 1$

adding $= 6326$

Hence mantissa = 0.6326

$\therefore \log_{10} 0.0004291$ = characteristic + mantissa

$= -4 + 0.6326$

$= \mathbf{\bar{4}.6326}$

Antilogarithms: The antilog of a number is that number whose logarithm is the given number.

Now to find out antilog of a given number, we start with the digits after the decimal point (mantissa) in the antilog table.

Procedure:

1. In case of negative logarithm, the mantissa must be converted into a positive mantissa before reference to antilog table.
2. The first two digits of mantissa are seen in the first column of the antilog table to locate the relevant row.
3. Read off the figure given before it in the column heading of the third digit.
4. Read off the figure before that row in the mean difference under column heading of the fourth digit.
5. Add the figure obtained in step (3) and (4). The number so obtained is called a desire number.
6. Finally, use the characteristic of the given number for placing the decimal point in result, obtained in step (5). These are three conditions:
   (a) If characteristic is o or positive: In this case the number of digits in the whole part of the desire number will be one more than the characteristic.
   e.g. If characteristic is 2 then adding 1, we get 3. Thus the point of decimal will be after 3 digits to right side in the number obtained from antilog.
   (b) If characteristic is (–1) or $\bar{1}$: In this case adding 1, we get 0. Thus the number will start from decimal point.
   (c) If characteristic is negative and more that (–1): In this case the number of zeros just after the decimal point will be one less than the characteristics.

***Example* 12**: Find antilog of (i) – 0.2682 (ii) $\bar{3}.5386$

Solution: (i) Here, the given number is negative. So, we have to convert the mantissa into positive. This is done by adding and subtracting 1 from this number, thus

$$\begin{aligned} -0.2682 &= -0.2682 + 1 - 1 \\ &= -1 + (1 - 0.2682) \\ &= -1 + 0.7318 \\ &= \bar{1}.7318 \end{aligned}$$

From antilog table,

The figure before .73 under 1 in main table	= 5383
The figure before .73 under 8 in mean difference	= 10
Adding	= 5393

Thus, desired number = 5393

The characteristic is –1. Therefore, the number must have no digit in the integral part. The number of zeros just after decimal point will (–1 + 1) i.e. 0.

Hence, antilog (–0.2682) = 0.5393

(ii)

The figure before .53 under 8 in main table	= 3551
The figure before .53 under 6 in mean difference	= 5
Adding	= 3456

Thus, desired number = 3456

Since the characteristic is $\bar{3}$ or (–3). Then number of zeros just after the decimal point will be (–3 + 1) = –2

Hence, antilog ($\bar{3}$.5386) = 0.003456.

***Example* 13:** Using logarithmic table, find the value of each of following

(i) $\dfrac{2.38\times 3.901}{4.83}$ (ii) $(0.8931)^{-4}$ (iii) $\sqrt{\dfrac{0.0148\times 0.274}{0.8735}}$

***Solution*:** (i) Let $x = \dfrac{2.38\times 3.901}{4.83}$

$\therefore \quad \log x = \log \dfrac{2.38\times 3.901}{4.83}$

$\Rightarrow \quad \log x = \log (2.38 \times 3.901) - \log 4.83$

$= \log 2.38 + \log 3.901 - \log 4.83$

$= 0.3766 + 0.5912 - 0.6839$

$\log x = 0.2839$

$x = \text{Antilog } (0.2839)$

$\mathbf{x = 1.923}$

(ii) Let $x = (0.8931)^{-4}$

$\log x = \log (0.8931)^{-4}$

$= -4 \log (0.8931)$

$= -4 \times (-1.9509)$

$= -4 \times (-1 + 0.9509)$

$= -4 \times -0.0491$

$\log x = 0.1964$

$x = \text{Antilog}\ (0.1964)$

$x = 1.571$

(iii) Let $x = \sqrt{\dfrac{0.0148 \times 0.278}{0.8735}}$

$$\log x = \log\left(\frac{0.0148 \times 0.278}{0.8735}\right)^{\frac{1}{2}}$$

$$= \frac{1}{2} \log\left(\frac{0.0148 \times 0.278}{0.8735}\right)$$

$$= \frac{1}{2}[\log (0.0148) + \log (0.278) - \log (0.8735)]$$

$$= \frac{1}{2}\ (\bar{2}.1703 + \bar{1}.4378 - \bar{1}.9412)$$

$$= \frac{1}{2}[(-2 + 0.1703) + (-1 + 0.4378) - (-1 + 9412)]$$

$$= \frac{1}{2}[-2 - 0.3331]$$

$= -1.1666$

$= -1 - 1 + 1 - 0.1666$

$= -2 + 0.8334$

$\log x = \bar{2}.8334$

$x = \text{Antilog}\ (\bar{2}.8334)$

$\mathbf{x = 0.06814}$

***Example* 14:** Prove that $7 \log_2 \dfrac{16}{15} + 5 \log_2 \dfrac{25}{24} + 3 \log_2 \dfrac{81}{80} = \log_2 2$

***Solution*:** L.H.S $= 7 \log \dfrac{16}{15} + 5 \log \dfrac{25}{24} + 3 \log \dfrac{81}{80}$

$$= 7 \log\left(\frac{2^4}{3\times 5}\right) + 5 \log\left(\frac{5^2}{2^3\times 3}\right) + 3 \log\left(\frac{3^4}{2^4\times 5}\right)$$

$= 7\,[\log 2^4 - \log (3\times5)] + 5\,[\log 5^2 - \log (2^3\times3)] + 3\,[\log 3^4 - \log (2^4\times5)]$

$= 7\,[4 \log 2 - \log 3 - \log 5)] + 5\,[2 \log 5 - 3 \log 2 - \log 3)] + 3\,[4 \log 3 - 4 \log 2 - \log 5]$

$= 28 \log 2 - 7 \log 3 - 7 \log 5 + 10 \log 5 - 15 \log 2 - 5 \log 3 + 12 \log 3 - 12 \log 2 - 3 \log 5$

$= \mathbf{\log_2 2}$ = R.H.S

***Example* 15:** Find the value of x if $7^x + 7^{x+2} = 250$

***Solution*:** Given $7^x + 7^{x+2} = 250$

$\Rightarrow$ $7^x + 7^x \cdot 7^2 = 250$

$7^x (1 + 49) = 250$

$7^x = \dfrac{250}{50}$

$7^x = 5$

$\log 7^x = \log 5$

$x \log 7 = \log 5$

$x \times 0.8451 = 0.6990$

$x = \dfrac{0.6990}{0.8451}$

$\mathbf{x = 0.8271}$

***Example* 16:** Find the value of x if

$$a^{3-x}\, b^{5x} = a^{x+5} \cdot b^{3x}$$

***Solution*:** Let $a^{3-x}\, b^{5x} = a^{x+5} \cdot b^{3x}$

Taking log on both the sides. we get

$\log (a^{3-x}\, b^{5x}) = \log (a^{x+5}\, b^{3x})$

$\log a^{3-x} + \log b^{5x} = \log a^{x+5} + \log b^{3x}$

$(3 - x) \log a + 5x \log b = (x + 5) \log a + 3x \log b$

$(3 - x) \log a - (x + 5) \log a + 5x \log b - 3x \log b = 0$

$\log a\,(3 - x - x - 5) + 2x \log b = 0$

$\log a\,(- 2 - 2x) + 2x \log b = 0$

$$-2 \log a - 2x \log a + 2x \log b = 0$$

$$x (\log b - \log a) = \log a$$

$$\mathbf{x = \frac{\log a}{\log b - \log a}}$$

Application of Logarithm in Pharmaceutical Computations

Calculation of Equilibrium Constants and Concentration: If the concentration of reaction and products are known at equilibrium in a reaction, the equilibrium constants can be calculated.

***Example* 17:** The equilibrium constant for the reaction

$$N_2(g) + 3H_2(g) \rightleftarrows 2NH_3(g)$$

at 715 K is 6.0×10^{-2}. If in a particular reaction there are 0.25 mol L^{-1} of H_2 and 0.06 mol L^{-1} of NH_3 present, calculate the concentration of N_2 at equilibrium.

***Solution*:** For the reaction, $N_2 (g) + 3H_2 (g) \rightleftarrows 2NH_3(g)$

We have $K = \frac{[NH_3]^2}{[H_2]^3 [N_2]}$ .....(i)

Given $K = 6.0 \times 10^{-2}$, $[H_2] = 0.25$ mol/l, $[NH_3] = 0.06$ mol/l

Putting in (i), we get

$$6.0 \times 10^{-2} = \frac{[0.06]^2}{[0.25]^3 [N_2]}$$

$$[N_2] = \frac{[0.06]^2}{[0.25]^3 \times 6 \times 10^{-2}}$$

Taking log on both sides

$$\log [N_2] = \log (0.06)^2 - [\log (0.25)^3 + \log (6 \times 10^{-2})]$$

$$= 2 \log (0.06) - 3 \log (0.25) - \log 6 + 2 \log 10$$

$$= 2(\bar{2}.07782) - 3(\bar{1}.3979) - (\bar{2}.7782)$$

$$= 2 (-2 + 0.7782) - 3 (-1 + 0.3979) - (-2 + 0.7782)$$

$$\log [N_2] = 0.545$$

$$[N_2] = \text{Antilog } (0.545)$$

$$\mathbf{[N_2] = 3.84 \quad Mol\ L^{-1}}$$

***Half Life Period*:** The half life period T of C-14 given the value of disintegration constant K from the equation $K = \frac{0.693}{T}$.

$$\text{The disintegration equation is } K = \frac{2.303}{t} \log \frac{a}{a-x}$$

***Example* 18:** The amount of C-14 isotopes in a piece of wood is found to be one-sixth of its amount present in a fresh piece of wood. Calculate the age of wood.

***Solution*:** Here, half life of C-14 = 5577 years = T

$$\text{By formula } K = \frac{0.693}{T}$$

$$K = \frac{0.693}{5577}$$

$$\text{The disintegration equation } K = \frac{2.303}{t} \log \frac{a}{a-x}$$

$$\frac{0.693}{5577} = \frac{2.303}{t} \times \log \frac{1}{\frac{1}{6}}$$

$$t = \frac{2.303 \times 5577 \times \log 6}{0.693}$$

$$= \frac{2.303 \times 5577 \times 0.7782}{0.693}$$

$$\mathbf{t = 14423 \text{ years}}$$

***Calculation of Boiling Point or Freezing Point*:** If the boiling point or the freezing point of a liquid at one pressure is known. It is easy to calculate it at another pressure by use of the Claperyon Claussius equation

$$\log_{10} \frac{P_2}{P_1} = \frac{\Delta H_\upsilon}{4.576} \left[\frac{1}{T_1} - \frac{1}{T_2} \right].$$

***Example* 19:** Calculate the vapour pressure of water at 90°C if its value at 100°C is 76 cm. The mean heat of vaporization of water in the temperature range 90°C and 100°C is 542 calories/gm.

***Solution*:** $\Delta H_\upsilon = 542 \text{ cal/gm} = 542 \times 18 \text{ cal/mole}$

$P_1 = 76$, $P_2 = ?$, $T_1 = 90 + 273 = 363°$ Abs, $T_2 = 100 + 273 = 373°$ Abs

According to formula

$$\log_{10}\frac{P_2}{P_1}=\frac{\Delta H_\upsilon}{4.576}\left[\frac{1}{T_1}-\frac{1}{T_2}\right].$$

$$\log_{10}\frac{P_2}{76}=\frac{542\times 18}{4.576}\left[\frac{1}{363}-\frac{1}{373}\right].$$

$$\log_{10} P_2-\log_{10} 76=\frac{542\times 18}{4.576}\left[\frac{373-363}{363\times 373}\right].$$

$$\log_{10} P_2-1.8808 = \log 542+\log 18+\log 10-\log 4.576-\log 363-\log 373$$

$$= 2.7340 + 1.2553 + 1 - 0.6605 - 2.5599 - 2.5717$$

$$= -.8028$$

$\log_{10} P_2 = -0.8028 + 1.8808$

$\log_{10} P_2 = 1.078$

P_2 = A.L. (1.078)

P_2 = 11.97 cm

PH equation

$$pH = pk_a + \log\frac{[salt]}{[acid]} \qquad \text{as } pk_a = -\log k_a.$$

***Example* 20:** 50 ml of 0.2 M acetic acid are mixed with 50 ml of 0.2 M sodium acetate solution. What will be the pH of the mixture. Ka = 1.85×10^{-5}.

***Solution*:** Total volume of mixture = 50 + 50 = 100 ml.

$$[Acid]=\frac{0.2}{100}$$

$$[salt]=\frac{0.2}{100}$$

We have $pH = pk_a + \log\frac{[salt]}{[acid]}$

$$= -\log(1.85\times 10^{-5}) + \log\frac{0.2/100}{0.2/100}$$

$$= -\log(1.85\times 10^{-5})$$

pH = 4.7325

Practice Questions

***Example* 1:** Solve the equation $\log_4 x + \log_4 (x-6) = 2$

***Solution*:** $\log_4 x + \log_4 (x-6) = 2 \Rightarrow \log_4 x(x-6) = 2$

$\Rightarrow$ $x^2 - 6x = 4^2$

$\Rightarrow$ $x^2 - 6x - 16 = 0$

$\Rightarrow$ $(x - 8)(x + 2) = 0$

$\Rightarrow$ $\mathbf{x = 8, -2}$

***Example* 2:** Evaluate $\log_4 40 + \log_4 2 - \log_4 20$

***Solution*:** $\log_4 40 + \log_4 2 - \log_4 20$

$$\log_4 \frac{40 \times 2}{20}$$

$= \mathbf{\log_4 4 = 1}$

***Example* 3:** Prove that $\dfrac{1}{\log_a abc} + \dfrac{1}{\log_b abc} + \dfrac{1}{\log_c abc} = 1$

***Solution*:**

$$\frac{1}{\log_a abc} + \frac{1}{\log_b abc} + \frac{1}{\log_c abc}$$

$= \log_{abc} a + \log_{abc} b + \log_{abc} c$

$= \mathbf{\log_{abc} abc = 1}$

***Example* 4:** If $a^x = b^y = c^z$ and $y^2 = xz$, prove that $\log_b a = \log_c b$

***Solution*:** If $a^x = c^z = b^y$

Taking log we get,

$\log_b a^x = \log_b c^z = \log_b b^y$

$\Rightarrow$ $x \log_b a = z \log_b c = y \log_b b$

$\Rightarrow$ $x \log_b a = z \log_b c = y$ $\quad (\because \log_b b = 1)$

Now $y^2 = x \log_b a.\ z \log_b c$

$xz = xz \log_b a.\ \log_b c$ $\quad (\because y^2 = xz$, given)

$1 = \log_b a\ .\ \log_b c$

$\Rightarrow$ $\mathbf{\log_b a = \log_c b}$

***Example* 5:** Show that $2 \log 8 - 3 \log 2 + \frac{1}{3} \log 64 = 5 \log 2$

***Solution*:**

$$2 \log 8 - 3 \log 2 + \frac{1}{3} \log 64$$

$$= 2 \log 2^3 - 3 \log 2 + \frac{1}{3} \log 2^6$$

$$= 6 \log 2 - 3 \log 2 + 2 \log 2$$

$$= \mathbf{5 \log 2}$$

***Example* 6:** Find the value of x, when

$$x = 2 \log 10 - \log 4 + \log (200) - \log\left(\frac{1}{2}\right)$$

***Solution*:**

$$x = 2 \log 10 - \log 4 + \log (200) - \log\left(\frac{1}{2}\right)$$

$$= \log 10^2 - \log 4 + \log 200 - \log\left(\frac{1}{2}\right)$$

$$= \log \frac{100 \times 200}{4 \times \frac{1}{2}}$$

$$= \log 10^4$$

$$= 4 \log 10$$

$$\mathbf{x = 4}$$

***Example* 7:** Prove that $2 \log 20 - \log 4 + 2 \log 10 = 4$

***Solution*:**

$$2 \log 20 - \log 4 + 2 \log 10$$

$$= \log 20^2 - \log 4 + \log 10^2$$

$$= \log 400 + \log 100 - \log 4$$

$$= \log \frac{400 \times 100}{4}$$

$$= \log 100 \times 100$$

$$= \log 10^4$$

$$= 4 \log 10$$

$$= \mathbf{4}$$

***Example* 8:** Prove that $3\log\left(\frac{1}{2}\right) + 2\log 2 + \log 20 = 1$

***Solution*:**

$$3\log\left(\frac{1}{2}\right) + 2\log 2 + \log 20$$

$$= \log\left(\frac{1}{2}\right)^3 + \log 2^2 + \log 20$$

$$= \log\frac{1}{8} + \log 4 + \log 20$$

$$= \log\left(\frac{1}{8}\times 4\times 20\right)$$

$$= \log 10$$

$$\mathbf{= 1}$$

***Example* 9:** Show that $3\log 4 + 2\log 5 - \frac{1}{3}\log 64 - \frac{1}{2}\log 16 = 2$

***Solution*:**

$$3\log 4 + 2\log 5 - \frac{1}{3}\log 64 - \frac{1}{2}\log 16$$

$$= \log 4^3 + \log 5^2 - \log\left(2^6\right)^{\frac{1}{3}} - \log\left(2^4\right)^{\frac{1}{2}}$$

$$= \log 64 + \log 25 - \log 4 - \log 4$$

$$= \log\frac{64\times 25}{4\times 4}$$

$$= \log 100$$

$$= \log 10^2$$

$$= 2\log 10$$

$$\mathbf{= 2}$$

***Example* 10:** Find the value 13.62 × 0.02453

***Solution*:** Let $x = 13.62 \times 0.2453$

$$\log x = \log 13.62 + \log(0.02453)$$

$$= 1.1341 + \bar{2}.3897$$

$$\mathbf{= \bar{1}.5238}$$

***Example* 11:** Find the value of .03752 ÷ 0.02432

***Solution*:** Let $x = 0.03752 \div 0.02432$

$$\log x = \log(0.03752) - \log(0.02432)$$

$$= \bar{2}.5742 - \bar{2}.3860$$

$$\log x = .1882$$

$$x = \text{Antilog}(.1882)$$

$$\mathbf{x = 1.543}$$

***Example* 12:** If log 64 = 1.8062, find log 8, log 2 and log 4.

***Solution*:**

$$\log 8 = \log\sqrt{64}$$

$$= \frac{1}{2}\log 64$$

$$= \frac{1}{2} \times 1.8062$$

$$\mathbf{= 0.9031}$$

$$\log 2 = \log(64)^{\frac{1}{6}}$$

$$= \frac{1}{6}\log 64$$

$$= \frac{1}{6} \times 1.8062$$

$$\mathbf{= 0.3010}$$

$$\log 4 = \log(64)^{\frac{1}{3}}$$

$$= \frac{1}{3}\log 64$$

$$= \frac{1}{3} \times 1.8062$$

$$\mathbf{= 0.6021}$$

***Example* 13:** If log 4 = 0.6021 and log 2 = 0.3010, find log 8, log 80, log 32.

***Solution*:**

$$\log 8 = \log (4 \times 2)$$
$$= \log 4 + \log 2$$
$$= 0.6021 + .3010$$
$$\mathbf{= 0.9031}$$
$$\log 80 = \log (8 \times 10)$$
$$= \log 8 + \log 10$$
$$= 0.9031 + 1$$
$$\mathbf{= 1.9031}$$
$$\log 32 = \log (8 \times 4)$$
$$= \log 8 + \log 4$$
$$= 0.9031 + 0.6021$$
$$\mathbf{= 1.5052}$$

***Example* 14:** Find the value of $\dfrac{472.9}{13.09 \times 86.27}$

***Solution*:** Let $x = \dfrac{472.9}{13.09 \times 86.27}$

$$\log x = \log 472.9 - [\log 13.09 + \log 86.27]$$
$$= 2.6747 - [1.1168 + 1.9359]$$
$$= \bar{1}.6220$$
$$x = \text{Antilog}\,(\bar{1}.6220)$$
$$\mathbf{x = 0.4188}$$

***Example* 15:** Find the value of 3.45 × 0.013 × 0.0008

***Solution*:** Let $x = 3.45 \times 0.013 \times 0.0008$

$$\log x = \log 3.45 + \log 0.013 + \log (0.0008)$$
$$= 0.5378 + \bar{2}.1139 + \bar{4}.9031$$
$$= 0.5378 - 2 + 0.1139 - 4 + 0.9031$$
$$= 1.5548 - 6$$
$$\log x = \bar{5}.5548$$
$$x = \text{Antilog}\,(\bar{5}.5548)$$
$$\mathbf{= 0.00003588}$$

***Example* 16:** Find the value of the following if log 2 = 0.3010, log 3 = 0.4771, log 5 = 0.6990, log 7 = 0.8454.

(i) $\log\frac{51}{26} + \log\frac{91}{119}$ (ii) log 105

***Solution*:** (i)

$$\log\frac{51}{26} + \log\frac{91}{119}$$

$$= \log\left(\frac{51}{26} \times \frac{91}{119}\right)$$

$$= \log\left(\frac{17\times3\times13\times7}{13\times2\times17\times7}\right)$$

$$= \log\left(\frac{3}{2}\right)$$

$$= \log 3 - \log 2$$

$$= 0.4771 - 0.3010$$

$$= \mathbf{0.1761}$$

(ii)

$$\log 105$$

$$= \log (3 \times 5 \times 7)$$

$$= \log 3 + \log 5 + \log 7$$

$$= 0.4771 + 0.6990 + 0.8451$$

$$= \mathbf{2.0212}$$

***Example* 17:** Prove that (i) $7\log\frac{16}{15} + 5\log\frac{25}{24} + 3\log\frac{81}{80} = \log 2$

(ii) $\frac{\log\sqrt{27} + \log\sqrt{8} - \log\sqrt{125}}{\log 6 - \log 5} = \frac{3}{2}$

(iii) $\log\frac{75}{16} - 2\log\frac{5}{9} + \log\frac{32}{243} = \log 2$

Solution: (i) L.H.S $= 7\log\frac{16}{15} + 5\log\frac{25}{24} + 3\log\frac{81}{80}$

$$= 7\log\left(\frac{2^4}{3\times 5}\right) + 5\log\left(\frac{5^2}{2^3\times 3}\right) + 3\log\left(\frac{3^4}{2^4\times 5}\right)$$

$= 7\,[\log 2^4 - \log(3\times 5)] + 5\,[\log 5^2 - \log(2^3\times 3)] + 3\,[\log 3^4 - \log(2^4\times 5)]$

$= 7\,[4\log 2 - \log 3 - \log 5] + 5\,[2\log 5 - 3\log 2 - \log 3] + 3\,[4\log 3 - 4\log 2 - \log 5]$

$= 28\log 2 - 7\log 3 - 7\log 5 + 10\log 5 - 15\log 2 - 5\log 3 + 12\log 3 - 12\log 2 - 3\log 5$

$= \log 2 =$ R.H.S.

(ii) LHS $= \dfrac{\log\sqrt{27} + \log\sqrt{8} - \log\sqrt{125}}{\log 6 - \log 5}$

$$= \frac{\log 3^{\frac{3}{2}} + \log 2^{\frac{3}{2}} - \log 5^{\frac{3}{2}}}{\log(2\times 3) - \log 5}$$

$$= \frac{\frac{3}{2}\log 3 + \frac{3}{2}\log 2 - \frac{3}{2}\log 5}{\log 2 + \log 3 - \log 5}$$

$$= \frac{\frac{3}{2}(\log 3 + \log 2 - \log 5)}{(\log 2 + \log 3 - \log 5)}$$

$= \dfrac{3}{2} =$ R.H.S

(iii) LHS $= \log\frac{75}{16} - 2\log\frac{5}{9} + \log\frac{32}{243}$

$$= \log\frac{75}{16} + \log\left(\frac{5}{9}\right)^{-2} + \log\frac{32}{243}$$

$$= \log\frac{75}{16} + \log\left(\frac{9}{5}\right)^{2} + \log\frac{32}{243}$$

$$= \log\frac{75}{16} + \log\frac{81}{25} + \log\frac{32}{243}$$

$$= \log\left(\frac{75}{16}\times\frac{81}{25}\times\frac{32}{243}\right)$$

$= \log 2 =$ R.H.S

***Example* 18:** Find the value of (i) $\log_4 32$ (ii) $\log_8 128$ (iii) $\log_5 64$ (iv) $\log_8 25$

***Solution*:** (i)

$$\log_4 32 = \log_2 32 \times \log_4 2$$
$$= \log_2 2^5 \times \frac{1}{\log_2 4}$$
$$= 5\log_2 2 \times \frac{1}{2\log_2 2}$$
$$= \frac{5}{2}$$

(ii)

$$\log_8 128 = \log_2 128 \times \log_8 2$$
$$= \log_2 2^7 \times \frac{1}{\log_2 8}$$
$$= 7\log_2 2 \times \frac{1}{3\log_2 2}$$
$$= \frac{7}{3}$$

(iii)

$$\log_5 64 = \log_{10} 64 \times \log_5 10$$
$$= \log_{10} 2^6 \times \frac{1}{\log_{10} 5}$$
$$= 6\,\frac{\log_{10} 2}{\log_{10} 5}$$
$$= 6 \times \frac{0.3010}{0.6990} = 2.584$$

(iv)

$$\log_8 25 = \log_{10} 25 \times \log_g 10$$
$$= \log_{10} 5^2 \times \frac{1}{\log_{10} 8}$$
$$= 2\log_{10} 5 \times \frac{1}{3\log_{10} 2}$$
$$= \frac{2 \times .6990}{3 \times .3010}$$
$$= \mathbf{1.55}$$

Exercises

1. If log (mn) = log m – log n, show that n = 1

[Hint: Given log (mn) = log m – log n

$$\log m + \log n = \log m - \log n$$
$$2 \log n = 0$$
$$\log n = 0$$
$$\log n = \log 1$$
$$\therefore n = 1$$

2. Find the values of (i) $7.2 \times 8.3 \times 0.94$ (ii) 5872×0.058
3. Find the values with the help of logarithms

   (i) $39.8 \div 19.9$ (ii) $80.17 \div 25.46$ (iii) $1.005 \div 0.039$

   (iv) 4^7 (v) $(0.00397)^4$ (vi) $(1.045)^7$
4. Find the values (i) $\sqrt[3]{12765.4}$ (ii) $\sqrt[7]{\frac{1}{1.235}}$ (iii) $\frac{81.37 \times 24.86}{315.5}$
5. Find the values (i) $\frac{\sqrt[3]{102.5} \times (0.0025)^4}{\sqrt[6]{0.7865}}$ (ii) $\frac{(6.45)^3 \times (0.00034)^{\frac{1}{3}}}{(9.37)^2 \times (8.93)^{\frac{1}{4}}}$
6. If log 2 = 0.3010 and log 3 = 0.4771, find the value of log 12 and log 54.
7. Find the value of $\log_{10}\left\{\frac{(16)^{\frac{1}{5}} \times (5)^2}{(108)^3}\right\}$, if $\log_{10} 2 = 0.3010$ and $\log_{10} 3 = 0.4771$
8. Prove that

   (i) $16 \log\frac{16}{15} + 12 \log\frac{25}{24} + 7 \log\frac{81}{80} + \log 2 = 1$

   (ii) $7 \log\frac{15}{16} + 7 \log\frac{8}{3} + 5 \log\frac{2}{5} + \log\frac{32}{25} + 3 \log 5 = 3$

   (iii) $\frac{\log\sqrt{27} + \log 8 + \log\sqrt{1000}}{\log 120} = \frac{3}{2}$

   (iv) $\frac{2\log 6 + 6\log 2}{4\log 2 + \log 27 - \log 9} = 2$

9. Prove that 2 log (1 + 2 + 4 + 7 + 14) = log 1 + log 2 + log 4 + log 7 + log 14.

10. Find the value $\log_2 3 \times \log_3 4 \times \log_4 5 \times \log_5 6 \times \log_6 7 \times \log_7 8$

11. Find the value of x, if

(i) $\log_2 x + \log_4 x \times \log_{16} x = \frac{21}{4}$

(ii) $\log x + \log (x - 3) = 1$

(iii) $(1 + x)^{10} = 1500$

12. If $a^2 + b^2 = 11ab$, prove that $\log \frac{a-b}{3} = \frac{\log a + \log b}{2}$

13. If $x^3 + y^3 = 0$ and $x + y \neq 0$, Prove that $\log(x + y) = \frac{1}{2}(\log x + \log y + \log 3)$

14. If $a^2 + b^2 = 23ab$, prove that $\log \frac{a+b}{5} = \frac{\log a + \log b}{2}$

15. Find the value of x, when log (x + 1) + log (x – 1) = log 11 + 2 log 3

16. Prove that

(i) log 108 = 3 log 3 + 2 log 2

(ii) $2 \log 5 + \log 8 - \frac{1}{2} \log 4 = 2$

(iii) $\log \frac{a^2}{bc} + \log \frac{b^2}{ca} + \log \frac{c^2}{ab} = 0$

Answers

2. (i) 56.16 (ii) 340.6
3. (i) 2 (ii) 3.149 (iii) 26. 93
   (iv) 16400 (v) 0.0000000002484 (vi) 0.7350
4. (i) 113 (ii) 0.9703 (iii) 6.143
5. (i) 0.0000000001903 (ii) 0.1234
6. 1.0791, 1.7323
7. – 4.4611
10. 3 log 2
11. (i) 8 (ii) 5, –2 (iii) 1.678
15. x = 10

CHAPTER 13

DATA ORGANIZATION

Graphical Representation of a frequency Distribution

When data of two items are compared with one another it is always easier to compare through graphs.

Generally the following types of graphs are used in representing frequency distributions:

1. Histograms
2. Frequency Polygon
3. Frequency curve
4. Cumulative frequency cure or the ogive

1. ***Histogram***: A histogram is a graph that represents the class frequency in a frequency distribution by vertical rectangles. For constructing a histogram, it is necessary to see that the class intervals are equal. The steps in constructing a histogram are:
    1. represent the class interval on base i.e., x-axis.
    2. represent the corresponding frequencies of class interval on y-axis.
    3. we draw rectangles with the area proportional to the frequencies of the respective class intervals.

***Example* 1:** Following are the ages of 360 patients getting medical treatment in a hospital on a day.

Ages	10-20	20-30	30-40	40-50	50-60	60-70
No. of Patients	90	50	60	80	50	30

Construct a histogram

***Solution*:** Taking class interval on x-axis and frequency on y-axis,

scale – Ages, 10 years = 1 cm on x-axis, patient 10 = 1 cm on y - axis.

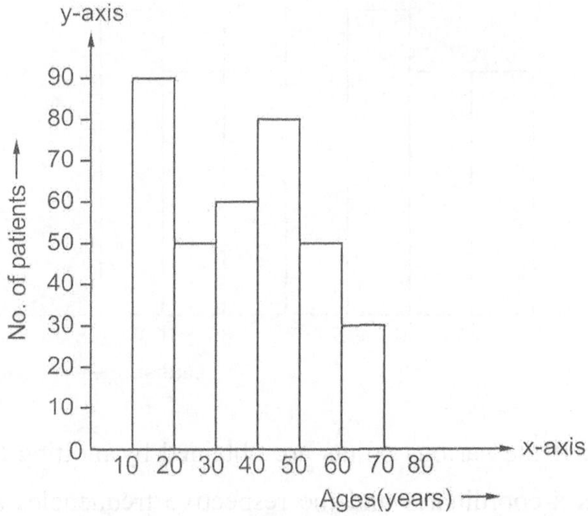

***Example* 2:** The water tax bills of 30 houses in a locality are as given below:

44, 84, 30, 95, 32, 96, 14, 112, 74, 88, 110, 102, 45, 75, 54, 74, 78, 66, 44, 35, 15, 20, 14, 40, 88, 76, 66, 112, 108

Construct a frequency distribution table and draw the histogram.

***Solution*:** Frequency distribution table

Class	**No.of Houses Tally Marks**	**Frequency**
10 – 25	1111	4
25 – 40	1111	4
40 – 55	~~1111~~	5
55 – 70	11	2
70 – 85	~~1111~~1	6
85 – 100	1111	4
100 – 115	~~1111~~	5
Total		**30**

Histogram

Taking class interval on x-axis and frequency on y-axis, scale 1 cm = 15 (class interval), 1 cm = 1 frequency

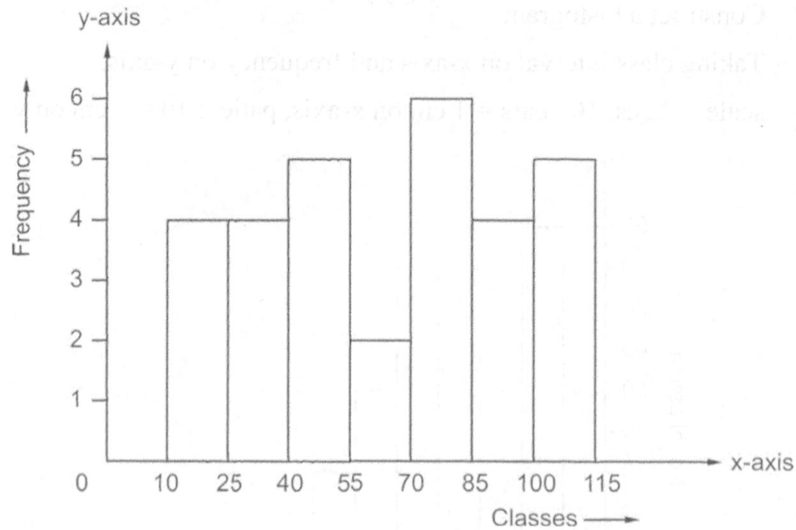

***Frequency Polygon*:** If the various points are obtained by plotting the central values of the class intervals as x-coordinates and the respective frequencies as the y-coordinates and these points are joined by straight lines taken in order, then form a polygon called frequency polygon. In a frequency polygon the variables or individuals of each class are assumed to be concentrated at the mid-point of the class interval.

***Example* 3:** Draw a frequency polygon for the data given below.

Class	0-10	10-20	20-30	30-40	40-50	50-60	60-70	70-80	80-90	90-100
Frequency	2	4	10	4	3	8	1	5	11	2

***Solution*:** Taking class interval 10 unit = 1 cm on x-axis, and frequency 1 unit = 1 cm on y-axis.

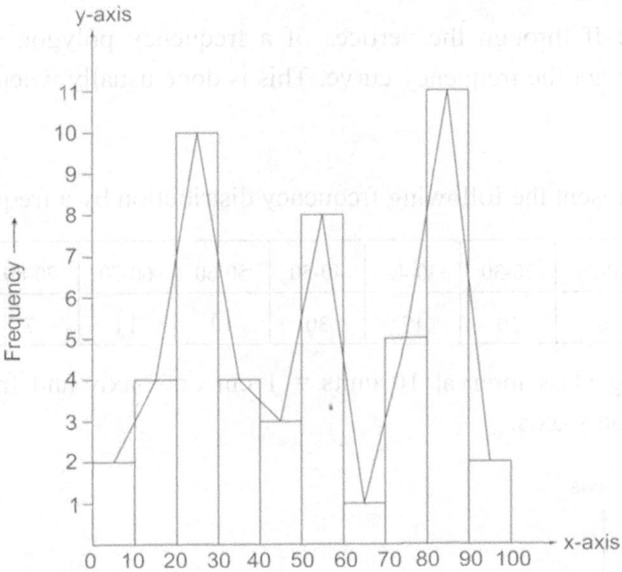

***Example* 4:** Represent the following data by a frequency polygon.

Marks upto	10	20	30	40	50
No. of Students	3	8	17	20	22

***Solution*:** To draw a frequency polygon, we convert the given data into frequency distribution as under.

Marks	0-10	10-20	20-30	30-40	40-50
Students	3	5	9	3	2

Taking marks 10 units = 1 cm on x-axis and no. of students 1 unit = 1 cm on y-axis.

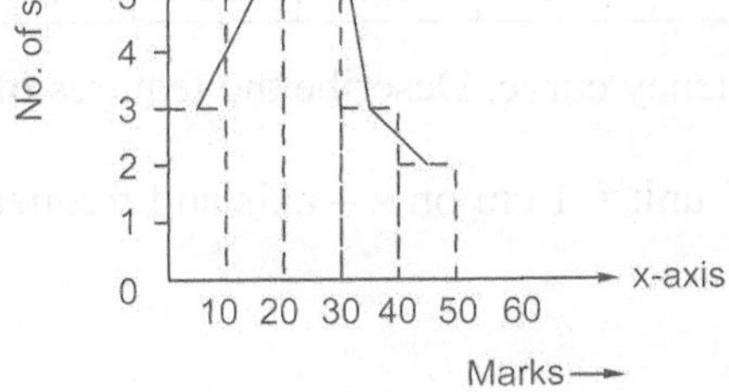

***Frequency Curve*:** If through the vertices of a frequency polygon a smooth freehand curve is drawn, we get the frequency curve. This is done usually when the class intervals are of small size.

***Example* 5:** Represent the following frequency distribution by a frequency curve.

Class	0-10	10-20	20-30	30-40	40-50	50-60	60-70	70-80	80-90	90-100
Frequency	2	6	10	17	30	10	14	7	4	1

***Solution*:** Taking class interval 10 units = 1 cm on x-axis and frequency 5 units = 1 cm on y-axis.

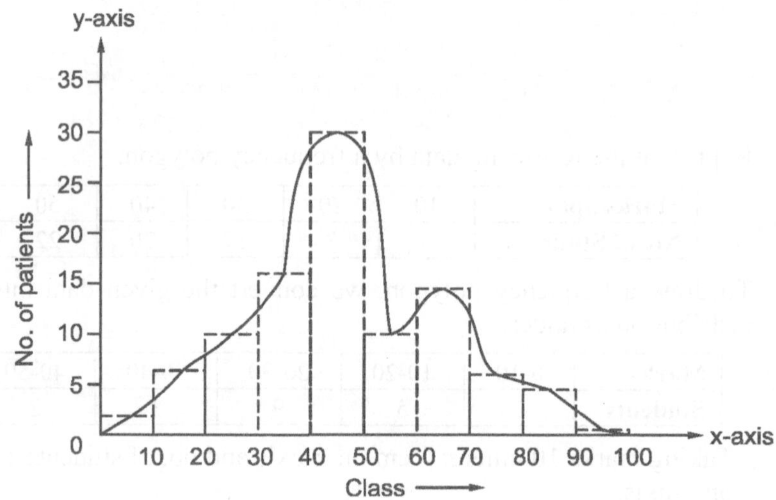

***Example* 6:** Suppose 1000 pieces of injection are inspected and the number of surface defects on each is recorded. If the number of defects is as below.

No. of Defects	0	1	2	3	4
Frequency	600	310	75	13	2

Draw a frequency curve. Describe the features of this curve.

***Solution*:** Taking class 1 unit = 1 cm on x – axis and frequency 100 units = 1 cm on y-axis.

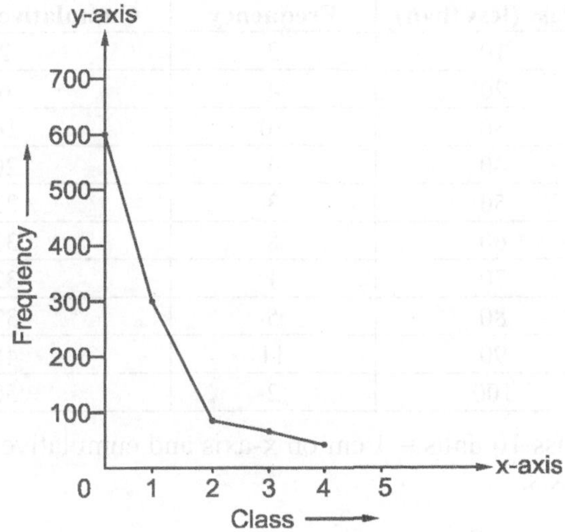

This graph shows

1. There are no surface defect for 600 pieces i.e., This is a minimum point.
2. If the number of pieces decreases then number of defects will increase.
3. The maximum number of defects are only 2 pieces.

Cumulative Frequency curve (Ogive): If from a cumulative frequency table the upper limits of the class taken as x-axis and the cumulative frequency as the y-axis and the points are plotted, then these points when joined by straight lines, we obtain less than type cumulative frequency polygon.

If more than cumulative frequency is plotted against the corresponding lower limits of each class and the points plotted are joined by straight lines, we obtain more than type cumulative frequency polygon. However, when the points plotted are joined by a freehand smooth curve, we obtain cumulative frequency curve.

***Example* 7**: Draw a less than type cumulative frequency polygon for the data given below.

Class	0-10	10-20	20-30	30-40	40-50	50-60	60-70	70-80	80-90	90-100
Frequency	2	4	10	4	3	8	1	5	11	2

Solution: Cumulative frequency table is

Class (less than)	Frequency	Cumulative frequency
10	2	2
20	4	6
30	10	16
40	4	20
50	3	23
60	8	31
70	1	32
80	5	37
90	11	48
100	2	50

Taking class 10 units = 1 cm on x-axis and cumulative frequency 5 units = 1 cm on y-axis.

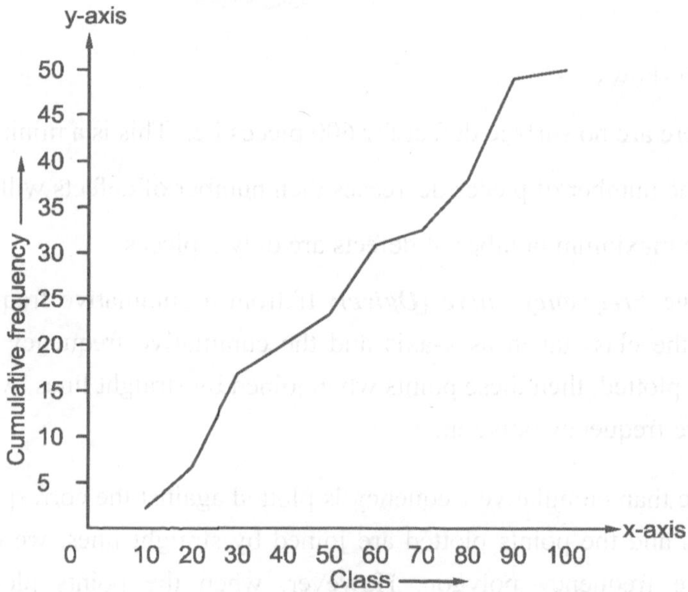

***Example* 8:** Following are the ages of 360 patients getting medical treatment in a hospital on a day.

Ages (in year)	10-20	20-30	30-40	40-50	50-60	60-70
No. of Patient	90	50	60	80	50	30

Construct the cumulative frequency table and cumulative frequency curve.

***Solution*:** Cumulative frequency table is

Ages (less than)	No. of Patient	Cumulative Frequency
20	90	90
30	50	140
40	60	200
50	80	280
60	50	330
70	30	360

Taking ages 10 units = 1 cm on x-axis and cumulative frequency 50 units = 1 cm on y-axis.

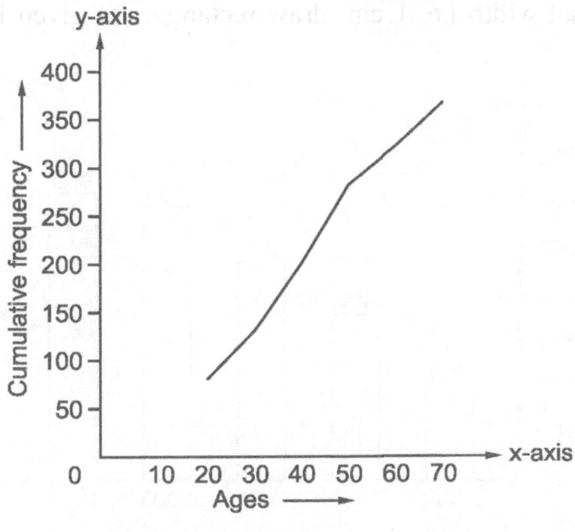

Diagrammatic Representation of Data: The data in different forms may be represented by graphs or diagrams. Diagrams have a more lasting effect on the brain. In case of time series data graphs make forecasting easier. Graphs help us in interpolation of values of the variables. The following are the main diagrams.

1. Bar diagram
2. Area diagram
3. Circle or Pie diagram
4. Sub divided Pie diagram
5. Prism and cubes diagram

1. ***Bar diagram***: When the comparision of simple magnitude of different items is done, the bar diagram are widely used. They are also called one dimensional diagrams. In bar diagrams, equal bases on a horizontal or vertical line are selected

and rectangles of equal width are constructed with the length proportional to the given data. The width of a bars is an arbitrary factor. The distance between two bars should be taken about one-half of the width of a bar.

***Example* 9:** Draw bar diagram to represent the following figures relating to the production of rice of a certain voltage.

Years	2000	2001	2002	2003	2004
Rice in Quintals	200	360	470	590	620

***Solution*:** Maximum value is 620 i.e., a scale of 1 cm = 100 quintals. Then length of different bars are represented by 2 cm, 3.6 cm, 4.7 cm, 5.9 cm and 6.2 cm with equal width i.e. 1 cm, draw rectangles of given length which is a bar diagram.

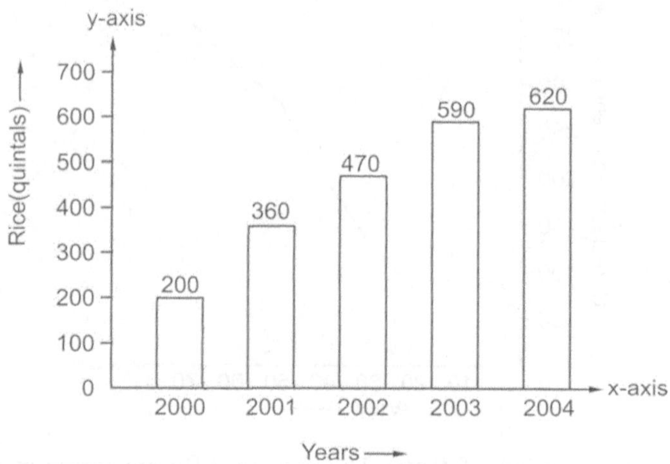

2. *Area Diagram*: When the difference between two quantities to be compared is large, bars would not show the comparision so clearly. In such cases circles or squares are used (2-D diagrams).

***Example* 10:** The following table gives the country of origin of feature films exhibits in India in 1984.

Country	India	America	Britain	Germany	Other Countries
No.of films	144	81	64	9	16

Represent them by Area Diagram.

***Solution*:** Here the ratio of greatest figure to the least is 144:9 $\Rightarrow$ 16 : 1, the bars would not show the comparision so clearly. Hence we make squares of sides of

12 cm, 9 cm, 8 cm, 3 cm and 4 cm. The area of the squares are proportional to the given figures.

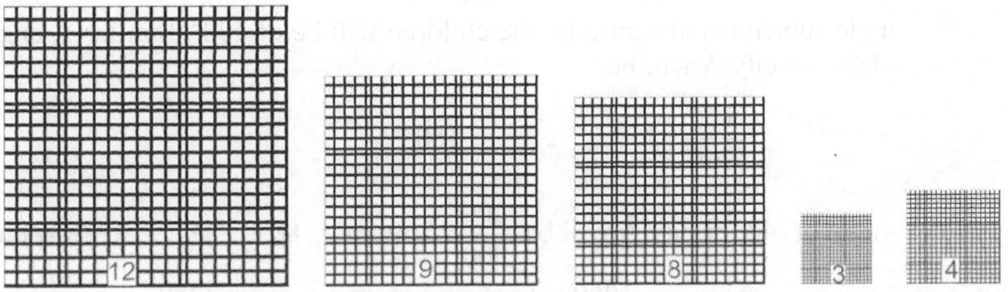

3. **Circle or Pie diagram:** When circles are drawn to represent its area equivalent to the figures, they are said to form pie-diagram or circle diagram. In case of circles, the square roots of magnitudes are proportional to the radii.

***Example* 11:** The circle diagram of above example 10 are as under.

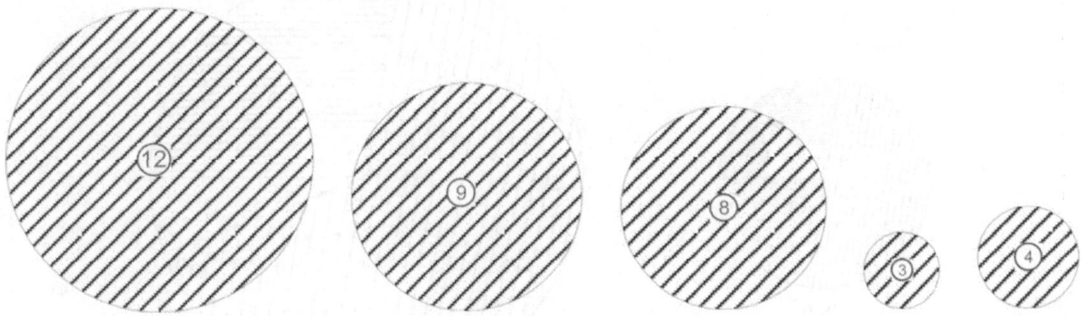

4. **Sub-divided pie-diagram:** Sub divided Pie-diagram are used when comparision of the component parts is done with another and the total. The total value is equated to 360° and then the angles corresponding to component parts are calculated.

***Example* 12:** Construct the Pie-diagram for the following data.

	Children	**Adults**	**Old**	**Total**
City A	250	450	200	900
City B	1000	2250	350	3600

Solution: We see the radii of two circles should be in the ratio

$$\sqrt{900} : \sqrt{3600} \quad \Rightarrow \quad 1:2$$

Draw two circles of radius 2 cm and 4 cm. The children of city A are 250 out of 900 and hence they must be $\frac{250}{900} \times 360$ out of 360 $\Rightarrow$ 100. Therefore, the angle subtended at centre by the children will be of 100°. Similarly, angle of adults for city A will be

$$\frac{450}{900} \times 360 \quad \Rightarrow \quad 180^{\circ}$$

Angle of old for city A will be $\frac{200}{900} \times 360 \Rightarrow 80^{\circ}$ and corresponding angles for city B are $\frac{1000}{3600} \times 360 = 100^{\circ}$ for children, $\frac{2250}{3600} \times 360 = 225^{\circ}$ for adults and $\frac{350}{3600} \times 360 = 35^{\circ}$ for old.

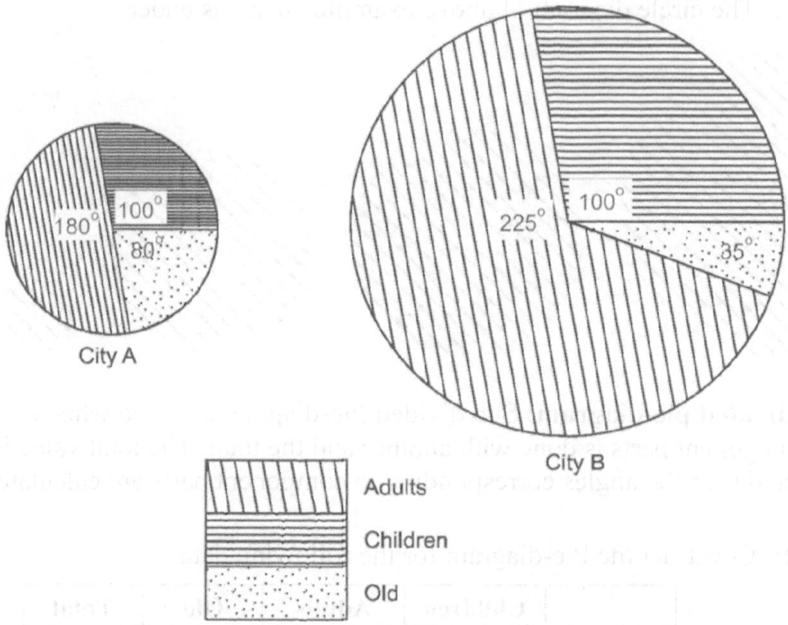

5. ***Prism and Cubes*:** When the ratio between the two quantities to be compared is very great so that even area diagrams are not suitable, the data can be represented by spheres, cubes or prisms. Cubes are in common use. Cubes are constructed on sides which are taken in the ratio of cube roots of the given quantities. It is also called three dimensional diagrams (3-D diagrams). For example,

If data given be 1, 216 and 1000.

In this we make three cubes of sides 1, $\sqrt[3]{216}$, $\sqrt[3]{1000}$ $\Rightarrow$ 1, 6, 10.

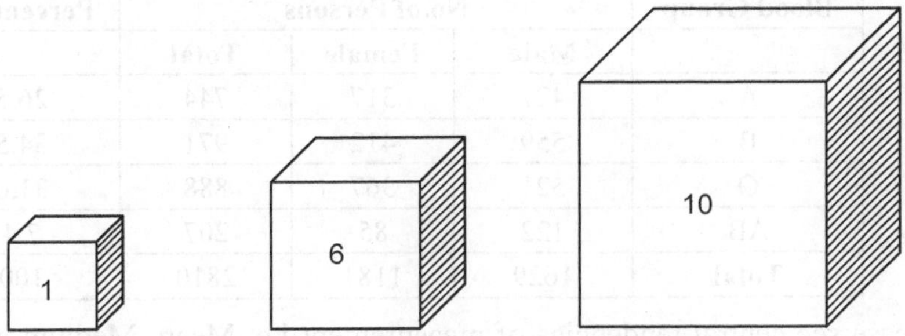

Exercise

1. Draw histogram and frequency polygon for the following table.

0-10	10-20	20-30	30-40	40-50
2	4	6	8	4

2. Represent the following distribution by a frequency polygon.

Scores	90-99	80-89	70-79	60-69	50-59	40-49	30-39	20-29
Frequency	2	12	22	20	14	3	1	1

3. The following table shows the frequency distribution for the number of students per teacher in 750 colleges and professional schools.

Students	x	1	4	7	10	13	16	19	22	25	28
Frequency	y	7	46	165	195	189	89	28	19	9	3

Draw a histogram and a frequency Polygon.

4. Draw a cumulative graph of the following distribution showing the monthly wages of a group of workmen.

Wages (in Rs.)	20	21	22	23	24	25	26	27	28
No. of workmen	8	10	11	16	20	25	15	9	6

5. Construct a Pie chart from the following data. (Distribution of blood group in minority community of Gujarat).

Blood Group	No.of Persons			Percentage
	Male	Female	Total	
A	427	317	744	26.5
B	559	412	971	34.5
O	521	367	888	31.6
AB	122	85	207	7.4
Total	1629	1181	2810	100

Remark: Please see central tendencies of measurement i.e. Mean, Medium and Mode in Part I.

Partition Values: These are the values of the variate which divide the total frequency into a number of equal parts. Some of them are quartiles, deciles and percentiles.

Quartiles: Quartiles are those values of the variate which divide the total frequency into four equal parts. Thus, Q_1 would stand for first quartile are called lower quartile and Q_3 is third quartile called upper quartile. The formula for computation are:

$$Q_1 = l + \frac{\frac{N}{4} - F}{f} \times i, \text{ (lower quartile)}$$

$$Q_2 = l + \frac{\frac{2N}{4} - F}{f} \times i$$

$$Q_3 = l + \frac{\frac{3N}{4} - F}{f} \times i, \text{ (upper quartile)}$$

$$Q_4 = l + \frac{\frac{4N}{4} - F}{f} \times i$$

Deciles: Deciles are those values of the variate which divide the total frequency into 10 equal parts. Thus.

The first decile $D_1 = l + \frac{\frac{N}{10} - F}{f} \times i$

The second decile $D_2 = l + \frac{\frac{2N}{10} - F}{f} \times i$

***Percentiles*:** Percentiles are the values of the variate which divides the total frequency into 100 equal parts. Thus

First percentile is $$P_1 = l + \frac{\frac{N}{100} - F}{f} \times i$$

Second percentile is $$P_2 = l + \frac{\frac{2N}{100} - F}{f} \times i$$

***Example* 1:** Calculate three quartiles, 7th decile and eighty second percentile

Salary (in Rs.)	0-10	10-20	20-30	30-40	40-50
No. of Servants	22	38	46	35	20

***Solution*:**

Class	**Frequency (f)**	**Cumulative frequency (F)**
0 – 10	22	22
10 – 20	38	60
20 – 30	46	106
30 – 40	35	141
40 – 50	20	161
Total	**161**	

For First quartile

$$Q_1 = l + \frac{\frac{N}{4} - F}{f} \times i$$

Here $$\frac{N}{4} = \frac{161}{4} = 40.25$$

$$\therefore \quad Q_1 = 10 + \frac{40.25 - 22}{38} \times 10$$

$$= 10 + 4.80 = \text{Rs. } 14.80$$

For second quartile,

$$\frac{2N}{4} = \frac{2 \times 161}{4} = 80.5$$

$$Q_2 = 20 + \frac{80.5 - 60}{46} \times 10$$

$$= 20 + 4.6 = \text{Rs. } 24.6$$

For third quartile,

$$\frac{3N}{4} = \frac{3 \times 161}{4} = 120.75$$

$$Q_3 = 30 + \frac{120.75 - 106}{35} \times 10$$

$$= 30 + 4.21 = \text{Rs. } 34.21$$

For 7th decile

$$\frac{7N}{10} = \frac{7 \times 161}{10} = 112.7$$

$$D_7 = l + \frac{\frac{7N}{10} - F}{f} \times i$$

$$= 30 + \frac{112.7 - 106}{35} \times 10$$

$$= 30 + 1.91 = \text{Rs. } 31.91$$

For 82nd percentile,

$$\frac{82N}{100} = \frac{82 \times 161}{100} = 132.02$$

$$P_{82} = l + \frac{\frac{82N}{100} - F}{f} \times i$$

$$= 30 + \frac{132.02 - 106}{35} \times 10$$

$$= 30 + 7.43$$

$$= \text{Rs. } 37.43$$

***Example* 2:** Calculate the lower and upper quartiles, fourth decile and 60th percentile of the following distribution.

Marks	5–10	10-15	15-20	20-25	25-30	30-35	35-40	40-45
No.of Students	5	6	15	10	5	4	2	2

***Solution*:** The cumulative frequency table is

Class	**Frequency**	**Cumulative frequency**
5-10	5	5
10-15	6	11
15-20	15	26
20-25	10	36
25-30	5	41
30-35	4	45
35-40	2	47
40-45	2	49

(i) For lower quartile Q_1, we have $\frac{N}{4}=\frac{49}{4}$ = 12.25 which lies in the class 15-20 then l = 15, F = 11, f = 15, i = 5

$$Q_1 = l + \frac{\frac{N}{4} - F}{f} \times i$$

$$= 15 + \frac{12.25 - 11}{15} \times 5 = 15 + 0.41 = 15.41$$

(ii) For upper quartile Q_3, we have $\frac{3N}{4}=\frac{3\times 49}{4}$ = 36.75

Which lies in class 25 – 30 then

l = 25, F = 36, f = 5, i = 5

$$Q_3 = l + \frac{\frac{3N}{4} - F}{f} \times i$$

$$= 25 + \frac{36.75 - 36}{5} \times 5 = 25 + 0.75 = 25.75$$

(iii) For 4th decile, we have $\frac{4N}{10} = \frac{4 \times 49}{10} = 19.6$ which lies in the class 15-20. Thus

$l = 15, f = 15, i = 5, F = 11$

$$\therefore \quad D_4 = l + \frac{\frac{4N}{10} - F}{f} \times i$$

$$= 15 + \frac{19.6 - 11}{15} \times 5 = 15 + 2.86 = 17.86$$

(iv) For 60th Percentile, we have $\frac{60N}{100} = \frac{60 \times 49}{100} = 29.4$ which lies in the class 20-25

then

$l = 20, F = 26, f = 10, i = 5$

$$\therefore \quad P_{60} = 20 + \frac{29.4 - 26}{10} \times 5$$

$$= 20 + 1.7 = 21.7$$

***Example* 3:** Calculate lower and upper quartile for the following data.

Class	0-4	4-6	6-8	8-12	12-18	18-20
Frequency	4	6	8	12	7	2

***Solution*:** The cumulative frequency table is

Class	**Frequency**	**Cumulative frequency (F)**
0-4	4	4
4-6	6	10
6-8	8	18
8-12	12	30
12-18	7	37
18-20	2	39
	N = 39	

(i) For lower quartile Q_1, we have $\frac{N}{4} = \frac{39}{4} = 9.75$ which lies in the class 4-6 then

$l = 4$, F = 4, f = 6, i = 2

$$Q_1 = l + \frac{\frac{N}{4} - F}{f} \times i$$

$$= 4 + \frac{9.75 - 4}{6} \times 2 = 4 + 1.92 = 5.92$$

(ii) For upper quartile Q_3 we have $\frac{3N}{4} = \frac{3 \times 39}{4} = 29.25$ which lies in the class 8-12.

Thus,

$l = 8$, F = 18, f = 12, i = 4

$$Q_3 = l + \frac{\frac{3N}{4} - F}{f} \times i$$

$$= 8 + \frac{29.25 - 18}{12} \times 4 = 8 + 3.75 = 11.75$$

1. Calculate 1st quartiles, 5th decile and 75th percentile of the following data

Marks	0–10	10-20	20-30	30-40	40-50
No. of Students	22	38	46	35	20

2. Calculate the lower and upper quartiles, 6th decile and 70th percentile of the following distribution

C.I	0-5	5-10	10-15	15-20	20-25	25-30	30-35	35-40	40-45
f	2	4	6	10	15	5	4	3	2

3. Find 30th percentile, 5th deciles and 3rd quartiles of the following distribution

Salary in Rs	20-30	30-40	40-50	50-60	60-70	70-80
No. of Emploies	4	5	10	12	8	6

CHAPTER 14

MEASURES OF DISPERSION

Definition of Dispersion

To explain the term 'Dispersion'. Let us consider an example of two player who earns the following numbers of game in 6 working days of a week.

	Mon	Tue	Wed	Thu	Fri	Sat	Total
I Player	20	20	25	25	30	30	150
II Player	15	20	25	25	30	35	150

Here, each player scored total 150 numbers in 6 days and the mean in both the cases is 25. Thus, there is no difference in mean but we notice that in the first case the score varies from 20 to 30 and in second case, the score varies from 15 to 35 i.e., the greatest deviation from the mean in first case is 5 and in second case is 10. Clearly it indicates a difference in the two series. Such variation is called dispersion.

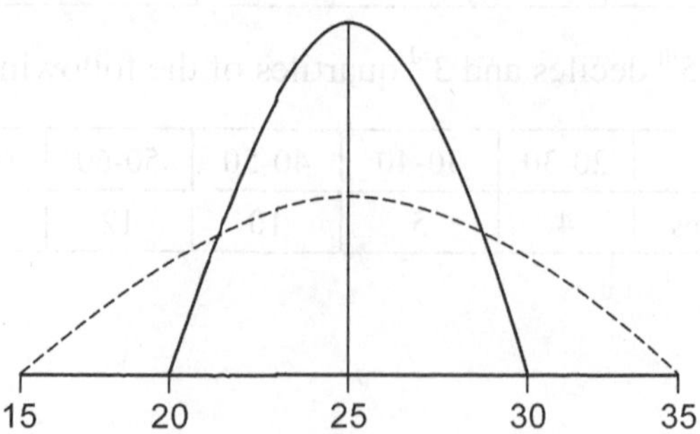

The word dispersion is used in two senses in statistics:

1. The scatteredness of the values of a variable due to variation among themselves is called dispersion.
2. The deviation from a measure of central tendency or any other fixed value are not uniform in their size. The scatteredness of these deviation is also referred to as dispersion.

The following are the measures of dispersion which are in common use:

1. Range
2. Quartile deviation or semi interquartile range
3. Average deviation
4. Standard deviation.

Range: Range is defined as the simplest possible measure of dispersion which is the difference between the greatest and the least values of the variable. It is also called absolute range.

$$\text{Range} = \text{Greatest value} - \text{Least value}$$

$$\text{Range coefficient of dispersion (relative range)} = \frac{\text{Difference between extreme values}}{\text{Sum of the extreme values}}$$

Merits, Demerits and uses of Range

Merits

1. Range can be easily understood.
2. Range is easy to calculate.
3. Range is absolute measure of dispersion.

Demerits

1. The range is subject to fluctuations of considerable magnitude from sample to sample.
2. It takes into account only the extreme items.
3. Range is based on all the observations.

Uses

1. Quality control
2. Weather forecasts
3. Fluctuation in the share prices

***Example* 1:** Find out the range and coefficient of range of the income of a certain person given in rupees for 12 months in a year.

139, 150, 151, 151, 157, 158, 160, 161, 162, 162, 173, 175

***Solution*:** First, we arrange the data in ascending order

139, 150, 151, 151, 157, 158, 160, 161, 162, 162, 173, 175

here, least value = 139, greatest value = 175

$$\therefore \quad \text{range} = \text{greatest value} - \text{least value}$$

$$= 175 - 139$$

$$= 36$$

$$\text{Coefficient of range} = \frac{\text{Difference between extreme values}}{\text{Sum of the Extreme values}}$$

$$= \frac{36}{175 + 139}$$

$$= \frac{36}{314}$$

$$= 0.115$$

***Example* 2:** Calculate range and the coefficient of range from the following data.

Marks	20-30	30-40	40-50	50-60	60-70
No. of Students	120	201	150	75	25

***Solution*:** Here, least value = 20

Greatest value = 70

Range = Greatest value – least value

$$= 70 - 20$$

$$= 50$$

$$\text{Coefficient of Range} = \frac{\text{Difference between extreme values}}{\text{Sum of the Extreme values}}$$

$$= \frac{50}{70 + 20}$$

$$= \frac{50}{90} = 0.55$$

Quartile Deviation: The difference between the upper and lower quartile is known as the Interquartile Range. Half of the interquartile range is called the Semi interquartile range or the quartile deviation. It is denoted by Q.D.

Let upper quartile be Q_3 and lower quartile be Q_1, then inter quartile range = $|Q_3 - Q_1|$

$$\text{Quartile Deviation (Q.D)} = \frac{Q_3 - Q_1}{2}$$

$$\text{Coefficient of Q.D} = \frac{Q_3 - Q_1}{Q_3 + Q_1}$$

Merits, Demerits and uses of Quartile Deviation

Merits

1. The quartile deviation is easily computed.
2. It is a better measure of dispersion than the range.
3. The quartile deviation is not affected by the extreme values.

Demerits

1. It does not take into account all the items.
2. It is affected by fluctuation of sampling.
3. It does not show the scatter around an average but rather a distance on a scale.

Uses

It can be used only for descriptive statistics.

***Example* 3:** Calculate range, inter quartile range and quartile deviation for the marks of 63 students in statistics given below.

Marks	0-10	10-20	20-30	30-40	40-50	50-60	60-70	70-80	80-90	90-100
No. of Students	5	7	10	16	11	7	3	2	2	0

Solution:

Marks	**Frequency (f)**	**Cumulative frequency (F)**
0 - 10	5	5
10 - 20	7	12
20 – 30	10	22
30 – 40	16	38
40 – 50	11	49

Contd....

50 – 60	7	56
60 – 70	3	59
70 – 80	2	61
80 – 90	2	63
90 – 100	0	63
Total	63	

Smallest value = 0

Greatest value = 100

Range = Greatest value – smallest value

= 100 – 0 = 100

$$Q_1 = l + \frac{\frac{N}{4} - F}{f} \times i$$

Here $\frac{N+1}{4} = \frac{63+1}{4} = 16$, l = 20, F = 12, f = 10, i = 10

$$Q_1 = 20 + \frac{15.75 - 12}{10} \times 10 = 23.75$$

$$Q_3 = l + \frac{\frac{3N}{4} - F}{f} \times i$$

Here $\frac{3(N+1)}{4} = 48$, $l = 40$, F = 38, f = 11, i = 10

$$Q_3 = 40 + \frac{47.25 - 38}{11} \times 10 = 48.4$$

$\therefore$ interquartile range = $Q_3 - Q_1$

$= 48.4 - 23.75$

$= 24.65$

$$\text{Quartile deviation} = \frac{Q_3 - Q_1}{2}$$

$$= \frac{24.64}{2} = 12.325$$

Mean deviation (Average Deviation): Arithmetic average of the group measured from an average (mean, median, or mode) taking all deviations as positive is known as mean deviation. Thus, mean deviation about any point A given by

$$\delta_A = \frac{1}{N}|\Sigma x - A| \qquad \text{or} \qquad \delta_A = \frac{\Sigma|x-\bar{x}|}{N}$$

$$\text{Mean deviation from median is } \delta M_d = \frac{|x - M_d|}{N}$$

$$\text{Mean deviation from mode is } \delta M_0 = \frac{|x - M_0|}{N}$$

$$\text{Coefficient of mean deviation} = \frac{\text{Mean Deviation from mean}}{\text{mean}}$$

$$\text{or} \qquad = \frac{\text{Mean Deviation from mode}}{\text{mode}}$$

$$\text{or} \qquad = \frac{\text{Mean Deviation from median}}{\text{median}}$$

Merits, Demerits and uses of Mean Deviation

Merits

1. Mean deviation is easy to understand and calculate.
2. Mean deviation about an arbitrary point is least when the point is median.
3. It is very less affected by the extreme values.

Demerits

1. Mean deviation sometime not useful for statistical references.
2. Mean deviation does not give accurate results.
3. It is not suitable for further algebraic treatments.

Uses

Mean deviation and its coefficient are used in studying economic problems such as distribution of income and wealth in a society.

***Example* 4:** Find the mean deviation from the arithmetic mean of the following distribution:

Marks	0 – 10	10 – 20	20 – 30	30 – 40	40 – 50
No.of students	5	8	15	16	6

Solution:

Class	Mid value (x)	Frequency (f)	fx	x – M	f|x – M|
0 – 10	5	5	25	–22	110
10 – 20	15	8	120	–12	96
20 – 30	25	15	375	–2	30
30 – 40	35	16	560	8	128
40 – 50	45	6	270	18	108
Total		50	1350		472

$$\text{Arithmetic mean } M = \frac{\sum fx}{\sum f} = \frac{1350}{50} = 27$$

$$\text{Mean deviation } \delta M = \frac{\sum f|x - M|}{\sum f}$$

$$= \frac{472}{50} = 9.44$$

***Example* 5 :** Calculate mean deviation from median of the data given below:

Age in year under	10	20	30	40	50	60	70	80
No.of person died	15	30	53	75	100	110	115	125

Solution:

$$\text{Median } M_d = l + \frac{\frac{N}{2} - F}{f} \times i$$

Here $\frac{N}{2} = \frac{125}{2} = 62.5$

Class interval is 30 – 40, $l = 30$, f = 22, F = 53, $i = 10$

Class	f	F	Mid value (x)	x – M_d	f $\|x - M_d\|$
0 – 10	15	15	5	–29.3	439.5
10 – 20	15	30	15	–19.3	289.5
20 – 30	23	53	25	–9.3	213.9
30 – 40	22	75	35	0.7	15.4
40 – 50	25	100	45	10.7	267.5
50 – 60	10	110	55	20.7	207.0
60 – 70	5	115	65	30.7	153.5
70 – 80	10	125	75	40.7	407.0
Total	125				1993. 3

$$M_d = 30 + \frac{62.5 - 53}{22} \times 10$$

$$= 30 + 4.32$$

$$= 34.32$$

Mean diviation about median

$$\delta M_d = \frac{\sum f|x - Md|}{\sum f}$$

$$= \frac{1993.3}{125} = 15.95$$

***Example* 6:** Calculate mean deviation about mode for the following distribution

Class	5-15	15-25	25-35	35-45	45-55	55-65
Frequency	12	19	26	23	14	6

***Solution*:**

Class	**f**	**Mid value (x)**	**x – M_0**	**f (x – M_0)**
5 – 15	12	10	–22	264
15 – 25	19	20	–12	228
25 – 35	26	30	–2	52
35 – 45	23	40	8	184
45 – 55	14	50	18	252
55 – 65	6	60	28	168
Total	100			1148

$$\text{Mode } M_0 = l + \frac{f - f_{-1}}{2f - f_{-1} - f_1} \times i$$

$$= 25 + \frac{26 - 19}{52 - 19 - 23} \times 10 = 32$$

Mean deviation about mode

$$\delta M_0 = \frac{\sum f|x - M_0|}{\sum f}$$

$$= \frac{1148}{100}$$

$$= 11.48$$

Standard Deviation: Standard deviation is the square root of the arithmetic mean of the square deviation. It is denoted by σ (sigma).

i.e., $$\sigma = \sqrt{\frac{\Sigma(x-M)^2}{n}} \quad \text{or} \quad \sigma = \sqrt{\frac{\Sigma f(x-M)^2}{\Sigma f}}$$

Method of Computing Standard Deviation

For individual series

(i) Direct method $$\sigma = \sqrt{\frac{\Sigma(x-M)^2}{n}}, \quad M = \frac{\Sigma x}{n}$$

(ii) Short cut method $$\sigma = \sqrt{\frac{\Sigma x^2}{n} - \left(\frac{\Sigma x}{n}\right)^2}$$

For Discrete series

(i) Direct method $$\sigma = \sqrt{\frac{\Sigma f(x-M)^2}{N}}$$

(ii) Short-cut method $$\sigma = \sqrt{\frac{\Sigma fu^2}{N} - \left(\frac{\Sigma fu}{N}\right)^2}$$

For grouped series

(i) direct method $$\sigma = \sqrt{\frac{\Sigma f(x-M)^2}{N}}$$

(ii) Short cut method $$\sigma = \sqrt{\frac{\Sigma fu^2}{N} - \left(\frac{\Sigma fu}{N}\right)^2}$$

(iii) Step-deviation method $$\sigma = i\sqrt{\frac{\Sigma fu^2}{N} - \left(\frac{\Sigma fu}{N}\right)^2}, \quad u = \frac{x-A}{i}$$

Variance: Variance is the square of standard deviation and is denoted by σ^2.

$$\text{Variance } (\sigma) = \frac{\Sigma(x-M)^2}{n}$$

Coefficient of Variance: Coefficient of variance is a relative measure of dispersion and denoted by C.V.

$$C.V = \frac{\sigma}{M} \times 100$$

Note: The series for which the variation is less is said to be more consistent.

***Standard Error of Means*:** When some observation in error then we can find the correct standard deviation by subtracting the uncorrect and adding the correct expressions. Hence,

$$\text{Corrected } \sigma = \sqrt{\frac{\text{corrected} \sum fx^2}{N} - (\text{corrected mean})^2}$$

***Example* 1:** For 100 observations, the mean and variance were obtained as 19 and 9 respectively. Later on it was discovered that the observation 12 was mis read as 21. Calculate the correct mean and variance of the actual data.

***Solution*:** Here $n = 100, \bar{x} = 19, \sigma^2 = 9$

Let x_1, be the uncorrect value and x_1^1 be the correct value of one observation. Thus $x_1 = 21,\ x_1^1 = 12$.

Therefore,

$$\begin{aligned}
\Sigma' x &= \text{correct sum of observations} \\
&= \Sigma x - x_1 + x_1^1 \\
&= n\bar{x} - x_1 + x_1^1 \\
&= 1900 - 21 + 12 = 1891
\end{aligned}$$

∴ Correct Mean

$$\bar{x}^1 = \frac{\Sigma' x}{n} = \frac{1891}{100} = 18.91$$

Again, correct

$$\begin{aligned}
\Sigma' x^2 &= \Sigma x^2 - x_1^2 + (x_1^1)^2 \\
&= n(\sigma^2 + \bar{x}^2) - x_1^2 + (x_1^1)^2 \\
&= 100(9 + 361) - (21)^2 + (12)^2 \\
&= 3700 - 441 + 144 = 36703
\end{aligned}$$

∴ Correct variance

$$\begin{aligned}
\sigma_1^2 &= \frac{\Sigma' x^2}{n} - (\bar{x}^1)^2 \\
&= \frac{36703}{100} - (18.91)^2 \\
&= 9.4419
\end{aligned}$$

***Fiducial (Confidence) Limits*:** Let μ be the mean and σ be the standard deviation of the population from which a random sample of size n is taken. Let $\bar{x}$ be the mean of the sample. If σ is unknown and not μ and μ lies within a range of possible values for which x is not significant at a specified level of probability. The relative deviation of x from μ is the ratio of $\bar{x}-\mu$ to the standard error of $\bar{x}$ i.e., $\frac{(\bar{x}-\mu)\sqrt{n}}{\sigma}$.

If $\bar{x}$ is not significant at 5% level of probability, this relative deviation must be less than that which is exactly numerically with a probability of 5%. Such a relative deviation is 1.96.

Hence the observed mean will not be significant at 5% level of probability

$$\left|\frac{\bar{x}-\mu}{\sigma}\sqrt{n}\right| < 1.96.$$

i.e., if $$\bar{x}-\frac{1.96\sigma}{\sqrt{n}} < \mu < \bar{x}+\frac{1.96\sigma}{\sqrt{n}}$$

The values $\bar{x} \pm \frac{1.96\sigma}{\sqrt{n}}$ are called 95% fiducial limits or confidence limits.

Merits, Demerits and uses of Standard Deviation

Merits

1. Standard deviation is based on all the observation.
2. Standard deviation represents true measurement of dispersion of a series.
3. It is extremely useful in correleation.

Demerits

1. It is difficult to compute.
2. It gives more weightage to extreme values.
3. It consumes more time and labour.

Uses

1. Standard deviation is widely used in biological studies.
2. It is most widely used in measure of dispersion.

***Example* 7:** Find the mean, median and standard deviation for the following.

Size of item	6	7	8	9	10	11	12
Frequency	3	6	9	13	8	5	4

***Solution*:** Cumulative frequency table is

x	f	C.f	fx	x – M	$(x – M)^2$	$f(x – M)^2$
6	3	3	18	–3	9	27
7	6	9	42	–2	4	24
8	9	18	72	–1	1	9
9	13	31	117	0	0	0
10	8	39	80	1	1	8
11	5	44	55	2	4	20
12	4	48	48	3	9	36
	N = 48		$\Sigma fx = 432$			$\Sigma f(x-M)^2 = 124$

(i) Mean $M = \frac{\Sigma fx}{N} = \frac{432}{48} = 9$

(ii) Median $M_d = \frac{N}{2} = \frac{48}{2} = 24$

∴ $M_d = 9$

(iii) Standard deviation $\sigma_M = \sqrt{\frac{\Sigma f(x-M)^2}{\Sigma f}}$

$$= \sqrt{\frac{124}{48}} = 1.6072 = 1.61$$

***Example* 8:** Calculate the mean and standard deviation from the following table.

Marks (more than)	0	10	20	30	40	50	60	70
No.of students	100	90	75	50	25	15	5	0

***Solution*:**

Class	f	Mid value (x)	$u = \frac{x - 35}{10}$	fu	u^2	fu^2
0 – 10	10	5	–3	–30	9	90
10 – 20	15	15	–2	–30	4	60
20 – 30	25	25	–1	–25	1	25
30 – 40	25	35	0	0	0	0
40 – 50	10	45	1	10	1	10
50 – 60	10	55	2	20	4	40
60 – 70	5	65	3	15	9	45
	N = 100			$\Sigma fu = -40$		$\Sigma fu^2 = 270$

Let us assume that a = 35, i = 10

(i) Arithmetic mean $m = a + \frac{\Sigma fu}{N} \times i$

$$= 35 + \frac{(-40)}{100} \times 10$$

$$= 35 - 4 = 31$$

(ii) Standard deviation $\sigma = \sqrt{\frac{\Sigma fu^2}{N} - \left(\frac{\Sigma fu}{N}\right)^2}$

$$= 10\sqrt{\frac{270}{100} - \left(\frac{-40}{100}\right)^2}$$

$$= 10\sqrt{2.7 - 0.16}$$

$$= 10 \times 1.5937$$

$$= 15.93$$

***Example* 9:** Calculate the mean, median and standard deviation from the following table.

Defect in size (mm)	1-10	11-20	21-20	31- 40	41-50	51-60
No. of items	3	6	26	31	16	8

***Solution*:**

Class	f	Mid value (x)	C.F	$u = \frac{x - 25.5}{10}$	u^2	fu	fu^2
0.5 – 10.5	3	5.5	3	–2	4	–6	12
10.5 = 20.5	6	15.5	9	–1	1	–6	6
20.5 – 30.5	26	25.5	35	0	0	0	0
30.5 – 40.5	31	35.5	66	1	1	31	31
40.5 = 50.5	16	45.5	82	2	4	32	64
50.6 – 60.5	8	55.5	90	3	9	24	72
	$\Sigma f = 90$					$\Sigma fu = 75$	$\Sigma fu2 = 185$

(i) Let us take assumed mean = 25.5, i = 10

Mean $M = a + \frac{\Sigma fu}{\Sigma f} \times i$

$$= 25.5 + \frac{75}{90} \times 10$$

$$= 25.5 + 8.33 = 33.83$$

(ii) Median class $= \frac{N^{th}}{2}$ term

$$= \frac{90}{2} = 45^{th} \text{ term,}$$

which lies in class interval 30.5 – 40.5

$\therefore$ $l = 30.5$, $F = 35$, $f = 31$, $i = 10$

$$M_d = l + \frac{\frac{N}{2} - F}{f} \times i$$

$$= 30.5 + \frac{\frac{90}{2} - 35}{31} \times 10$$

$$= 30.5 + 3.22$$

$$= 33.72$$

(iii) Standard Deviation $\sigma = i\sqrt{\frac{\Sigma fu^2}{N} - \left(\frac{\Sigma fu}{N}\right)^2}$

$$= 10\sqrt{\frac{185}{90} - \left(\frac{75}{90}\right)^2}$$

$$= 10\sqrt{2.0555 - 0.6944}$$

$$= 10 \times 1.166$$

$$\sigma = 11.66$$

***Example* 10:** The team A and B made the following goals in the football match.

No.of Goals	No.of Games A	No.of Games B
0	27	17
1	9	9
2	8	6
3	5	5
4	4	3

Calculate Standard deviation for team A and team B.

***Solution*:**

x	f_A	$f_A.x$	$(x - M_A)^2$	$f_A(x - M_A)^2$	f_B	$f_B.x$	$(x-M_B)^2$	$f_B(x-M_B)^2$
0	27	0	1.21	32.67	17	0	1.44	24.48
1	9	9	0.01	0.09	9	9	0.04	0.36
2	8	16	0.81	6.48	6	12	0.64	3.84
3	5	15	3.61	18.05	5	15	3.24	16.20
4	4	16	8.41	33.64	3	12	7.84	23.52
Total	$\Sigma f_A = 53$	$\Sigma xf_A = 56$		90.93	$\Sigma f_B = 40$	$\Sigma xf_B = 48$		68.4

(i) Standard deviation from team 'A'

$$\text{Mean } (M_A) = \frac{\Sigma xf_A}{\Sigma f_A}$$

$$= \frac{56}{53} = 1.06 = 1.1 \text{ (approximate)}$$

$$\therefore \quad \sigma_A = \sqrt{\frac{\Sigma f_A (x - M_A)^2}{\Sigma f_A}}$$

$$= \sqrt{\frac{90.93}{53}} = 1.3098$$

(ii) Standard deviation from team 'B'

$$\text{Mean } M_B = \frac{\Sigma x f_B}{\Sigma f_B}$$

$$= \frac{48}{40} = 1.2$$

$$\sigma_B = \sqrt{\frac{\Sigma f_B (x - M_B)^2}{\Sigma f_B}}$$

$$= \sqrt{\frac{68.4}{40}} = 1.3077$$

***Example* 11:** For the distribution

x	5	6	7	8	9	10	11	12	13	14	15
f	18	25	34	47	68	90	80	62	35	27	11

Find the mean, median and standard deviation

***Solution*:**

x	f	C.F	fx	(x-10)	$(x-10)^2$	$f(x-10)^2$
5	18	18	190	–5	25	450
6	25	43	150	–4	16	400
7	34	77	238	–3	9	306
8	47	124	376	–2	4	188
9	68	192	612	–1	1	68
10	90	282	900	0	0	0
11	80	362	880	1	1	80
12	62	424	744	2	4	248
13	35	459	455	3	9	315
14	27	486	378	4	16	432
15	11	497	165	5	25	275
	497		5088			2762

(i) Mean $= \frac{\Sigma fx}{\Sigma f} = \frac{5088}{497} = 10.23 = 10$ (aapproximate)

(ii) Median Class $= \left(\frac{N+1}{2}\right)^{th}$ term

$= \left(\frac{497+1}{2}\right)^{th}$ term

$= 249^{th}$ term

$M_d = 10$

(iii) Standard Deviation $\sigma = \sqrt{\frac{\Sigma f(x-M)^2}{\Sigma f}}$

$= \sqrt{\frac{2762}{497}}$

$= \sqrt{5.56}$

$= 2.3$ (approx)

***Example* 12:** Find the standard deviation for the following observations 192, 288, 236, 229, 184, 260, 348, 291, 330, 243.

***Solution*:**

X	x – M	(x – M)^2
192	–68	4624
288	28	784
236	–24	576
229	–31	961
184	–76	5776
260	0	0
348	88	7744
291	31	961
330	70	4900
243	–17	289
Σx = 2601		26615

$$\text{Mean } M = \frac{\Sigma x}{n} = \frac{2601}{10} = 260.1$$

$$= 260 \text{ (approx)}$$

$$\text{Standard deviation } \sigma = \sqrt{\frac{\Sigma(x-M)^2}{n}}$$

$$= \sqrt{\frac{26615}{10}}$$

$$= \sqrt{2661.5}$$

$$= 51.6 \text{ (approx)}$$

***Example* 13:** The details of runs gained by two batsman A and B in different innings are as follows:

A	24	79	31	114	14	2	68	1	110	7
B	5	18	42	53	9	47	52	17	81	56

(i) Which of two batsman is better run scorer ?

(ii) Which of the two batsman has more consistency in the number of runs ?

***Solution*:**

(i) To compute the coefficient of variation of runs of batsman A.

x	$x-M_A$	$(x-M_A)^2$
24	– 21	441
79	34	1156
31	– 14	196
114	69	4761
14	– 31	961
2	– 43	1849
68	23	529
1	– 44	1936
110	65	4225
7	– 38	1444
$\Sigma x = 450$		17498

$$\therefore \quad \text{Mean } M_A = \frac{\Sigma x}{n} = \frac{450}{10} = 45 \text{ runs}$$

Standard deviation of batsman A is

$$\sigma_A = \sqrt{\frac{\Sigma(x - M_A)^2}{n}}$$

$$= \sqrt{\frac{17498}{10}} = 41.83 \text{ runs}$$

Coefficient of variation $V_A = \frac{\sigma_A}{M_A} \times 100$

$$= \frac{41.83}{45} \times 100 = 92.96\%$$

(ii) To compute the coefficient of variation of runs of batsman B.

x	**x − M_B**	**(x − M_B)²**
5	−33	1089
18	−20	400
42	4	16
53	15	225
9	−29	841
47	9	81
52	14	196
17	−21	441
81	43	1849
56	18	324
Σx = 380		5462

Mean $M_B = \frac{\Sigma x}{n} = \frac{380}{10} = 38 \text{ runs}$

Standard deviation of batsman B is

$$\sigma_B = \sqrt{\frac{\Sigma(x - M_B)^2}{n}}$$

$$= \sqrt{\frac{5462}{10}} = 23.37 \text{ runs}$$

∴ Coefficient of variance

$$V_B = \frac{\sigma_B}{M_B} \times 100$$

$$= \frac{23.37}{38} \times 100$$

$$= 61.5\%$$

Hence,

(i) Since $M_A > M_B$ ∴ batsman A is better run scorer

(ii) Since $V_B < V_A$ ∴ batsman B has more consistency in the number of runs.

Combining Standard Deviation of two sets: Let x and y be two groups containing n_1 and n_2 numbers. Let $\bar{x}_1$ and $\bar{x}_2$ be their means and standard deviation be σ_1 and σ_2 respectively. Then the combined standard deviation of two groups x and y are denoted by σ or σ_{12} and is defined as

$$\sigma_{12} = \sqrt{\frac{n_1\sigma_1^2 + n_2\sigma_2^2}{n_1 + n_2} + \frac{n_1 n_2}{(n_1 - n_2)^2}(\bar{x}_1 - \bar{x}_2)^2}$$

'or'

$$\sigma_{12} = \sqrt{\frac{n_1(\sigma_1^2 + d_1^2) + n_2(\sigma_2^2 + d_2^2)}{n_1 + n_2}}$$

where $d_1 = \bar{x}_1 - \bar{x}$ and $d_2 = \bar{x}_2 - \bar{x}$

and $\bar{x} = \dfrac{n_1\bar{x}_1 + n_2\bar{x}_2}{n_1 + n_2}$ is the combined mean

***Example* 14:** The daily average wages of a group of 70 workers is Rs. 13.5 and the standard deviation is Rs. 1.4. The daily wages of another group of 80 workers is Rs. 15 and variance is Rs. 4. Find (i) Daily average wages (ii) Variance and (iii) Coefficient of variation for all the 150 workers.

Solution: Here $n_1 = 70$, $\bar{x}_1 = 13.5$, $\sigma_1 = 1.4$

$n_2 = 80$, $\bar{x}_2 = 1.5$ $\sigma_2 = \sqrt{4} = 2$

Total number of workers $n = n_1 + n_2 = 70 + 80 = 150$

(i) Daily average wages of 150 workers are

$$\bar{x}_1 = \frac{n_1\bar{x}_1 + n_2\bar{x}_2}{n_1 + n_2}$$

$$= \frac{70 \times 13.5 + 80 \times 15}{150}$$

$$= \frac{94.5 + 1200}{150} = 14.3$$

(ii) $d_1 = \bar{x}_1 - \bar{x} = 13.5 - 14.3 = -0.8 \quad \Rightarrow d_1^2 = 0.64$

$d_2 = \bar{x}_2 - \bar{x} = 15 - 14.3 = 0.7 \quad \Rightarrow d_2^2 = 0.49$

∴ Combined variance

$$\sigma^2 = \sqrt{\frac{n_1(\sigma_1^2 + d_1^2) + n_2(\sigma_2^2 + d_2^2)}{n_1 + n_2}}$$

$$= \frac{70((1.4)^2 + 0.64) + 80(4 + 0.49)}{150}$$

$$= \frac{182 + 359.2}{150}$$

$$= 3.608$$

(iii) Combined Standard Deviation $\sigma = \sqrt{3.608}$

$\sigma = 1.9$ (approx)

∴ Combined coefficient of variance

$$V = \frac{\sigma}{\bar{x}} \times 100$$

$$= \frac{1.9}{14.3} \times 100 = 13.28\%$$

***Example* 15:** A distribution consists of three components with frequency 200, 250, and 300 having means 26, 11 and 16 and standard deviation 3, 4 and 5 respectively. Find the mean, variance of the combined distribution.

***Solution*:** Here $n_1 = 200 \quad \bar{x}_1 = 26 \quad \sigma_1 = 3$

$n_2 = 250, \quad \bar{x}_2 = 11 \quad \sigma_2 = 4$

$n_3 = 300 \quad \bar{x}_3 = 16 \quad \sigma_3 = 5$

(i) Combined mean $= \dfrac{n_1\bar{x}_1 + n_2\bar{x}_2 + n_3\bar{x}_3}{n_1 + n_2 + n_3}$

$$= \frac{200\times 26 + 250\times 11 + 300\times 16}{200 + 250 + 300}$$

$$= \frac{5200 + 2750 + 4800}{750} = \frac{12750}{750} = 17$$

(ii) $d_1 = \bar{x}_1 - \bar{x} = 26 - 17 = 9 \qquad \Rightarrow d_1^2 = 81$

$d_2 = \bar{x}_2 - \bar{x} = 11 - 17 = -6 \qquad \Rightarrow d_2^2 = 36$

$d_3 = \bar{x}_3 - \bar{x} = 11 - 17 = -6 \quad 16 - 17 = -1 \qquad \Rightarrow d_3^2 = 1$

$\therefore$ Combined variance

$$\sigma^2 = \frac{n_1(\sigma_1^2 + d_1^2) + n_2(\sigma_2^2 + d_2^2) + n_3(\sigma_3^2 + d_3^2)}{n_1 + n_2 + n_3}$$

$$= \frac{200(9+81) + 250(16+36) + 300(25+1)}{200 + 250 + 300}$$

$$= \frac{18000 + 13000 + 7800}{750}$$

$$= \frac{38800}{750}$$

$$= 51.73$$

***Example* 16:** Fill up the missing values in the following table :

	First Group	**Second Group**	**Combined Group**
No.of observations	100	150	-
Mean	-	55	51
Standard Deviation	7	-	$\sqrt{130}$

***Solution*:** Here, $n_1 = 100 \qquad \bar{x}_1 = a \qquad \sigma_1 = 7$

$n_2 = 150, \qquad \bar{x}_2 = 55 \qquad \sigma_2 = b$

$n = c \qquad \bar{x} = 51 \qquad \sigma = \sqrt{130}$

∴ Combined mean

$$\bar{x} = \frac{n_1\bar{x}_1 + n_2\bar{x}_2}{n_1 + n_2}$$

$$51 = \frac{100\,a + 150 \times 55}{100 + 150}$$

$$51 = \frac{100\,a + 8250}{250}$$

$$51 \times 250 = 100\,a + 8250$$

$$100\,a = 12750 - 8250$$

$$100\,a = 4500$$

$$a = \frac{4500}{100} = 45$$

∴ $\bar{x}_1 = 45$

and $n = n_1 + n_2$

$= 100 + 150$

$n = 250$

Now, $d_1 = \bar{x}_1 - \bar{x} = 45 - 51 = -6 \quad \Rightarrow d_1^2 = 36$

$d_2 = \bar{x}_2 - \bar{x} = 55 - 51 = 4 \quad \Rightarrow d_2^2 = 16$

∴ Combined variance

$$\sigma^2 = \frac{n_1(\sigma_1^2 + d_1^2) + n_2(\sigma_2^2 + d_2^2)}{n_1 + n_2}$$

$$\left(\sqrt{130}\right)^2 = \frac{100(49 + 36) + 150(b^2 + 16)}{100 + 150}$$

$$130 \times 250 = 8500 + 150\,b^2 + 2400$$

$$32500 = 10900 + 150\,b^2$$

$$150\,b^2 = 32500 - 10900$$

$$= 21600$$

$$b^2 = \frac{21600}{150} = 144$$

∴ $b = \sigma_2 = 12$

Exercises

1. Find the standard deviation for the following distribution.

x	2	3	4	5	9	10	12	13	15
f	25	37	44	59	68	43	31	32	12

2. Calculate the mean, median, mode, standard deviation for the following.

Mid value	15	20	25	30	35	40	45	50	55
Frequency	2	22	19	14	3	4	6	1	1

3. For the following data find the standard deviation and coefficient of variation.

Marks	0-10	10-20	20-30	30-40	40-50	50-60	60-70	70-80
No. of persons	5	10	20	40	30	20	10	4

4. Calculate the standard deviation of the following two series and find out which series shows greater variation.

A	192	288	236	229	184	260	348	291	330	243
B	83	87	93	109	124	126	126	101	102	108

5. The scores of two golfers for 10 rounds each are

A	55	54	52	53	56	58	52	50	51	49
B	108	107	105	105	107	104	103	104	106	101

By calculating the coefficient of variation in each case, find which player may be regarded as the more consistant player?

6. Find the mean and standard deviation

$n_1 = 10,\ \bar{x} = 5,\ \sigma_1 = 2\ ;\ n_2 = 20,\ \bar{x}_2 = 4.8,\ \sigma_2 = 1.5$

7. The mean height of 200 men is 65.50 inches and the standard deviation is 0.08 inch. Similarly the mean height of a set of another 400 men is 67.25 inches and the standard deviation is 0.06 inch. Find the mean height of 600 men and the standard deviation.

8. The means of two h samples of sizes 50 and 100 are 54.4 and 50.3 and standard deviation are 8 and 7 respectively. Obtain the mean and standard deviation of the sample of size 150 obtained after combination of two samples.
9. A distribution consists of three components with frequencies 200, 250 and 300 having means of 25, 10 and 15, standard deviation of 3, 4 and 5 respectively. Show that mean of the combined distribution is 16 and its standard deviation 7.2 approximately.

Answers

1. 3.76
2. 27.85, 25.65, 21.84, 8.85
3. 1.56, 39.2
4. 51.6, 14.96, B
5. $M_A = 53$, $\sigma_A = 2.65$, $V_A = 4.99\%$, $M_B = 105$, $\sigma_B = 2.0$, $V_B = 1.90\%$, B is more consistent
6. 4.87, 81.69
7. 66.67 inch, 0.8277 inc
8. 51.56, 7.55

QUESTION BANK

1. (a) Form a differential equation by eliminating arbitrary constants

$y = a \sin x + b \cos x \quad \Rightarrow \quad \frac{d^2y}{dx^2} + y = 0$

(b) If $\frac{dy}{dx} = e^{x+y}$ and it is given that for x = 1, y = 1, find y when x = –1.

(c) Solve $(1 + x^2)\frac{dy}{dx} + 2xy = 4x^2$

2. (a) Find the general solution of

$\frac{d^2y}{dx^2} - 5\frac{dy}{dx} + 6y = e^{4x} \Rightarrow c_1 e^{\frac{5}{2}x} + c_2 e^{\frac{1}{2}x} + \frac{1}{2} e$

(b) Solve $\frac{d^2y}{dx^2} - 4y = 2\sin\left(\frac{x}{2}\right) \Rightarrow c_1 \cos 2x - c_2 \sin 2x - \frac{5}{2} \sin$

3. The water tax bills of 30 houses in a locality are as given below:

44, 84, 30, 95, 32, 34, 96, 14, 112, 74, 88,

110, 102, 45, 75, 54, 74, 78, 66, 44, 35, 15,

20, 14, 40, 88, 76, 66, 112, 108

Construct a frequency distribution table and draw the histogram.

4. The teams A and B made the following goals in the football match:

No. of Goals	No. of Games A	No. of Games B
0	27	17
1	9	9
2	8	6
3	5	5
4	4	3

Calculate standard deviation for team A and team B.

5. The probability that a injection manufactured by a company will be defective is 1/10. If 12 such injections are manufactured, find the probability that:
   (a) Exactly two will be defective
   (b) At least two will be defective
   (c) None will be defective
6. Find the general solution of simultaneous equation.

$$\frac{d^2y}{dt^2} + x + y = 0$$

$$\frac{d^2x}{dt^2} - 3x - 4y = 0$$

7. Blood serum cholesterol levels of 10 persons are 240, 260, 290, 245, 255, 288, 273, 263, 277, 251. Calculate standard deviation with the help of assumed mean.
8. The incidence of occupational disease in an industry is such that the workman have a 10% chance of suffering from it. What is the probability that in a group of 7, five or more will suffer from it?
9. Find the correlation coefficient between the sales and expenses of the following 10 firms.

Firms	Sales	Expenses
1	50	11
2	50	13
3	55	14
4	60	16
5	65	16
6	65	15
7	65	15
8	60	14
9	60	13
10	50	13

10. (a) (i) write the rules according to which numbers are rounded off.
   (ii) Round off the numbers to four significant figures
   1.6381, 29.0568, 0.859279, 3.14159
   (b) Explain the terms related with frequency Histogram, frequency polygon and ogive.

11. For the distribution

x	f
5	18
6	25
7	34
8	47
9	68
10	90
11	80
12	62
13	35
14	27
15	11

find the mean, median and standard deviation.

12. (a) **Five men in a company of 20 are B. Pharma. If 3 men are picked out of 20 at random, what is the probability that at least one is a B.Pharma?**

(b) **Fit a Poisson Distribution to the following**

x	y
0	192
1	100
2	24
3	3
4	1

13. (a) **Form the differential equation of all circles having radius a and centre as origin.**

(b) Solve $(x + y + 1)\dfrac{dy}{dx} = 1$

(c) Solve $x\dfrac{dy}{dx} - 2y = 2x$

14. (a) Solve the equation

$$\frac{d^2y}{dx^2} - 5\frac{dy}{dx} + 6y = 2e^{2x}$$

(b) Solve the equation

$$\frac{d^2y}{dx^2} + y = x^2$$

15. A bacterial population B is known to have a rate of growth proportional to B itself. If between noon and 2 p.m. the population triples, at what time, no controls being exerted, should B become 100 times? What it was at noon?

16. (a) (i) If $x = 0.178693$ and $y = 0.178439$ correct to 6 significant digit.

Then $z = x - y$ is correct to how many significant digit?

(ii) Write the rules according to which numbers are rounded off.

(b) (i) What are the characteristics of histogram?

(ii) Explain the terms frequency curve and ogive.

17. (a) Tickets are numbered from 1 to 100. They are well shuffled and a ticket is drawn at random. What is the probability that the drawn ticket has :

(i) an even number?

(ii) A number which is greater than 75?

(b) Find the standard deviation for the following observations.

192, 288, 236, 229, 184, 260, 348, 291, 330, 243

18. Obtain the line of regression of y on x for the data given below.

x	y
1.53	33.50
1.78	36.50
2.60	40.00
2.95	45.80
3.42	53.50

19. (a) Define ordinary differential equation and form the differential equation by eliminating the constants :

$$y = e^x (A \cos x + B \sin x)$$

(b) Solve the following differential equations

(i) $x(1 - x^2)\dfrac{dy}{dx} + (2x^2 - 1)\,y = x^3$

(ii) $\dfrac{dy}{dx} = e^{2x + 3y}$

20. (a) Solve

$$\frac{d^3y}{dx^3} - 7\frac{dy}{dx} - 6y = 0$$

(b) Find the general solution of

$$\frac{d^2y}{dx^2} + 4y = \cos 2x$$

21. Out of 800 families with 4 children each, how many families would be expected to have.

(i) 2 boys and 2 girls

(ii) No girl

(iii) At most 2 girls

Assume equal probabilities for boys and girls.

22. Blood Serum Cholestrol level of 10 persons are

240, 260, 290, 245, 255, 288, 273, 263, 277, 251

Calculate standard deviation with the help of assumed mean.

23. Find the correlation coefficient between the sales and expenses of the following 10 firms.

Firms	Sales	Expenses
1	50	11
2	50	13
3	55	14
4	60	16
5	65	16
6	65	15
7	65	15
8	60	14
9	60	13
10	50	13

24. (a) Form a differential equation by eliminating arbitrary constants

$$y = a\,e^{kx} + B\,e^{-kx}$$

(b) Solve

$$(1 - x^2)(1 - y)\,dx = xy(1 + y)\,dy$$

(c) Solve

$$\frac{dy}{dx} + y\tan x = \sec x$$

25. (a) Solve the equation

$$\frac{d^2y}{dx^2} - 2\frac{dy}{dx} + y = e^x$$

(b) Solve the equation

$$\frac{d^2y}{dt^2} + 5\frac{dy}{dt} + 6y = \sin t$$

26. Radium decomposes at a rate proportional to the quantity of radium present. Suppose that it is found that in 25 year's approximately 1.1% of a certain quantity of radium has decomposed. Determine approximately how long will it take for one-half of the original amount of radium to decompose?

27. (a) (i) Explain the term significant digit:

(ii) What is the approx. maximum error if a number is rounded off to 3 places of decimal?

(b) Number of marks obtained by 30 students in statistics in class according to roll no. are:

1, 0, 10, 2, 3, 0, 5, 7, 2, 2

4, 3, 3, 2, 6, 7, 6, 7, 5, 4

8, 8, 1, 2, 3, 4, 5, 6, 7, 9

Prepare a frequency table of each variate.

28. (a) Eight employees in a pharmaceutical company 30 are B.Pharma. If 5 employees are selected from at random, what is the probability that:

(i) they are all B. Pharma

(ii) atleast 1 is B.Pharma

(b) Calculate the mean and standard deviation of following:

Size of item	Frequency
6	3
7	6
8	9
9	13
10	8
11	5
12	4

29. Find a relation of the type R = aV +b, when some value of R and V obtained from an experiment are:

V	R
60	109
65	114
70	118
75	123
80	127
85	130
90	133

30. (a) Find the order and degree of the differential equation:

$$\left\{\frac{d^2y}{dx^2}+\frac{dy}{dx}+5\right\}^3=7$$

(b) Verify that $y = (ax + 3)^3$ is a solution of the differential equation $\frac{d^3y}{dx^3}=6a^3$.

(c) Solve the differential equation

$(1 - x^2)\, dy = (1 - y^2)\, dx$

(d) Solve

$$\frac{dy}{dx}+y\cot x-\sec x=0$$

31. (a) Solve the equation

$$\frac{d^2y}{dx^2}+2\frac{dy}{dx}+y=e^{3x}$$

(b) Solve the equation

$$\frac{d^2y}{dx^2}+3\frac{dy}{dx}+2y=\cos x$$

32. (a) The value of π is 3. 1415927:

(i) Round off to four places of decimal.

(ii) Write the value of π to five significant digits.

(b) Explain Editing of data, Classification of data and Tabulation of data.

33. (a) Find standard deviation

Class	f
0 – 10	2
10 – 20	3
20 – 30	5
30 – 40	7
40 – 50	9

(b) In a Pharmaceutical company there are 40 employees working and 8 are B.Pharma. If 6 employees are selected at random, find the probability that :

(i) all are B. Pharma

(ii) at least 3 are B. Pharma

34. (a) Find the coefficient of correlation of the following data:

x	y
10	9
6	4
9	6
10	9
12	11
13	13
11	8
9	4

(b) Find range and coefficient of range for the following series:

Class	**f**
10 – 20	2
20 – 30	5
30 – 40	7
40 – 50	9
50 – 60	15

35. (a) Explain kurtosis and how flatness is measured.

(b) Explain ANOVA

(c) There are 500 workers in a factory and their mean weight is 151 lbs and S.D. is 15 lb. Assuming weights are normally distributed, find how many workers weigh between 120 and 156 lbs.

Table 1 Four Place Common Logarithms

Log10 N or log N

											Proportional Parts								
N	0	1	2	3	4	5	6	7	8	9	1	2	3	4	5	6	7	8	9
10	0000	0043	0086	0128	0170	0212	0253	0294	0334	0374	4	8	12	17	21	25	30	33	37
11	0414	0453	0492	0531	0569	0607	0645	0682	0719	0755	4	8	11	15	19	23	26	30	31
12	0792	0828	0864	0899	0934	0969	1004	1038	1072	1106	3	7	10	14	17	21	24	28	31
13	1139	1173	1206	1239	1271	1303	1335	1367	1399	1430	3	6	10	13	16	19	23	26	29
14	1461	1492	1523	1553	1584	1614	1644	1673	1703	1732	3	6	9	12	15	18	21	24	27
15	1761	1790	1818	1847	1875	1903	1931	1959	1987	2014	3	6	8	11	14	17	20	22	25
16	2041	2068	2095	2122	2148	2175	2201	2227	2253	2279	3	5	8	11	13	16	18	21	24
17	2304	2330	2355	2380	2405	2430	2455	2480	2504	2529	2	5	7	10	12	15	17	20	23
18	2553	2577	2601	2625	2648	2672	2695	2718	2742	2765	2	5	7	9	12	11	16	19	21
19	2788	2810	2833	2856	2878	2900	2923	2945	2967	2989	2	4	7	9	11	13	16	18	20
20	3010	3032	3054	3075	3096	3118	3139	3160	3181	3201	2	4	6	8	11	13	15	17	19
21	3222	3243	3263	3284	3304	3324	3348	3365	3385	3404	2	4	6	8	10	12	14	16	18
22	3424	3444	3464	3483	3502	3522	3541	3560	3579	3598	2	4	6	8	10	12	14	15	17
23	3617	3636	3655	3674	3692	3711	3729	3747	3766	3784	2	4	6	7	9	11	13	15	17
24	3802	3820	3839	3856	3874	3892	3909	3927	3945	3962	2	4	5	7	9	11	12	14	16
25	3979	3997	4014	4031	4048	4065	4082	4099	4116	4133	2	3	5	7	9	10	12	14	15
26	4150	4166	4183	4200	4216	4232	4249	4265	4281	4298	2	3	5	7	8	10	11	13	15
27	4314	4330	4346	4362	4378	4393	4409	4425	4440	4456	2	3	5	6	8	9	11	13	14
28	4472	4487	4502	4518	4533	4548	4564	4579	4594	4609	2	3	5	6	8	9	11	12	14
29	4624	4639	4654	4669	4683	4698	4713	4728	4742	4757	1	3	4	6	7	9	10	12	13
30	4771	4786	4800	4814	4829	4843	4857	4871	4886	4900	1	3	4	6	7	9	10	11	13
31	4914	4928	4942	4955	4969	4983	4997	5011	5024	5038	1	3	4	6	7	8	10	11	12
32	5051	5065	5079	5092	5105	5119	5132	5145	5159	5172	1	3	4	5	7	8	9	11	12
33	5185	5198	5211	5224	5237	5250	5263	5276	5289	5302	1	3	4	5	6	8	9	10	12
34	5315	5328	5340	5353	5366	5378	5391	5403	5416	5428	1	3	4	5	6	8	9	10	11
35	5441	5453	5465	5478	5490	5502	5514	5527	5539	5551	1	2	4	5	6	7	8	10	11
36	5563	5575	5587	5599	5611	5623	5635	5647	5658	5670	1	2	4	5	6	7	8	10	11
37	5682	5694	5705	5717	5729	5740	5752	5763	5775	5786	1	2	3	5	6	7	8	9	10
38	5798	5809	5821	5832	5843	5355	5866	5877	5888	5899	1	2	3	5	6	7	8	9	10
39	5911	5922	5933	5944	5955	5966	5977	5988	5999	6010	1	2	3	4	5	7	8	9	10
40	6021	6031	6042	6053	6064	6075	6085	6096	6107	6117	1	2	3	4	5	6	8	9	10
41	6128	6138	6149	6160	6170	6180	6191	6201	6212	6222	1	2	3	4	5	6	7	8	9
42	6232	6243	6253	6263	6274	6284	6294	6304	6314	6325	1	2	3	4	5	6	7	8	9
43	6335	6345	6355	6365	6375	6385	6395	6405	6415	6425	1	2	3	4	5	6	7	8	9
44	6435	6444	6454	6464	6474	6484	6493	6503	6513	6522	1	2	3	4	5	6	7	8	9
45	6532	6542	6551	6561	6571	6580	6590	6599	6609	6618	1	2	3	4	5	6	7	8	9
46	6628	6637	6646	6656	6665	6675	6684	6693	6702	6712	1	2	3	4	5	6	7	7	8
47	6721	6730	6739	6749	6758	6767	6776	6785	6794	6803	1	2	3	4	5	5	6	7	8
48	6812	6821	6830	6839	6848	6857	6866	6875	6884	6893	1	2	3	4	4	5	6	7	8
49	6902	6911	6920	6928	6937	6946	6955	6964	6972	6981	1	2	3	4	4	5	6	7	8
50	6990	6998	7007	7016	7024	7033	7042	7050	7059	7067	1	2	3	3	4	5	6	7	8
51	7076	7084	7093	7101	7110	7118	7126	7135	7143	7152	1	2	3	3	4	5	6	7	8
52	7160	7168	7177	7185	7193	7202	7210	7218	7226	7235	1	2	2	3	4	5	6	7	7
53	7243	7251	7259	7267	7275	7284	7292	7300	7308	7316	1	2	2	3	4	5	6	6	7
54	7324	7332	7340	7348	7356	7364	7372	7380	7388	7396	1	2	2	3	4	5	6	6	7
N	0	1	2	3	4	5	6	7	8	9	1	2	3	4	5	6	7	8	9

Table 1 (Contd.)

Four Place Common Logarithms

$\log_{10} N$ or log N

N	0	1	2	3	4	5	6	7	8	9	Proportional Parts 1	2	3	4	5	6	7	8	9
55	7404	7412	7419	7427	7435	7443	7451	7459	7466	7474	1	2	2	3	4	5	5	6	7
56	7482	7490	7497	7505	7513	7520	7528	7536	7543	7551	1	2	2	3	4	5	5	6	7
57	7559	7566	7574	7582	7589	7597	7604	7612	7619	7627	1	2	2	3	4	5	5	6	7
58	7634	7642	7649	7657	7664	7672	7679	7686	7694	7701	1	1	2	3	4	4	5	6	7
59	7709	7716	7723	7731	7738	7745	7752	7760	7767	7774	1	1	2	3	4	4	5	6	7
60	7783	7789	7796	7803	7810	7818	7825	7832	7839	7846	1	1	2	3	4	4	5	6	6
61	7853	7860	7868	7875	7882	7889	7896	7903	7910	7917	1	1	2	3	4	4	5	6	6
62	7924	7931	7938	7945	7952	7959	7966	7973	7980	7987	1	1	2	3	3	4	5	6	6
63	7993	8000	8007	8014	8021	8028	8035	8041	8048	8055	1	1	2	3	3	4	5	5	6
64	8062	8069	8075	8082	8089	8096	8102	8109	8116	8122	1	1	2	3	3	4	5	5	6
65	8129	8136	8142	8149	8156	8162	8169	8176	8182	8189	1	1	2	3	3	4	5	5	6
66	8195	8202	8209	8215	8222	8228	8235	8241	8248	8254	1	1	2	3	3	4	5	5	6
67	8261	8267	8274	8280	8287	8293	8299	8306	8312	8319	1	1	2	3	3	4	5	5	6
68	8325	8331	8338	8344	8351	8357	8363	8370	8376	8382	1	1	2	3	3	4	4	5	6
69	8388	8395	8401	8407	8414	8420	8426	8432	8439	8445	1	1	2	2	3	4	4	5	6
70	8451	8457	8463	8470	8476	8482	8488	8494	8500	8506	1	1	2	2	3	4	4	5	6
71	8513	8519	8525	8531	8537	8543	8549	8555	8561	8567	1	1	2	2	3	4	4	5	5
72	8573	8579	8585	8591	8597	8603	8609	8615	8621	8627	1	1	2	2	3	4	4	5	5
73	8633	8639	8645	8651	8657	8663	8669	8675	8681	8686	1	1	2	2	3	4	4	5	5
74	8692	8698	8704	8710	8716	8722	8727	8733	8739	8745	1	1	2	2	3	4	4	5	5
75	8751	8756	8762	8768	8774	8779	8785	8791	8797	8802	1	1	2	2	3	3	4	5	5
76	8808	8814	8820	8825	8831	8837	8842	8848	8854	8859	1	1	2	2	3	3	4	5	5
77	8865	8871	8876	8882	8887	8893	8899	8904	8910	8915	1	1	2	2	3	3	4	4	5
78	8921	8927	8932	8938	8943	8949	8954	8960	8965	8971	1	1	2	2	3	3	4	4	5
79	8976	8982	8987	8993	8998	9004	9009	9015	9020	9025	1	1	2	2	3	3	4	4	5
80	9031	9036	9042	9047	9053	9058	9063	9069	9074	9079	1	1	2	2	3	3	4	4	5
81	9085	9090	9096	9101	9106	9112	9117	9122	9128	9133	1	1	2	2	3	3	4	4	5
82	9138	9143	9149	9154	9159	9165	9170	9175	9180	9186	1	1	2	2	3	3	4	4	5
83	9191	9196	9201	9206	9212	9217	9222	9227	9232	9238	1	1	2	2	3	3	4	4	5
84	9243	9348	9253	9258	9263	9269	9274	9279	9284	9289	1	1	2	2	3	3	4	4	5
85	9294	9299	9304	9309	9315	9320	9325	9330	9335	9340	1	1	2	2	3	3	4	4	5
86	9345	9350	9355	9360	9365	9370	9375	9380	9385	9390	1	1	2	2	3	3	4	4	5
87	9395	9400	9405	9410	9415	9420	9425	9430	9435	9440	0	1	1	2	2	3	3	4	4
88	9445	9450	9455	9460	9465	9469	9474	9479	9484	9489	0	1	1	2	2	3	3	4	4
89	9494	9499	9504	9509	9513	9518	9523	9528	9533	9538	0	1	1	2	2	3	3	4	4
90	9542	9547	9552	9557	9562	9566	9571	9576	9581	9586	0	1	1	2	2	3	3	4	4
91	9590	9595	9600	9605	9609	9614	9619	9624	9628	9633	0	1	1	2	2	3	3	4	4
92	9638	9643	9647	9652	9657	9661	9666	9671	9675	9680	0	1	1	2	2	3	3	4	4
93	9685	9689	9694	9699	9703	9708	9713	9717	9722	9727	0	1	1	2	2	3	3	4	4
94	9731	9736	9741	9745	9750	9754	9759	9763	9768	9773	0	1	1	2	2	3	3	4	4
95	9777	9782	9786	9791	9795	9800	9805	9809	9814	9818	0	1	1	2	2	3	3	4	4
96	9823	9827	9832	9836	9841	9845	9850	9854	9859	9863	0	1	1	2	2	3	3	4	4
97	9868	9872	9877	9881	9886	9890	9894	9899	9903	9908	0	1	1	2	2	3	3	4	4
98	9912	9917	9921	9926	9930	9934	9939	9943	9948	9952	0	1	1	2	2	3	3	4	4
99	9956	9961	9965	9969	9974	9978	9983	9987	9991	9996	0	1	1	2	2	3	3	3	4
N	0	1	2	3	4	5	6	7	8	9	1	2	3	4	5	6	7	8	9

Table 2

FOUR PLACE COMMON ANTILOGARITHMS

10^p or antilong p

p	0	1	2	3	4	5	6	7	8	9	PP 1	PP 2	PP 3	PP 4	PP 5	PP 6	PP 7	PP 8	PP 9
·00	1000	1002	1005	1007	1009	1012	1014	1016	1019	1021	0	0	1	1	1	1	2	2	2
·01	1023	1026	1028	1030	1033	1035	1038	1040	1042	1045	0	0	1	1	1	1	2	2	2
·02	1047	1050	1052	1054	1057	1059	1062	1064	1067	1069	0	0	1	1	1	1	2	2	2
·03	1072	1074	1076	1079	1081	1084	1086	1089	1091	1094	0	0	1	1	1	1	2	2	2
·04	1096	1099	1102	1104	1107	1109	1112	1114	1117	1119	0	1	1	1	1	2	2	2	2
·05	1122	1125	1127	1130	1132	1135	1138	1140	1143	1146	0	1	1	1	1	2	2	2	2
·06	1148	1151	1153	1156	1159	1161	1164	1167	1169	1172	0	1	1	1	1	2	2	2	2
·07	1175	1178	1180	1183	1186	1189	1191	1194	1197	1199	0	1	1	1	1	2	2	2	2
·08	1202	1205	1206	1211	1213	1216	1219	1222	1225	1227	0	1	1	1	1	2	2	2	3
·09	1230	1233	1236	1239	1242	1245	1247	1250	1253	1256	0	1	1	1	1	2	2	2	3
·10	1259	1262	1265	1268	1271	1274	1276	1279	1282	1285	0	1	1	1	1	2	2	2	3
·11	1288	1291	1294	1297	1300	1303	1306	1309	1312	1315	0	1	1	1	2	2	2	2	3
·12	1318	1321	1324	1327	1330	1334	1337	1340	1343	1346	0	1	1	1	2	2	2	2	3
·13	1349	1352	1355	1358	1361	1365	1368	1371	1374	1377	0	1	1	1	2	2	2	3	3
·14	1380	1384	1387	1390	1393	1396	1400	1403	1406	1409	0	1	1	1	2	2	2	3	3
·15	1413	1416	1419	1422	1426	1429	1432	1435	1439	1442	0	1	1	1	2	2	2	3	3
·16	1445	1449	1452	1455	1459	1462	1466	1469	1472	1476	0	1	1	1	2	2	2	3	3
·17	1479	1483	1486	1489	1493	1496	1500	1503	1507	1510	0	1	1	1	2	2	2	3	3
·18	1514	1517	1521	1524	1528	1531	1535	1538	1542	1545	0	1	1	1	2	2	2	3	3
·19	1549	1552	1556	1560	1563	1567	1570	1574	1578	1581	0	1	1	1	2	2	3	3	3
·20	1585	1589	1592	1596	1600	1603	1607	1611	1614	1618	0	1	1	1	2	2	3	3	3
·21	1622	1626	1629	1633	1637	1641	1644	1648	1652	1656	0	1	1	2	2	2	3	3	3
·22	1660	1663	1667	1671	1675	1679	1683	1687	1690	1694	0	1	1	2	2	2	3	3	3
·23	1698	1702	1706	1710	1714	1718	1722	1726	1730	1734	0	1	1	2	2	2	3	3	4
·24	1738	1742	1746	1750	1754	1758	1762	1766	1770	1774	0	1	1	2	2	2	3	3	4
·25	1778	1782	1786	1791	1795	1799	1803	1807	1811	1816	0	1	1	2	2	2	3	3	4
·26	1820	1824	1828	1832	1837	1841	1845	1849	1854	1858	0	1	1	2	2	3	3	3	4
·27	1862	1866	1871	1875	1879	1884	1888	1892	1897	1901	0	1	1	2	2	3	3	3	4
·28	1905	1910	1914	1919	1923	1928	1932	1936	1941	1945	0	1	1	2	2	3	3	4	4
·29	1950	1954	1959	1963	1968	1972	1977	1982	1986	1991	0	1	1	2	2	3	3	4	4
·30	1995	2000	2004	2009	2014	2018	2023	2028	2032	2037	0	1	1	2	2	3	3	4	4
·31	2042	2046	2051	2056	2061	2065	2070	2075	2080	2084	0	1	1	2	2	3	3	4	4
·32	2089	2094	2099	2104	2109	2113	2118	2123	2128	2133	0	1	1	2	2	3	3	4	4
33	2138	2143	2148	2153	2158	2163	2168	2173	2178	2183	0	1	1	2	2	3	3	4	4
·34	2188	2193	2198	2203	2208	2213	2218	2223	2228	2234	1	1	2	2	3	3	4	4	5
·35	2239	2244	2249	2254	2259	2265	2270	2275	2280	2286	1	1	2	2	3	3	4	4	5
·36	2291	2296	2301	2307	2312	2317	2323	2328	2333	2339	1	1	2	2	3	3	4	4	5
·37	2344	2350	2355	2360	2366	2371	2377	2382	2388	2393	1	1	2	2	3	3	4	4	5
·38	2399	2404	2410	2415	2421	2427	2432	2438	2443	2449	1	1	2	2	3	3	4	4	5
·39	2455	2460	2466	2472	2477	2483	2489	2495	2500	2506	1	1	2	2	3	3	4	5	5
·40	2512	2518	2523	2529	2535	2541	2547	2553	2559	2564	1	1	2	2	3	4	4	5	5
·41	2570	2576	2582	2588	2594	2600	2606	2612	2618	2624	1	1	2	2	3	4	4	5	5
·42	2630	2636	2642	2649	2655	2661	2667	2673	2679	2685	1	1	2	2	3	4	4	5	6
·43	2692	2698	2704	2710	2716	2723	2729	2735	2742	2748	1	1	2	3	3	4	4	5	6
·44	2754	2761	2767	2773	2780	2786	2793	2799	2805	2812	1	1	2	3	3	4	4	5	6
·45	2818	2825	2831	2838	2844	2851	2858	2864	2871	2877	1	1	2	3	3	4	5	5	6
·46	2884	2891	2897	2904	2911	2917	2924	2931	2938	2944	1	1	2	3	3	4	5	5	6
·47	2951	2958	2965	2972	2979	2985	2992	2999	3006	3013	1	1	2	3	3	4	5	5	6
·48	3020	3027	3034	3041	3048	3055	3062	3069	3076	3083	1	1	2	3	4	4	5	6	6
·49	3090	3097	3105	3112	3119	3126	3133	3141	3148	3155	1	1	2	3	4	4	5	6	6
p	0	1	2	3	4	5	6	7	8	9	1	2	3	4	5	6	7	8	9

(PP = Proportional Parts)

Table 2 (Contd.)

FOUR PLACE COMMON ANTILOGARITHMAS

10^p or antilog p

											Proportional Parts								
p	0	1	2	3	4	5	6	7	8	9	1	2	3	4	5	6	7	8	9
·50	3162	3170	3177	3184	3192	3199	3206	3214	3221	3228	1	1	2	3	4	4	5	6	7
·51	3236	3243	3251	3258	3266	3273	3281	3289	3296	3304	1	2	2	3	4	5	5	6	7
·52	3311	3319	3327	3334	3342	3350	3357	3365	3373	3381	1	2	2	3	4	5	5	6	7
·53	3388	3396	3404	3412	3420	3428	3436	3443	3451	3459	1	2	2	3	4	5	6	6	7
·54	3467	3475	3483	3491	3499	3508	3516	3524	3532	3540	1	2	2	3	4	5	6	6	7
·55	3548	3556	3565	3573	3581	3589	3597	3606	3614	3622	1	2	2	3	4	5	6	7	7
·56	3631	3639	3648	3656	3664	3673	3681	3690	3698	3707	1	2	3	3	4	5	6	7	8
·57	3715	3724	3733	3741	3750	3758	3767	3776	3784	3793	1	2	3	3	4	5	6	7	8
·58	3802	3811	3819	3828	3837	3846	3855	3864	3873	3882	1	2	3	4	4	5	6	7	8
·59	3890	3899	3908	3917	3926	3936	3945	3954	3963	3972	1	2	3	4	5	5	6	7	8
·60	3981	3990	3999	4009	4018	4027	4036	4046	4055	4064	1	2	3	4	5	6	6	7	8
·61	4074	4083	4093	4102	4111	4121	4130	4140	4150	4159	1	2	3	4	5	6	7	8	9
·62	4169	4178	4188	4198	4207	4217	4227	4236	4246	4256	1	2	3	4	5	6	7	8	9
·63	4266	4276	4285	4295	4305	4315	4325	4335	4345	4355	1	2	3	4	5	6	7	8	9
·64	4365	4375	4385	4395	4406	4416	4426	4436	4446	4457	1	2	3	4	5	6	7	8	9
·65	4467	4477	4487	4498	4508	4519	4529	4539	4550	4560	1	2	3	4	5	6	7	8	9
·66	4571	4581	4592	4603	4613	4624	4634	4645	4656	4667	1	2	3	4	5	6	7	9	10
·67	4677	4688	4699	4710	4721	4732	4742	4753	4764	4775	1	2	3	4	5	7	8	9	10
·68	4786	4797	4808	4819	4831	4842	4853	4864	4875	4887	1	2	3	4	6	7	8	9	10
·69	4898	4909	4920	4932	4943	4955	4966	4977	4989	5000	1	2	3	5	6	7	8	9	10
·70	5012	5023	5035	5047	5058	5070	5082	5093	5105	5117	1	2	4	5	6	7	8	9	11
·71	5129	5140	5152	5164	5176	5188	5200	5212	5224	5236	1	2	4	5	6	7	8	10	11
·72	5248	5260	5272	5284	5297	5309	5321	5333	5346	5358	1	2	4	5	6	7	9	10	11
·73	5370	5383	5395	5408	5420	5433	5445	5458	5470	5483	1	3	4	5	6	8	9	10	11
·74	5495	5508	5521	5534	5546	5559	5572	5585	5598	5610	1	3	4	5	6	8	9	10	12
·75	5623	5636	5649	5662	5675	5689	5702	5715	5728	5741	1	3	4	5	7	8	9	10	12
·76	5754	5768	5781	5794	5808	5821	5834	5848	5861	5875	1	3	4	5	7	8	9	11	12
·77	5888	5902	5916	5929	5943	5957	5970	5984	5998	6012	1	3	4	5	7	8	10	11	12
·78	6026	6039	6053	6067	6081	6095	6109	6124	6138	6152	1	3	4	6	7	8	10	11	13
·79	6166	6180	6194	6209	6223	6237	6252	6266	6281	6295	1	3	4	6	7	9	10	11	13
·80	6310	6324	6339	6353	6368	6383	6397	6412	6427	6442	1	3	4	6	7	9	10	12	13
·81	6457	6471	6486	6501	6516	6531	6546	6561	6577	6592	2	3	5	6	8	9	11	12	14
·82	6607	6622	6637	6653	6668	6683	6699	6714	6730	6745	2	3	5	6	8	9	11	12	14
·83	6761	6776	6792	6808	6823	6839	6855	6871	6887	6902	2	3	5	6	8	9	11	13	14
·84	6918	6934	6950	6966	6982	6998	7015	7031	7047	7063	2	3	5	6	8	10	11	13	15
·85	7079	7096	7112	7129	7145	7161	7178	7194	7211	7228	2	3	5	7	8	10	12	13	15
·86	7244	7261	7278	7295	7311	7328	7345	7362	7379	7396	2	3	5	7	8	10	12	13	15
·87	7413	7430	7447	7464	7482	7499	7516	7534	7551	7568	2	3	5	7	9	10	12	14	16
·88	7586	7603	7621	7638	7656	7674	7691	7709	7727	7745	2	4	5	7	9	11	12	14	16
·89	7762	7780	7798	7816	7834	7852	7870	7889	7907	7925	2	4	5	7	9	11	13	14	16
·90	7943	7962	7980	7998	8017	8035	8054	8072	8091	8110	2	4	6	7	9	11	13	15	17
·91	8128	8147	8166	8185	8204	8222	8241	8260	8279	8299	2	4	6	8	9	11	13	15	17
·92	8318	8337	8356	8375	8395	8414	8433	8453	8472	8492	2	4	6	8	10	12	14	15	17
·93	8511	8531	8551	8570	8590	8610	8630	8650	8670	8690	2	4	6	8	10	12	14	16	18
·94	8710	8730	8750	8770	8790	8810	8831	8851	8872	8892	2	4	6	8	10	12	14	16	18
·95	8913	8933	8954	8974	8995	9016	9036	9057	9078	9099	2	4	6	8	10	12	15	17	19
·96	9120	9141	9162	9183	9204	9226	9247	9268	9290	9311	2	4	6	8	11	13	15	17	19
·97	9333	9354	9376	9397	9419	9441	9462	9484	9506	9528	2	4	7	9	11	13	15	17	20
·98	9550	9572	9594	9616	9638	9661	9683	9705	9727	9750	2	4	7	9	11	13	16	18	20
·99	9772	9795	9817	9840	·863	9886	9908	9931	9954	9977	2	5	7	9	11	14	16	18	20
p	0	1	2	3	4	5	6	7	8	9	1	2	3	4	5	6	7	8	9

Table 3 Inverse Quantities, Powers, Roots, Logarithms

x	$1/x$	x^2	x^3	$\sqrt{x}$	$\sqrt[3]{x}$	$\log x$	x	$1/x$	x^2	x^3	$\sqrt{x}$	$\sqrt[3]{x}$	$\log x$
1.0	1.000	1.000	1.000	1.000	1.000	0.0000	5.5	0.182	30.25	166.4	2.345	1.765	1.7047
1.1	0.909	1.210	1.331	1.049	1.032	0.0953	5.6	0.179	31.36	175.6	2.366	1.776	1.7228
1.2	0.833	1.440	1.728	1.095	1.063	0.1823	5.7	0.175	32.49	185.2	2.387	1.786	1.7405
1.3	0.769	1.690	2.197	1.140	1.091	0.2624	5.8	0.172	33.64	195.1	2.408	1.797	1.7579
1.4	0.714	1.960	2.744	1.183	1.119	0.3365	5.9	0.169	34.81	205.4	2.429	1.807	1.7750
1.5	0.667	2.250	3.375	1.225	1.145	0.4055	6.0	0.167	36.00	216.0	2.449	1.817	1.7918
1.6	0.625	2.560	4.096	1.265	1.170	0.4700	6.1	0.164	37.21	227.0	2.470	1.827	1.8083
1.7	0.588	2.890	4.913	1.304	1.193	0.5306	6.2	0.161	38.44	238.3	2.490	1.837	1.8245
1.8	0.556	3.240	5.832	1.342	1.216	0.5878	6.3	0.159	39.69	250.0	2.510	1.847	1.8405
1.9	0.526	3.610	6.859	1.378	1.239	0.6419	6.4	0.156	40.96	262.1	2.530	1.857	1.8563
2.0	0.500	4.000	8.000	1.414	1.260	0.6931	6.5	0.154	42.25	274.6	2.550	1.866	1.8718
2.1	0.476	4.410	9.261	1.449	1.281	0.7419	6.6	0.151	43.56	287.5	2.569	1.876	1.8871
2.2	0.454	4.840	10.65	1.483	1.301	0.7885	6.7	0.149	44.89	300.8	2.588	1.885	1.9021
2.3	0.435	5.290	12.17	1.517	1.320	0.8329	6.8	0.147	46.24	314.4	2.608	1.895	1.9169
2.4	0.417	5.760	13.82	1.549	1.339	0.8755	6.9	0.145	47.61	328.5	2.627	1.904	1.9315
2.5	0.400	6.250	15.62	1.581	1.357	0.9163	7.0	0.143	49.00	343.0	2.646	1.913	1.9459
2.6	0.385	6.760	17.58	1.612	1.375	0.9555	7.1	0.141	50.41	357.9	2.665	1.922	1.9601
2.7	0.370	7.290	19.68	1.643	1.392	0.9933	7.2	0.139	51.84	373.2	2.683	1.931	1.9741
2.8	0.357	7.840	21.95	1.673	1.409	1.0296	7.3	0.137	53.29	389.0	2.702	1.940	1.9879
2.9	0.345	8.410	24.39	1.703	1.426	1.0647	7.4	0.135	54.76	405.2	2.720	1.949	2.0015
3.0	0.333	9.000	27.00	1.732	1.442	1.0986	7.5	0.133	56.25	421.9	2.739	1.957	2.0149
3.1	0.323	9.610	29.79	1.761	1.458	1.1314	7.6	0.132	57.76	439.0	2.757	1.966	2.0281
3.2	0.312	10.24	32.77	1.789	1.474	1.1632	7.7	0.130	59.29	456.5	2.775	1.975	2.0412
3.3	0.303	10.89	35.94	1.817	1.489	1.1939	7.8	0.128	60.84	474.6	2.793	1.983	2.0541
3.4	0.294	11.56	39.30	1.844	1.504	1.2238	7.9	0.127	62.41	493.0	2.811	1.992	2.0669
3.5	0.286	12.25	42.88	1.871	1.518	1.2528	8.0	0.125	64.00	512.0	2.828	2.000	2.0794
3.6	0.278	12.96	46.66	1.897	1.533	1.2809	8.1	0.123	65.61	531.4	2.846	2.008	2.0919
3.7	0.270	13.69	50.65	1.924	1.547	1.3083	8.2	0.122	67.24	551.4	2.864	2.017	2.1041
3.8	0.263	14.44	54.87	1.949	1.560	1.3350	8.3	0.120	68.89	571.8	2.881	2.025	2.1163
3.9	0.256	15.21	59.32	1.975	1.574	1.3610	8.4	0.119	70.56	592.7	2.898	2.033	2.1282
4.0	0.250	16.00	64.00	2.000	1.587	1.3863	8.5	0.118	72.25	614.1	2.915	2.041	2.1401
4.1	0.244	16.81	68.92	2.025	1.601	1.4110	8.6	0.116	73.96	636.1	2.933	2.049	2.1518
4.2	0.238	17.64	74.09	2.049	1.613	1.4351	8.7	0.115	75.69	658.5	2.950	2.057	2.1633
4.3	0.233	18.49	79.51	2.074	1.626	1.4586	8.8	0.114	77.44	681.5	2.966	2.065	2.1748
4.4	0.227	19.36	85.18	2.098	1.639	1.4816	8.9	0.112	79.21	705.0	2.983	2.072	2.1861
4.5	0.222	20.25	91.12	2.121	1.651	1.5041	9.0	0.111	81.00	729.0	3.000	2.080	2.1972
4.6	0.217	21.16	97.34	2.145	1.663	1.5261	9.1	0.110	82.81	753.6	3.017	2.088	2.2083
4.7	0.213	22.09	103.8	2.168	1.675	1.5476	9.2	0.109	84.64	778.7	3.033	2.095	2.2192
4.8	0.208	23.04	110.6	2.191	1.687	1.5686	9.3	0.108	86.49	804.4	3.050	2.103	2.2300
4.9	0.204	24.01	117.6	2.214	1.698	1.5892	9.4	0.106	88.36	830.6	3.066	2.110	2.2107
5.0	0.200	25.00	125.0	2.236	1.710	1.6094	9.5	0.105	90.25	857.4	3.082	2.118	2.2513
5.1	0.196	26.01	132.7	2.258	1.721	1.6292	9.6	0.104	92.16	884.7	3.098	2.125	2.2618
5.2	0.192	27.04	140.6	2.280	1.732	1.6487	9.7	0.103	94.09	912.7	3.114	2.133	2.2721
5.3	0.189	28.09	148.9	2.302	1.744	1.6677	9.8	0.102	96.04	941.2	3.130	2.140	2.2824
5.4	0.185	29.16	157.5	2.324	1.754	1.6864	9.9	0.101	98.01	970.3	3.146	2.147	2.2925
							10.0	0.100	100.00	1000.0	3.162	2.154	2.3026

Table 4 Trigonometric Functions

x^0	*x (radians)*	sin *x*	tan *x*	cot *x*	cos *x*		
0	0.0000	0.0000	0.0000	∞	1.0000	1.5708	90
1	0.0175	0.0175	0.0175	57.29	0.9998	1.5533	89
2	0.0349	0.0349	0.0349	28.64	0.9994	1.5359	88
3	0.0524	0.0523	0.0524	19.08	0.9986	1.5184	87
4	0.0698	0.0698	0.0699	14.30	0.9976	1.5010	86
5	0.0873	0.0872	0.0875	11.43	0.9962	1.4835	85
6	0.1047	0.1045	0.1051	9.514	0.9945	1.4661	84
7	0.1222	0.1219	0.1228	8.144	0.9925	1.4486	83
8	0.1396	0.1392	0.1405	7.115	0.9903	1.4312	82
9	0.1571	0.1564	0.1584	6.314	0.9877	1.4137	81
10	0.1745	0.1736	0.1763	5.671	0.9848	1.3963	80
11	0.1920	0.1908	0.1944	5.145	0.9816	1.3788	79
12	0.2094	0.2079	0.2126	4.705	0.9781	1.3614	78
13	0.2269	0.2250	0.2309	4.331	0.9744	1.3439	77
14	0.2443	0.2419	0.2493	4.011	0.9703	1.3265	76
15	0.2618	0.2588	0.2679	3.732	0.9659	1.3090	75
16	0.2793	0.2756	0.2867	3.487	0.9613	1.2915	74
17	0.2967	0.2924	0.3057	3.271	0.9563	1.2741	73
18	0.3142	0.3090	0.3249	3.078	0.9511	1.2566	72
19	0.3316	0.3256	0.3443	2.904	0.9455	1.2392	71
20	0.3491	0.3420	0.3640	2.747	0.9397	1.2217	70
21	0.3665	0.3584	0.3839	2.605	0.9336	1.2043	69
22	0.3840	0.3746	0.4040	2.475	0.9272	1.1868	68
23	0.4014	0.3907	0.4245	2.356	0.9205	1.1694	67
24	0.4189	0.4067	0.4452	2.246	0.9135	1.1519	66
25	0.4363	0.4226	0.4663	2.145	0.9063	1.1345	65
26	0.4538	0.4384	0.4877	2.050	0.8988	1.1170	64
27	0.4712	0.4540	0.5095	1.963	0.8910	1.0996	63
28	0.4887	0.4695	0.5317	1.881	0.8829	1.0821	62
29	0.5061	0.4848	0.5543	1.804	0.8746	1.0647	61
30	0.5236	0.5000	0.5774	1.732	0.8660	1.0472	60
31	0.5411	0.5150	0.6009	1.6643	0.8572	1.0297	59
32	0.5585	0.5299	0.6249	1.6003	0.8480	1.0123	58
33	0.5760	0.5446	0.6494	1.5399	0.8387	0.9948	57
34	0.5934	0.5592	0.6745	1.4826	0.8290	0.9774	56
35	0.6109	0.5736	0.7002	1.4281	0.8192	0.9599	55
36	0.6283	0.5878	0.7265	1.3764	0.8090	0.9425	54
37	0.6458	0.6018	0.7536	1.3270	0.7986	0.9250	53
38	0.6632	0.6157	0.7813	1.2799	0.7880	0.9076	52
39	0.6807	0.6293	0.8098	1.2349	0.7771	0.8901	51
40	0.6981	0.6428	0.8391	1.1918	0.7660	0.8727	50
41	0.7156	0.6561	0.8693	1.1504	0.7547	0.8552	49
42	0.7330	0.6691	0.9004	1.1106	0.7431	0.8378	48
43	0.7505	0.6820	0.9325	1.0724	0.7314	0.8203	47
44	0.7679	0.6947	0.9657	1.0355	0.7193	0.8029	46
45	0.7854	0.7071	1.0000	1.0000	0.7071	0.7854	45
		cos *x*	cot *x*	tan *x*	sin *x*	*x* radians)	x^0

Table 5 Exponential, Hyperbolic and Trigonometric Functions

x	e^x	e^{-x}	sinh x	cosh x	tanh x	sin x	cos x
0.0	1.0000	1.0000	0.0000	1.0000	0.0000	0.0000	1.0000
0.1	1.1052	0.9048	0.1002	1.0050	0.0997	0.0998	0.9950
0.2	1.2214	0.8187	0.2013	1.0201	0.1974	0.1987	0.9801
0.3	1.3499	0.7408	0.3045	1.0453	0.2913	0.2955	0.9553
0.4	1.4918	0.6703	0.4108	1.0811	0.3799	0.3894	0.9211
0.5	1.6487	0.6065	0.5211	1.1276	0.4621	0.4794	0.8776
0.6	1.8221	0.5488	0.6367	1.1855	0.5370	0.5646	0.8253
0.7	2.0138	0.4966	0.7586	1.2552	0.6044	0.6442	0.7648
0.8	2.2255	0.4493	0.8881	1.3374	0.6640	0.7174	0.6967
0.9	2.4596	0.4066	1.0265	1.4331	0.7163	0.7833	0.6216
1.0	2.7183	0.3679	1.1752	1.5431	0.7616	0.8415	0.5403
1.1	3.0042	0.3329	1.3356	1.6685	0.8005	0.8912	0.4536
1.2	3.3201	0.3012	1.5095	1.8107	0.8337	0.9320	0.3624
1.3	3.6603	0.2725	1.6984	1.9709	0.8617	0.9636	0.2675
1.4	4.0552	0.2466	1.9043	2.1509	0.8854	0.9854	0.1700
1.5	4.4817	0.2231	2.1293	2.3524	0.9051	0.9975	0.0707
1.6	4.9530	0.2019	2.3756	2.5775	0.9217	0.9996	− 0.0292
1.7	5.4739	0.1827	2.6456	2.8283	0.9354	0.9917	− 0.1288
1.8	6.0496	0.1653	2.9422	3.1075	0.9468	0.9738	− 0.2272
1.9	6.6859	0.1496	3.2682	3.4177	0.9562	0.9463	− 0.3233
2.0	7.3891	0.1353	3.6269	3.7622	0.9640	0.9093	− 0.4161
2.1	8.1662	0.1225	4.0219	4.1443	0.9704	0.8632	− 0.5048
2.2	9.0250	0.1108	4.4571	4.5679	0.9757	0.8085	− 0.5885
2.3	9.9742	0.1003	4.9370	5.0372	0.9801	0.7457	− 0.6663
2.4	11.0232	0.0907	5.4662	5.5569	0.9837	0.6755	− 0.7374
2.5	12.1825	0.0821	6.0502	6.1323	0.9866	0.5985	− 0.8011
2.6	13.4637	0.0743	6.6947	6.7690	0.9890	0.5155	− 0.8569
2.7	14.8797	0.0672	7.4063	7.4735	0.9910	0.4274	− 0.9041
2.8	16.4446	0.0608	8.1919	8.2527	0.9926	0.3350	− 0.9422
2.9	18.1741	0.0550	9.0596	9.1146	0.9940	0.2392	− 0.9710
3.0	20.0855	0.0498	10.0179	10.0677	0.9950	0.1411	− 0.9900
3.1	22.1979	0.0450	11.0764	11.1215	0.9959	0.0416	− 0.9991
3.2	24.5325	0.0408	12.2459	12.2366	0.9967	− 0.0584	− 0.9983
3.3	27.1126	0.0369	13.5379	13.5748	0.9973	− 0.1577	− 0.9875
3.4	29.9641	0.0334	14.9654	14.9987	0.9978	− 0.2555	− 0.9668
3.5	33.1154	0.0302	16.5426	16.5728	0.9982	− 0.3508	− 0.9365

Table 6 Bessel Functions

x	$J_0(x)$	$J_1(x)$	x	$J_0(x)$	$J_1(x)$
0·0	1.0000	0.0000	3·0	– 0.2601	0.3991
0·1	0.9975	0.0499	3·1	– 0.2921	0.3009
0·2	0.9900	0.0995	3·2	– 0.3202	0.2613
0·3	0.9776	0.1483	3·3	– 0.3443	0.2207
0·4	0.9604	0.1960	3·4	– 0.3643	0.1792
0·5	0.9385	0.2423	3·5	– 0.3801	0.1374
0·6	0.9120	0.2867	3·6	– 0.3918	0.0955
0·7	0.8812	0.3290	3·7	– 0.3992	0.0538
0·8	0.8463	0.3688	3·8	– 0.4026	0.0128
0·9	0.8075	0.4059	3·9	– 0.4018	– 0.0272
1·0	0.7652	0.4401	4·0	– 0.3971	– 0.0660
1·1	0.7196	0.4709	4·1	– 0.3887	– 0.1033
1·2	0.6711	0.4983	4·2	– 0.3766	– 0.1386
1·3	0.6201	0.5220	4·3	– 0.3610	– 0.1719
1·4	0.5669	0.5419	4·4	– 0.3423	– 0.2028
1·5	0.5118	0.5579	4·5	– 0.3205	– 0.2311
1·6	0.4554	0.5699	4·6	– 0.2961	– 0.2566
1·7	0.3980	0.5778	4·7	– 0.2693	– 0.2791
1·8	0.3400	0.5815	4·8	– 0.2404	– 0.2985
1·9	0.2818	0.5812	4·9	– 0.2097	– 0.3147
2·0	0.2239	0.5767	5·0	– 0.1776	– 0.3276
2·1	0.1666	0.5683	5·1	– 0.1443	– 0.3371
2·2	0.1104	0.5560	5·2	– 0.1103	– 0.3432
2·3	0.0555	0.5399	5·3	– 0.0758	– 0.3460
2·4	0.0025	0.5202	5·4	– 0.0412	– 0.3453
2·5	– 0.0484	0.4971	5·5	– 0.0068	– 0.3414
2·6	– 0.0968	0.4708	5·6	0.0270	– 0.3343
2·7	– 0.1424	0.4416	5·7	0.0599	– 0.3241
2·8	– 0.1850	0.4097	5·8	0.0917	– 0.3110
2·9	– 0.2243	0.3754	5·9	0.1220	– 0.2951

Table 7 Area under the Normal Curve

0 z

z	0	1	2	3	4	5	6	7	8	9
0.0	0.0000	0.0040	0.0080	0.0120	0.0160	0.0199	0.0239	0.0279	0.0319	0.0359
0.1	0.0398	0.0438	0.0478	0.0517	0.0557	0.0596	0.0636	0.0675	0.0714	0.0754
0.2	0.0793	0.0832	0.0871	0.0910	0.0948	0.0987	0.1026	0.1064	0.1103	0.1141
0.3	0.1179	0.1217	0.1255	0.1293	0.1331	0.1368	0.1406	0.1443	0.1480	0.1517
0.4	0.1554	0.1591	0.1628	0.1664	0.1700	0.1736	0.1772	0.1808	0.1844	0.1879
0.5	0.1915	0.1950	0.1985	0.2019	0.2054	0.2088	0.2123	0.2157	0.2190	0.2224
0.6	0.2258	0.2291	0.2324	0.2357	0.2389	0.2422	0.2454	0.2486	0.2518	0.2549
0.7	0.2580	0.2612	0.2642	0.2673	0.2704	0.2734	0.2764	0.2794	0.2823	0.2852
0.8	0.2881	0.2910	0.2939	0.2967	0.2996	0.3023	0.3051	0.3078	0.3106	0.3133
0.9	0.3159	0.3186	0.3212	0.3238	0.3264	0.3289	0.3315	0.3340	0.3365	0.3389
1.0	0.3413	0.3438	0.3461	0.3485	0.3508	0.3531	0.3554	0.3577	0.3599	0.3621
1.1	0.3643	0.3665	0.3686	0.3708	0.3729	0.3749	0.3770	0.3790	0.3810	0.3830
1.2	0.3849	0.3869	0.3888	0.3907	0.3925	0.3944	0.3962	0.3980	0.3997	0.4015
1.3	0.4032	0.4049	0.4066	0.4082	0.4099	0.4115	0.4131	0.4147	0.4162	0.4177
1.4	0.4192	0.4207	0.4222	0.4236	0.4251	0.4265	0.4279	0.4292	0.4306	0.4319
1.5	0.4332	0.4345	0.4357	0.4370	0.4382	0.4394	0.4406	0.4418	0.4429	0.4441
1.6	0.4452	0.4463	0.4474	0.4484	0.4495	0.4505	0.4515	0.4525	0.4535	0.4545
1.7	0.4554	0.4564	0.4573	0.4582	0.4591	0.4599	0.4608	0.4616	0.4625	0.4633
1.8	0.4641	0.4649	0.4656	0.4664	0.4671	0.4678	0.4686	0.4693	0.4699	0.4706
1.9	0.4713	0.4719	0.4726	0.4732	0.4738	0.4744	0.4750	0.4756	0.4761	0.4767
2.0	0.4772	0.4778	0.4783	0.4788	0.4793	0.4798	0.4803	0.4808	0.4812	0.4817
2.1	0.4821	0.4826	0.4830	0.4834	0.4838	0.4842	0.4846	0.4850	0.4854	0.4857
2.2	0.4861	0.4864	0.4868	0.4871	0.4875	0.4878	0.4881	0.4884	0.4887	0.4890
2.3	0.4893	0.4896	0.4898	0.4901	0.4904	0.4906	0.4909	0.4911	0.4913	0.4916
2.4	0.4918	0.4920	0.4922	0.4925	0.4927	0.4929	0.4931	0.4932	0.4934	0.4936
2.5	0.4938	0.4940	0.4941	0.4943	0.4945	0.4946	0.4948	0.4949	0.4951	0.4952
2.6	0.4953	0.4955	0.4956	0.4957	0.4959	0.4960	0.4961	0.4962	0.4963	0.4964
2.7	0.4965	0.4966	0.4967	0.4968	0.4969	0.4970	0.4971	0.4972	0.4973	0.4974
2.8	0.4974	0.4975	0.4973	0.4977	0.4977	0.4978	0.4979	0.4979	0.4980	0.4981
2.9	0.4981	0.4982	0.4982	0.4983	0.4984	0.4984	0.4985	0.4985	0.4986	0.4986
3.0	0.4987	0.4987	0.4987	0.4988	0.4988	0.4989	0.4989	0.4989	0.4990	0.4990
3.1	0.4990	0.4991	0.4991	0.4991	0.4992	0.4992	0.4992	0.4992	0.4993	0.4993

Table 8 Values of (t) with probability p and degrees of freedom v

P / v	0·50	0·10	0·05	0·02	0·01
1	1·000	6·34	12·71	31·82	63·66
2	0·816	2·92	4·30	6·96	9·92
3	0·765	2·35	3·18	4·54	5·84
4	0·741	2·13	2·78	3·75	4·60
5	0·727	2·02	2·57	3·36	4·03
6	0·718	1·94	2·45	3·14	3·71
7	0·711	1·90	2·36	3·00	3·50
8	0·706	1·86	2·31	2·90	3·36
9	0·703	1·83	2·26	2·82	3·25
10	0·700	1·81	2·23	2·76	3·17
11	0·697	1·80	2·20	2·72	3·11
12	0·695	1·78	2·18	2·68	3·06
13	0·694	1·77	2·16	2·65	3·01
14	0·692	1·76	2·14	2·62	2·98
15	0·691	1·75	2·13	2·60	2·95
16	0·690	1·75	2·12	2·58	2·92
17	0·689	1·74	2·11	2·57	2·90
18	0·688	1·73	2·10	2·55	2·88
19	0·688	1·73	2·09	2·54	2·86
20	0·687	1·72	2·09	2·53	2·84
21	0·686	1·72	2·08	2·52	2·83
22	0·686	1·72	2·07	2·51	2·82
23	0·685	1·71	2·07	2·50	2·81
24	0·685	1·71	2·06	2·49	2·80
25	0·684	1·71	2·06	2·48	2·79
26	0·684	1·71	2·06	2·48	2·78
27	0·684	1·70	2·05	2·47	2·77
28	0·683	1·70	2·05	2·47	2·76
29	0·683	1·70	2·04	2·46	2·76
30	0·683	1·70	2·04	2·46	2·75

Table 9 Values of x^2 with probability p and df v

P / v	0·99	0·95	0·50	0·30	0·20	0·10	0·05	0·01
1	0·0002	0·004	0·46	1·07	1·64	2·71	3·84	6·64
2	0·020	0·103	1·39	2·41	3·22	4·60	5·99	9·21
3	0·115	0·35	2·37	3·66	4·64	6·25	7·82	11·34
4	0·30	0·71	3·36	4·88	5·99	7·78	9·49	13·28
5	0·55	1·14	4·35	6·06	7·29	9·24	11·07	15·09
6	0·87	1·64	5·35	7·23	8·56	10·64	12·59	16·81
7	1·24	2·17	6·35	8·38	9·80	12·02	14·07	18·48
8	1·65	2·73	7·34	9·52	11·03	13·36	15·51	20·09
9	2·09	3·32	8·34	10·66	12·24	14·68	16·92	21·67
10	2·56	3·94	9·34	11·78	13·44	15·99	18·31	23·21
11	3·05	4·58	10·34	12·90	14·63	17·28	19·68	24·72
12	3·57	5·23	11·34	14·01	15·81	18·55	21·03	26·22
13	4·11	5·89	12·34	15·12	16·98	19·81	22·36	27·69
14	4·66	6·57	13·34	16·22	18·15	21·06	23·68	29·14
15	5·23	7·26	14·34	17·32	19·31	22·31	25·00	30·58
16	5·81	7·96	15·34	18·42	20·46	23·54	26·30	32·00
17	6·41	8·67	16·34	19·51	21·62	24·77	27·59	33·41
18	7·02	9·39	17·34	20·60	22·76	25·99	28·87	34·80
19	7·63	10·12	18·34	21·69	23·90	27·20	30·14	36·19
20	8·26	10·85	19·34	22·78	25·04	28·41	31·41	37·57
21	8·90	11·59	20·34	23·86	26·17	29·62	32·67	38·93
22	9·54	12·34	21·34	24·94	27·30	30·81	33·92	40·29
23	10·20	13·09	22·34	26·02	28·43	32·01	35·17	41·64
24	10·86	13·85	23·34	27·10	29·55	33·20	36·42	42·98
25	11·52	14·61	24·34	28·17	30·68	34·68	37·65	44·31
26	12·20	15·38	25·34	29·25	31·80	35·56	38·88	45·64
27	12·88	16·15	26·34	30·32	32·91	36·74	40·11	46·96
28	13·56	16·93	27·34	31·39	34·03	37·92	41·34	48·28
29	14·26	17·71	28·34	32·46	35·14	39·09	42·56	49·59
30	14·95	18·49	29·34	33·53	36·25	40·26	43·77	50·89

Table 10 5% and 1% points of F

v_2 \ v_1	1	2	3	4	5	6	8	12	24	∞
2	18.51	19.00	19.16	19.25	19.30	19.32	19.37	19.41	19.45	19.50
	98.49	99.00	99.17	99.25	99.30	99.33	99.36	99.42	99.46	99.50
3	10.13	9.55	9.28	9.12	9.01	8.94	8.84	8.74	8.64	8.53
	34.12	30.82	29.46	28.71	28.24	27.91	27.49	27.05	26.60	26.12
4	7.71	6.94	6.59	6.39	6.26	6.16	6.04	5.91	5.77	5.63
	21.20	18.00	16.69	15.98	15.52	15.21	14.80	14.37	13.93	13.46
5	6.61	5.79	5.41	5.19	5.05	4.95	4.82	4.68	4.53	4.36
	16.26	13.27	12.06	11.39	10.97	10.67	10.27	9.89	9.47	9.02
6	5.99	5.14	4.76	4.53	4.39	4.28	4.15	4.00	3.84	3.67
	13.74	10.92	9.78	9.15	8.75	8.47	8.10	7.72	7.31	6.88
7	5.59	4.74	4.35	4.12	3.97	3.87	3.73	3.57	3.41	3.23
	12.25	9.55	8.45	7.85	7.46	7.19	6.84	6.47	6.07	5.65
8	5.32	4.46	4.07	3.84	3.69	3.58	3.44	3.28	3.12	2.93
	11.26	8.65	7.59	7.01	6.63	6.37	6.03	5.67	5.28	4.86
9	5.12	4.26	3.86	3.63	3.48	3.37	3.23	3.07	2.90	2.71
	10.56	8.02	6.99	6.42	6.06	5.80	5.47	5.11	4.73	4.31
10	4.96	4.10	3.71	3.48	3.33	3.22	3.07	2.91	2.74	2.54
	10.04	7.56	6.55	5.99	5.64	5.39	5.06	4.71	4.33	3.91
12	4.75	3.88	3.49	3.26	3.11	3.00	2.85	2.69	2.50	2.30
	9.33	6.93	5.95	5.41	5.06	4.82	4.50	4.16	3.78	3.36
14	4.60	3.74	3.34	3.11	2.96	2.85	2.70	2.53	2.35	2.13
	8.86	6.51	5.56	5.03	4.69	4.46	4.14	3.80	3.43	3.00
16	4.49	3.63	3.24	3.01	2.85	2.74	2.59	2.42	2.24	2.01
	8.53	6.23	5.29	4.77	4.44	4.20	3.89	3.55	3.18	2.75
18	4.41	3.55	3.16	2.93	2.77	2.66	2.51	2.34	2.15	1.92
	8.28	6.01	5.09	4.58	4.25	4.01	3.71	3.37	3.01	2.57
20	4.35	3.49	3.10	2.87	2.71	2.60	2.45	2.28	2.08	1.84
	8.10	5.85	4.94	4.43	4.10	3.87	3.56	3.23	2.86	2.42
25	4.24	3.38	2.99	2.76	2.60	2.49	2.34	2.16	1.96	1.71
	7.77	5.57	4.68	4.18	3.86	3.63	3.32	2.99	2.62	2.17
30	4.17	3.32	2.92	2.69	2.53	2.42	2.27	2.09	1.89	1.62
	7.56	5.39	4.51	4.02	3.70	3.47	3.17	2.84	2.47	2.01
40	4.08	3.23	2.84	2.61	2.45	2.34	2.18	2.00	1.79	1.51
	7.31	5.18	4.31	3.83	3.51	3.29	2.99	2.66	2.29	1.81
60	4.00	3.15	2.76	2.52	2.37	2.25	2.10	1.92	1.70	1.39
	7.08	4.98	4.13	3.65	3.34	3.12	2.82	2.50	2.12	1.60